Electronics – Circuits and Systems

Electronics – Circuits and Systems

Owen Bishop

OXFORD AUCKLAND BOSTON JOHANNESBURG MELBOURNE NEW DELHI

Newnes
An imprint of Butterworth-Heinemann
Linacre House, Jordan Hill, Oxford OX2 8DP
225 Wildwood Avenue, Woburn, MA 01801-2041
A division of Reed Educacational and Professonal Publishing Ltd

 A member of the Reed Elsevier plc group

First published 1999
Reprinted 2000 (twice)

British Library Cataloguing in Publication Data
A catalogue record for this book is available from the British Library

Library of Congress Cataloguing in Publication Data
A catalogue record for this book is available from the Library of Congress

ISBN 0 7506 4195 9

Printed and bound in Great Britain

Contents

Preface

This book is written for a wide range of pre-degree courses in electronics. The contents have been carefully matched to current UK syllabuses at Level 3 / A-level, but the topics covered, depth of coverage, and student activities have been designed so that the resulting book will be a student-focused text suitable for the majority of courses at pre-degree level around the world. The only prior knowledge assumed is basic maths and science.

The UK courses covered by this text are:

- Advanced GNVQ Optional Units in Electronics and Microelectronics from Edexcel and AQA (City and Guilds)
- BTEC National units in Electronics
- A-level and AS syllabuses from OCR, AQA (NEAB) and WJEC.

The book is essentially practical in its approach, encouraging students to assemble and test real circuits in the laboratory. In response to the requirements of certain GNVQ syllabuses, the book shows how circuit behaviour may be studied with a computer, using circuit simulator software.

The book is suitable for class use, and also for self-instruction. The main text is backed up by boxed-off discussions and summaries, which the student may read or ignore, as appropriate. There are frequent 'Test Your Knowledge' questions in the margins with answers given in the Supplement. Another feature of the book is the placing of short 'memos' in the margins. These are intended to remind the student of facts recently encountered but probably not yet learnt. They also provide definitions of terms, particularly of some of the useful jargon associated with electronics and computing.

Each chapter ends with a batch of examination-type questions, and in most instances with a selection of multiple choice questions. Answers to the multiple choice questions appear in the Supplement.

Owen Bishop

Practical circuits and systems

Circuit ideas

As well as being a textbook, this is a sourcebook of circuit ideas for laboratory work and as the basis of practical electronic projects.

All circuits in this book have been tested on the workbench or on computer, using a circuit simulator. All circuit diagrams are complete with component values, so the student will have no difficulty in building circuits that will work.

Testing circuits

The circuit diagrams in this book provide full information about the types of components used and their values. Try to assemble as many as you can of these circuits and get them working. Check that they behave in the same ways as described in the text. Try altering some of the values slightly, predict what should happen, and then test the circuit to check that it does.

There are two ways of building a test circuit:

- Use a breadboarding system to build the circuit temporarily from individual components or circuit modules.
- Use a computer to run a circuit simulator. 'Build' the circuit on the simulator, save it as a file and then run tests on it.

The simulator technique is usually quicker and cheaper than breadboarding. It is easier to modify the circuit, and quicker to run the tests and to plot results. There is no danger of accidentally burning out components.

Conventions used in this book

Units are printed in roman (upright) type V, A, s, S, μF

Values are printed in italic (sloping) type and include:

Fixed values	V_{CC}, R_1
Varying values	v_{GS}, g_m, i_D
Small *changes* in values	v_{gs}, i_d

Resistors are numbered, R1, R2, ... The *resistance* of a resistor R1 is represented by the symbol R_1. The same applies to capacitors (C1, C2 ...) and inductors (L1, L2, ...).

Significant figures

When working the numerical problems in this book give the answers to three significant figures, unless otherwise indicated.

Units in calculations

Usually the units being used in a calculation are obvious but, where they are not so obvious, they are stated in square brackets. Sometimes we show one unit divided by or multiplied by another.

Example

On p. 23 we have stated:

$$R_1 = 14.4/2.63\ [\mathrm{V}/\mu\mathrm{A}] = 5.48\ \mathrm{M}\Omega$$

A voltage measured in volts is being divided by a current measured in *micro*amperes. Mathematically, this equation should be written:

$$R_1 = 14.4/(2.63 \times 10^{-6}) = 5.48 \times 10^{6}$$

Doing this makes the equation difficult to understand and to remember. To avoid this problem we quote the *units* instead of powers of 10. When the result is being worked out on a calculator it is easy to key in the values (14.4, 2.63) and follow each by keypresses for 'EXP -6' or other exponents where required. The result, in Engineering or Scientific format, tells us its units. In this example the display shows 5.475285171^{06}. We round this to 3 significant figures, '5.48', and the '^{06}' index informs us that the result is in megohms.

1 FET amplifiers

Summary

A common-source amplifier based on a MOSFET (MOS field effect transistor) is taken as a typical transistor amplifier and its features are described. It is contrasted with the MOSFET common-drain amplifier (voltage follower) and a common-source amplifier based on a JFET. We also look at how a MOSFET can be used as a transistor switch.

Common-source amplifier

The amplifier in Fig. 1.1 is built around a metal-oxide semiconductor field effect transistor, a MOSFET for short. The type illustrated in the figure is the most often-used one, *an n-channel enhancement MOSFET.*

Given a fixed supply voltage $V_{DD,}$ the size of the current i_D flowing through the transistor depends on v_{GS} the voltage difference between the gate (g) and the source (s). Current flows in at the drain (d) and out at the source (s). Fig. 1.2 shows how v_{GS} and i_D are related. Below a

Source and drain: The names refer to the flow of electrons into the transistor at the source and out through the drain. The current i_D shown in Fig. 1.1 is *conventional* current, which is taken to flow in the opposite direction.

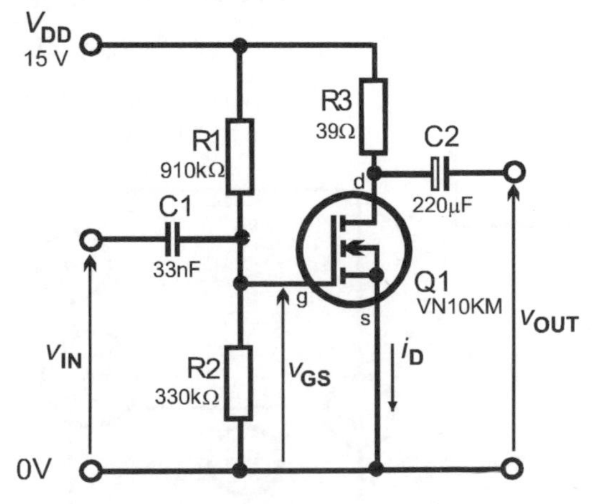

Figure 1.1 *The source terminal is common to both the input and output sides of this MOSFET amplifier, so it is known as a common-source amplifier.*

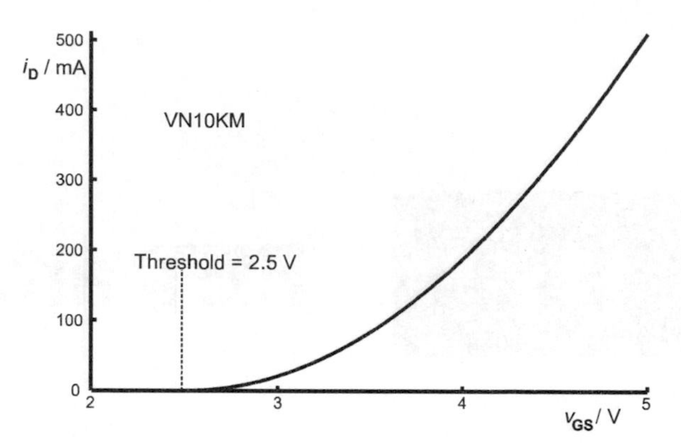

Figure 1.2 *As the gate-source voltage is increased above the threshold, the drain current increases more and more rapidly.*

certain voltage, known as the *threshold* voltage, there is no current through the transistor. As voltage is increased above the threshold, current begins to flow, the current increasing with increasing voltage.

Transconductance

The amount by which i_D increases for a given small increase in v_{GS} is known as the *transconductance* of the transistor. Its symbol is g_m and its unit is the siemens (amperes per volt). We measure g_m by using the circuit of Fig 1.3. The drain voltage V_{DD} is held constant while the gate voltage v_{GS} is varied over a range of values. The meter registers the drain current i_D for each value of v_{GS}. Fig. 1.2 shows typical results.

Test your knowledge 1.1

1 When v_{GS} is increased by 0.2 V the increase of i_D through a given transistor is found to be 25 mA. Calculate g_m.

2 Fig. 1.2 shows that when v_{GS} = 3 V then g_m = 80 mS. What increase in v_{GS} is needed to increase i_D by 15 mA?

Example

From the data of Fig 1.2, find g_m when v_{GS} is 4 V. By measurement on the graph, as v_{GS} rises from 3.75 V to 4.25 V (an increase of 0.5 V), i_D rises by 120 mA.

$$g_m = 120/0.5 \text{ [mA/V]} = 240 \text{ mS}$$

The typical transconductance quoted for the VN10KM in data sheets is 200 mS.

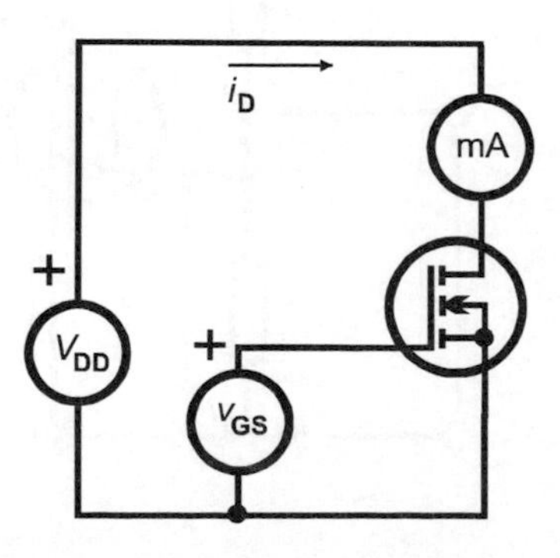

Figure 1.3 *This is the test circuit used to obtain the data for the graph in Fig. 1.2.*

Fig. 1.2 is not a straight-line graph. It slopes more steeply as v_{GS} increases. This means that g_m is not constant. It increases with increasing v_{GS}. The effect of this is to introduce distortion into the signal. The only way to avoid this is to keep the signal amplitude small, so that the transistor is operating over only a small part of the curve in Fig. 1.2. The smaller part of the curve is closer to a straight line and there is less distortion.

Biasing

In Fig 1.1 the gate of the transistor is held at a fixed (quiescent) voltage so that the transistor is passing a current, even when there is no signal. A pair of resistors R1 and R2 acting as a potential divider biases it.

Example

Given that V_{DD} = 15 V, calculate suitable values for R1 and R2. Because MOSFETs require hardly any gate current, we can use high-value resistors in the potential divider. If we decide on 330 kΩ for R2, then:

$$R_1 = \left(\frac{15}{4} \times 330\,\text{k}\right) - 330\,\text{k} = 907.5\text{k}$$

The nearest E24 value is 910 kΩ.

Test your knowledge 1.2

Find a pair of standard resistor values totalling about 1 MΩ for biasing a transistor to +3.8 V when the supply voltage is 18 V.

Two resistors in parallel: Their total resistance is:

$$(R_1 \times R_2)/(R_1 + R_2)$$

We have chosen biasing resistors of fairly high value because we want the amplifier to have high input resistance. This is an advantage because it does not draw large currents from any signal source to which it may be connected.

Example

R_1 and R_2 in parallel provide a DC path to ground. Their parallel resistance is:

$$R_T = (910 \times 330)/(910 + 330) = 242\ \text{k}\Omega$$

Coupling

Capacitor C1 couples the amplifier to a previous circuit or device (such as a microphone). Sometimes it is possible to use a directly wired connection instead of a capacitor. But usually making a direct connection to another circuit will pull the voltage at the gate of Q1 too high or too low for the transistor to operate properly. With a capacitor, the quiescent voltages on either side of the capacitor can differ widely, yet signals can pass freely across the capacitor from the signal source to the amplifier.

Potential divider

The same current *i* flows through both resistors in Fig. 1.4.

Ohm's law applies:

$$i = v_{OUT}/R_2 \quad \text{and} \quad i = v_{IN}\,(R_1 + R_2)$$

Combining these equations:

$$v_{OUT}/R_2 = v_{IN}\,(R_1 + R_2)$$
$$\Rightarrow \quad v_{OUT} = v_{IN} \times R_2/(R_1 + R_2)$$

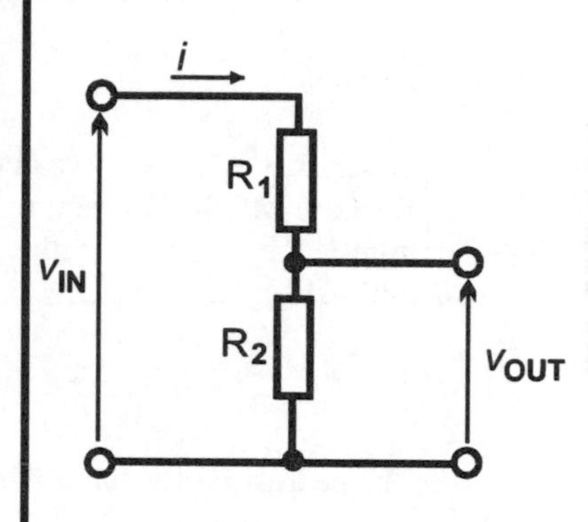

Figure 1.4 *The voltage output of this potential divider circuit is less than the voltage input.*

The calculation assumes that no current flows out of the divider at the junction of R1 and R2.

Test your knowledge 1.3

A pair of biasing resistors have values 470 kΩ and 750 kΩ. What is their total parallel resistance?

C1 and the biasing resistors form a high-pass filter, which might limit the ability of the amplifier to handle low frequencies. We must select a value for C1 so that the filter passes all signals in the frequency range intended for this amplifier.

Example

Suppose we want to pass signals of 20 Hz and above. In other words, the –3 dB point of the amplifier is to be at 20 Hz. The input resistance of R_1 and R_2 in parallel is 242 kΩ. Now calculate the capacitance required:

$$C = \frac{1}{2\pi fR} = \frac{1}{2\pi \times 20 \times 242 \times 10^3} = 32.9 \times 10^{-9}$$

The nearest available value is 33 nF.

Test your knowledge 1.4

If the amplifier was intended to pass signals of frequency 10 kHz and higher, what should be the value of C1?

Sinusoids

A sinusoidal signal is one that, when plotted to show how the voltage or current varies in time, has the shape of a sine curve (Fig. 1.5). Its amplitude and frequency define a sinusoid. Its period is equal to 1/frequency.

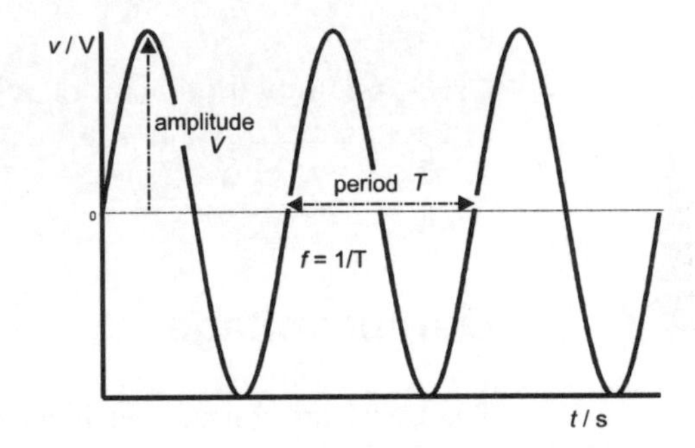

Figure 1.5 *When plotted as a graph, a voltage sinusoid has the same shape as a sine curve.*

Highpass filter

When a resistor and capacitor are connected as in Fig. 1.6, signals of high frequency pass more easily than signals of other frequencies.

The cut-off frequency or half-power frequency of the filter is found from:

$$f_C = 1/(2\pi RC)$$

Signals of this frequency are reduced to half power as they pass through the filter. Signals of higher frequency are reduced by a very small amount or not at all. Signals of lower frequency are reduced to less than half power. This frequency is also known as the *–3 dB point.*

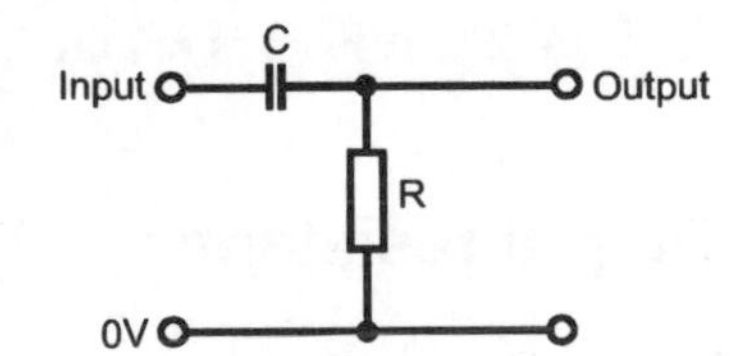

Figure 1.6 *A capacitor and resistor connected as in this network form a highpass filter.*

If the resistor and capacitor are exchanged, the network becomes a lowpass filter.

Activity 1.1 Filters

Set up a highpass filter (Fig. 1.5) on a breadboard, or on a simulator. Calculate f_c. Apply a sinusoidal signal, frequency f_c, and amplitude 1V, from a signal generator. Use an oscilloscope to measure the amplitude of the signal after it has passed through the filter. What is the ratio of the *power* of the output signal to the *power* of the input signal? What is the ratio of their voltage amplitudes?

Repeat for various combinations of resistance and capacitance, giving a range of half-power frequencies below and above f_c. Repeat, but with a lowpass filter.

Output voltage

The transistor converts an input voltage (v_{GS}) into an output current (i_D). This current is converted into an output voltage by passing it through the drain resistor R3. By Ohm's Law:

$$v_{OUT} = i_D R_3$$

Test your knowledge 1.5

If the quiescent i_D is 100 mA and V_{DD} is 12 V, what is the best value for the drain resistor?

Ideally the no-signal (quiescent) output of the amplifier should be halfway between 0V and V_{DD}. This allows the output to swing widely in either direction without distortion. The value of R3 is chosen to obtain this.

Example

Given that the gate is held at 4 V, Fig 1.2 shows that i_D is then equal to 200 mA. We must arrange that this current flowing through R3 causes a voltage drop of approximately $V_{DD}/2$. The voltage drop required is 7.5 V:

$$R_3 = \frac{7.5}{0.2} = 37.5$$

A 39Ω resistor is required.

Output resistance

Current flowing to the output must pass from the positive line and through R3. In other words, the output resistance of the amplifier is equal to the value of the drain resistor, R_3. In this amplifier R3 has a very low resistance. This is an advantage because it means that the amplifier is able to supply a reasonably large current to any circuit or device (such as a speaker) to which it is connected. In other words, it has low output resistance. If the circuit to which it is supplying current takes a large current, there is a relatively low drop of voltage across R3. It can supply a large *current* without a resultant fall in output *voltage*.

There is a highpass filter at the output formed by R3 and C2. Although this is 'the other way up' compared with the highpass filter in Fig 1.5, it still acts as a highpass filter. In this filter the low frequency signals are absorbed into the positive supply line instead of the 0 V line, but the effect is just the same. To comply with the specification of this amplifier, the highpass filter must have its cut-off frequency (–3 dB point) at 20 Hz.

Example

Using the same equation as above:

$$C = \frac{1}{2\pi f_C R} = \frac{1}{2\pi \times 20 \times 39} = 204 \times 10^{-6}$$

A 220 μF electrolytic or tantalum capacitor is required.

Testing the amplifier

When a sinusoidal signal of amplitude 100 mV and frequency 1 kHz is fed to the amplifier, the output is as plotted in Fig 1.7. The output signal is a sinusoid of the same frequency, but with an amplitude of approximately 900 mV. The voltage gain of the amplifier is:

Example

Calculate the expected voltage gain, given g_m = 240 mS and R_3 = 39 Ω.

If v_{IN} increases by an amount v_{in} 100 mV, the corresponding increase in i_D is:

$$i_d = g_m \times v_{in} = 0.240 \times 0.1 = 0.024$$

If the current through R3 increases by 0.024 A, the voltage across R3 increases by:

$$v_{out} = i_d \times R_3 = 0.024 \times 39 = 0.936$$

The calculation shows that when v_{IN} increases, the voltage across R3 *increases* too. Since the positive end of R3 is connected directly to the +15V line, the voltage at the other end of R3 must *fall* by 0.936 V. As input increases, output decreases, and the other way about. This is an *inverting amplifier* with the output 180° out of phase with the

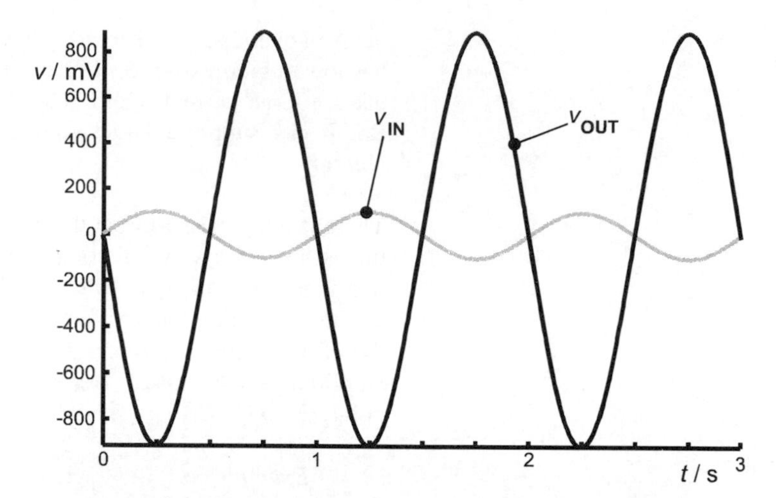

Figure 1.7 *The output (black curve) from the common-source amplifier has the same frequency as the input (grey curve), a greater amplitude, and is 180° out of phase.*

input. The expected voltage gain is –0.936/0.100 = –9.36. This is based on a fairly rough estimate of g_m from measurements taken on Fig. 1.2. The result compares well with the test gain of –9.04 shown in Fig 1.7.

Frequency response

The frequency response of the amplifier is obtained by inputting sinusoidal signals of constant amplitude but different frequencies and measuring the output amplitude. In the test results shown in Fig 1.8, the frequency is varied over the range from 1 Hz to 100 MHz. The graph plots the output amplitude at each frequency given constant input amplitude of 100 mV. The graph, known as a *Bode Plot*, is plotted on logarithmic scales so that a wide range of values can be shown on a single diagram. The scale along the frequency axis is graduated in intervals of 10 *times*, that is 1 Hz, 10 Hz, 100 Hz, 1000 Hz (= 1 kHz), and so on. The scale along the amplitude axis is graduated in decibels. On this scale, 0 dB is the maximum amplitude obtained in the tests, that is 904 mV. The amplifier maintains its output at this amplitude for all frequencies between about 100 Hz and 10 MHz. Amplitude falls off at frequencies below 100 Hz, because of the action of the highpass filters at the input and output. The point marked on the curve at the half-power value (–3 dB) occurs when the frequency is 20 Hz. This is as expected, because we calculated the values of C1 and C2 to have this effect.

The amplitude falls off at very high frequencies, above about 100 MHz, because of the effects of capacitance within the transistor.

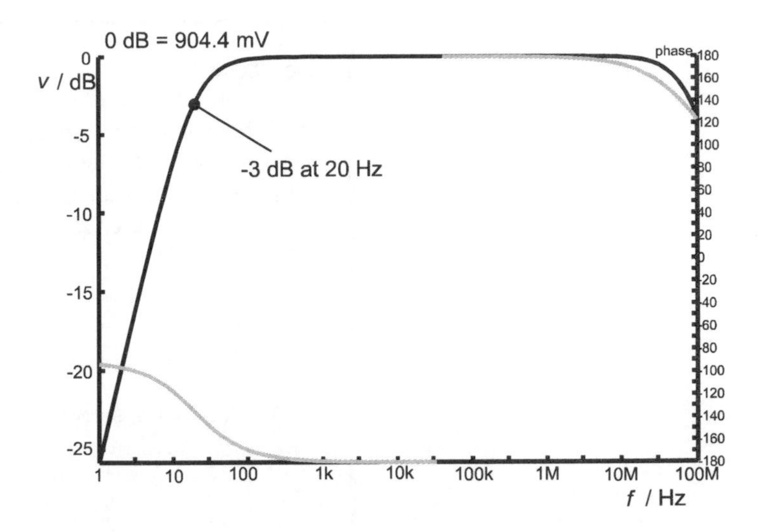

Figure 1.8 *This Bode Plot shows that the output amplitude (black) of the common-source amplifier falls off at low and high frequencies. The plot also shows how phase (grey) varies with frequency.*

Phase

Fig 1.8 also provides information on the variation of phase with frequency. For most of the range, the input and output signals are exactly 180° out of phase, the amplifier acting as an inverting amplifier. Phase differences depart from 180° at the lowest and highest frequencies because of the effects of the capacitances of C1, C2 and the transistor.

Activity 1.2 MOSFET c-s amplifier

Using Fig. 1.1 as a guide, build a single-stage MOSFET common-source amplifier. Connect a signal generator to its input and an oscilloscope to its output. Find the gain of the amplifier at 1 kHz, as explained under the heading 'Testing the Amplifier'. Then try varying the frequency of the input signal in steps over the range 10 Hz to 100 MHz. Keep the input amplitude constant at 100 mV. Measure the output amplitude at each frequency. Plot a graph of output amplitude against frequency (use a logarithmic scale).

Try altering R1 and R2 to see the effect of biasing. For example, make R_2 = 100 kΩ. Then make R_2 = 1MΩ.

Try altering C1 or C2 to see the effect, if any, on the frequency response of the amplifier.

Decibels

In electronics, the decibel is most often used to express the *ratio* between two powers, two voltages or two currents. If, for example, a signal of power P_1 is amplified to power P_2, the amplification in decibels is calculated from:

$$n = 10 \times \log_{10} (P_2/P_1)$$

Example

A 50 mW signal is amplified to 300 mW:

$$\begin{aligned} n &= 10 \times \log_{10} (300/50) \\ &= 10 \times \log_{10} 6 = 7.78 \end{aligned}$$

The power amplification is 7.78 dB.

Power is proportional to voltage squared or to current squared, so we multiply by 20 when calculating n from voltage or currents.

Example

A signal of amplitude 60 mV is amplified to 500 mW:

$$\begin{aligned} n &= 20 \times \log_{10} (500/60) \\ &= 20 \times \log_{10} 8.333 = 18.4 \end{aligned}$$

The voltage amplification is 18.4 dB.

If there is power loss, the decibel value is negative. In particular, a *power loss* of half is:

$$n = 10 \log_{10} 0.5 = -3.01$$

This –3 dB point is often used to define the 'cut-off point' in amplifiers and filters. The lower and upper –3 dB point are also used to define the limits of their bandwidth.

Similarly a *voltage* or *current* reduction of one half is equivalent to –6 dB.

Remember:

Use the factor 10 when dealing with *power*, and the factor 20 when dealing with *voltage* or *current*.

Test your knowledge 1.6

1 An amplifier increases the power of a signal from 5 mW to 37.5 mW. Express this in decibels.

2 In the amplifier of Fig. 1.1 a 2 kHz signal of 100 mV amplitude is amplified to 904.4 mV. Express this in decibels. What is the output amplitude of a 20 Hz 100 mV signal?

Common-drain amplifier

The MOSFET amplifier of Fig. 1.9 is known as a common-drain amplifier because the drain terminal is common to both the input and output circuits. There is a resistor between the source and the 0 V line and the output is taken from the source terminal. Biasing and coupling arrangements are the same as in Fig. 1.1 except that R_3 is bigger, for a reason that will be explained later. This means that the value of C2 needs to be re-calculated if we still want to provide a highpass filter with cut-off point at 20 Hz.

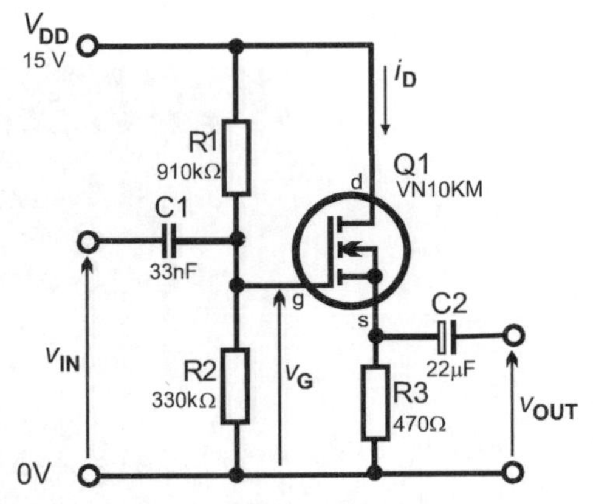

Figure 1.9 *A common-drain MOSFET amplifier has unity voltage gain but high current gain. It is also known as a voltage follower.*

Voltage gain

When the output voltage is v_{OUT}, the drain current is given by:

$$i_D = v_{OUT} / R_3$$

From the definition of transconductance we know that:

$$i_D = g_m v_{GS}$$

It can be seen in Fig 1.9 that the voltage at the gate of Q1 equals v_{IN} and the voltage at the source equals v_{OUT}. Therefore the gate-source voltage, v_{GS}, is equal to their difference, and:

$$i_D = g_m (v_{IN} - v_{OUT})$$

$$\Rightarrow \quad v_{OUT} / R_3 = g_m (v_{IN} - v_{OUT})$$

$$\Rightarrow \quad \text{voltage gain} = \frac{v_{OUT}}{v_{IN}} = \frac{R_3 g_m}{1 + R_3 g_m}$$

It is clear from this equation that the voltage gain is less than 1. If R_3 is appreciably greater than $1/g_m$, the '+1' may be ignored and the equation approximates to:

$$\text{Voltage gain} = 1$$

Because this amplifier has unity voltage gain, it is called a *voltage follower*. Its usefulness depends on its high input resistance (which may be several hundred kilohms) and its low output resistance (which may be only a few hundred ohms). It is used as a buffer when we need to connect a circuit or device that has high output resistance to one that has low input resistance. For example, the voltage signal from a microphone (high output resistance) may be wholly or partly lost if the microphone is coupled to an audio amplifier with relatively low input resistance.

Activity 1.3 MOSFET transconductance

Set up the circuit of Fig. 1.3 to measure the transconductance of the MOSFET you used in Activity 1.2. Use a power pack set at 15 V for V_{DD} and a variable power pack for v_{GS}. Fig. 1.2 shows a suitable range of input values.

Plot a graph similar to Fig 1.2 and from this calculate two or three values for g_m at different values of v_{GS}.

Activity 1.4 MOSFET c-d amplifier

Using Fig. 1.9 as a guide, build a single-stage MOSFET common-drain amplifier. Use the component values quoted in the figure. Connect a signal generator to its input and an oscilloscope to its output. Find the gain of the amplifier at 1 kHz. Then try varying the frequency of the input signal in steps over the range 10 Hz to 100 MHz. Keep the input amplitude constant at 100 mV. Measure the output amplitude at each frequency.

Plot a graph of output amplitude against frequency. It is best to plot the frequency on a logarithmic scale, as in Fig. 1.8. There is no need to convert the output amplitudes to the decibel scale.

Try increasing the amplitude of the input signal and observe the output signal to find out what is the largest input amplitude that this circuit can follow without distortion.

Try altering R1 and R2 to see the effect of biasing. For example, make R_2 = 100 kΩ. Then make R_2 = 1MΩ.

Try altering C1 or C2 to see the effect, if any, on the frequency response of the amplifier.

Matching output to input

In Fig. 1.10(a), v_S represents the piezo-electric or magnetic transducer in the microphone which converts sound energy into electrical energy. R_{OUT} represents the electrical resistance of the microphone unit. The microphone is connected to an audio amplifier that has input resistance R_L. If R_L is less than R_{OUT}, the voltage drop across R_L is less than that across R_{OUT}. In other words, most of the signal voltage is 'lost' in R_{OUT} and only a fraction of it appears across R_L, ready to be amplified. Only a reduced signal can eventually appear from the amplifier.

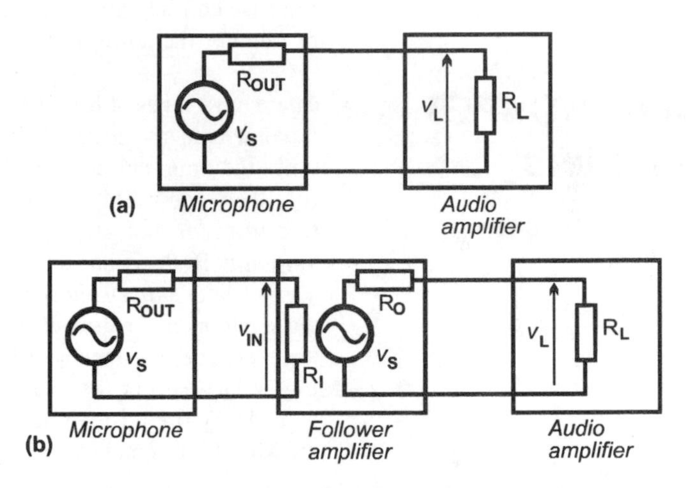

Figure 1.10 *In (a) the microphone is not able to supply enough current to drive the audio amplifier. In (b) we use a voltage follower to match the output of the microphone to the input of the audio amplifier.*

In Fig. 1.10(b), a follower amplifier (like Fig. 1.9) is connected as a buffer between the microphone and the audio amplifier. The follower amplifier has high input resistance R_I and low output resistance R_O. Because R_I is much greater than R_{OUT}, most of v_S appears across R_I and only a small part of it is 'lost' across R_{OUT}. We now have transferred most of the original signal to the follower amplifier. Here its voltage is not amplified, so we again represent the signal by v_S. But now the resistance through which the signal has to pass is R_O, which is very much *lower* than R_L. This being so, only a small portion of v_S is 'lost' across R_O and most of it appears across R_L. Most of the signal from the microphone has now been transferred through the follower amplifier to the audio amplifier, where it can then be amplified further and perhaps fed to a speaker.

This technique for using a follower amplifier to transfer the maximum signal between one circuit and another is known as *impedance matching*.

Current and power gain

Although the voltage gain of a follower amplifier is only about 1, it has a high current gain. The input current to the insulated gate is only a few picoamperes, while the output current is measured in milliamperes or even amperes. So current gain of voltage amplifiers (and other FET amplifiers) is very high. It is not possible to put an exact figure on the gain because the current entering the gate is a leakage current, and may vary widely from transistor to transistor.

Because current gain is high and power depends on current multiplied by voltage, the power gain of MOSFET amplifiers is very high.

Other MOSFET amplifiers

There are 4 types of MOSFET. In the *n-channel* types the drain current is conducted as electrons through n-type material. In the *p-channel* types the drain current is conducted as holes through p-type material. In either type of MOSFET there is only one kind of charge carrier (electrons *or* holes), which gives these transistors the description *unipolar*. Both n-channel and p-channel MOSFETs are made in two sorts. The *enhancement* sort has no ready-made conduction channel in the semiconductor. It is formed when we apply a charge to the gate. In the n-channel enhancement MOSFET (the sort most often used, as in Fig. 1.1) we form the channel by applying a positive voltage to the gate. The more positive the gate, the wider the channel and the better the MOSFET conducts.

In the p-channel enhancement MOSFET, we apply a negative voltage to the gate to form the channel. The more negative the gate, the wider the channel and the bigger the drain current. Amplifiers based on p-channel enhancement MOSFETs have circuits similar to Figs. 1.1 and 1.9, except that the polarities are reversed. The drain is held negative of the source, and the gate is made negative to turn the transistor on.

Depletion MOSFETs are manufactured with a conductive channel (n-type only, since p-type are not made) already in existence. This is *reduced* in width as the gate is made more negative, so *reducing* the drain current.

JFET amplifier

JFETs are another type of unipolar transistor. They are made in n-channel and p-channel versions, both of which are depletion devices. In the remainder of this chapter we describe an amplifier based on an n-channel JFET.

Fig. 1.11 is a common-source JFET amplifier similar to that of Fig. 1.1 and operating in much the same way, except for the biasing of the gate.

In JFETs the gate is insulated from the substrate of the transistor by being reverse-biased. It is held negative of the source, causing a depletion region to form between the gate and the substrate, in the same way that it is formed in a reverse-biased diode. The usual way of biasing the gate is to connect a single high-value resistor (R1) between the gate and the 0V line. This holds the gate very close to 0V. We then make the source a few volts *positive* of the gate by wiring a resistor (R3) between the source and the 0V line. Relative to the source, the gate is negative.

The threshold voltage of the 2N3819 is –4 V, so a suitable quiescent gate voltage is in the region of –2 V. The gate can be biased at any voltage between its threshold and about +0.5 V. Above +0.5 V the junction at the gate becomes forward-biased, the depletion region breaks down and the gate is no longer insulated from the body of the transistor.

Example

In Fig. 1.11 the quiescent current through the transistor is 0.5 mA. The voltage drop across R3 is:

$$v_{GS} = -i_D R_3 = -0.0005 \times 3900 = -1.95 \text{ V}$$

In practice, the voltage across R3 varies with the signal, making v_{GS} vary too. As the signal increases, v_{GS} increases, which tends to partly turn off the transistor and thus to reduce the signal. This is *negative feedback*, which reduces the gain of the amplifier. A large-value capacitor C3 is used to stabilise the voltage at the source, so reducing the feedback and increasing the gain. Looking at this another way, the

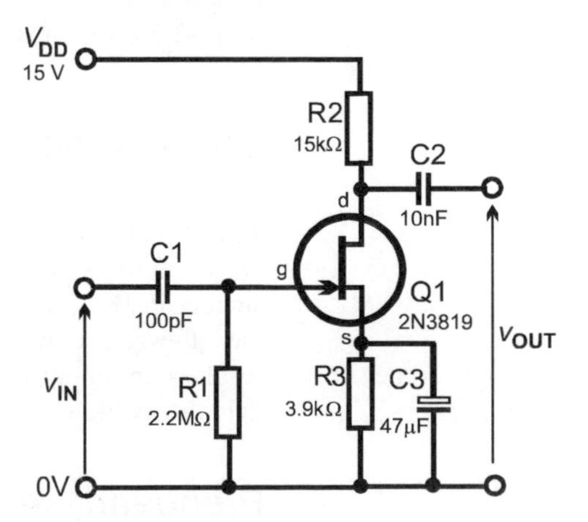

Figure 1.11 *A JFET common-source amplifier is biased by a single pull-down resistor.*

capacitor passes the signal through from the source to the 0V line, leaving only the quiescent voltage at the source. For this reason, the capacitor is called a *by-pass* capacitor.

Transconductance

The transconductance of JFETs is generally much lower than that of MOSFETs. For the 2N3819, g_m = 0.4 mS (compared with 240 mS for the MOSFET). We calculate the expected voltage gain as for the MOSFET common-source amplifier.

Example

If v_{IN} increases by an amount v_{in} = 100 mV, the corresponding increase in i_D is:

$$i_d = g_m \times v_{in} = 0.0004 \times 0.1 = 40\ \mu A$$

If the current through R2 increases by 40 μA, the voltage across R2 increases by:

$$v_{out} = i_d \times R_2 = 0.0004 \times 15\,000 = 0.6$$

The voltage at the drain *falls* by 0.6 V.

Voltage gain = –0.6/0.1 = –6

This calculation assumes that there is no change in g_m due to variations in v_{GS}. As explained above, the gain is reduced by negative feedback if we do not have a bypass capacitor across R3. Tests confirm that the gain is –6 when the bypass capacitor is present, but we find that gain is only –2.5 if the capacitor is absent.

Output resistance

The output resistance of a common-source amplifier depends on the value of the drain resistor, R2. Drain currents are, on the whole, smaller in JFET amplifiers. So we need a larger drain resistor to bring the quiescent output voltage to about half the supply voltage. The result is an output resistance that is reasonably low, but not as low as that obtainable with a MOSFET amplifier.

Frequency response

The frequency response of the JFET amplifier is similar to that of the MOSFET amplifier, except that amplitude begins to fall off above

Bandwidth: The range of frequencies over which the output of an amplifier is within 3dB of its maximum.

about 500 kHz instead of extending as far as 100 MHz. In other words, the JFET amplifier does not have such a high *bandwidth* as the MOSFET amplifier. This is because of the greater input capacitance of JFETs when compared with MOSFETs. The input capacitance of the 2N3819, for example, is 8 nF, compared with only 48 pF for the VN10KM.

Summing up, the JFET common-source amplifier shares with the MOSFET version the advantage of a very high input resistance. Its voltage gain is not as high, and its output falls off sooner at high frequencies. In a switching circuit, the very low on resistance of MOSFETs is usually a big advantage.

Activity 1.5 JFET c-s amplifier

Using Fig. 1.11 as a guide, build a single-stage JFET common-source amplifier. Use the component values quoted in the figure. Try varying the frequency of the input signal in steps over the range 10 Hz to 100 MHz. Keep the input amplitude constant at 100 mV. Measure the output amplitude at each frequency.

Plot a graph of output amplitude against frequency.

Try reducing the bypass capacitor C3 to 10 μF and then to 1 μF and observe the effect on signal gain. Finally remove and test again.

Activity 1.6 JFET transconductance

Set up the circuit of Fig. 1.3 to measure the transconductance of the JFET you used in Activity 1.5. Use a power pack set at 15 V for V_{DD} and a variable power pack for v_{GS}. Connect the variable power pack with its *negative* terminal to the gate of the transistor and its *positive* terminal to the 0 V rail (the rail to which the source is connected). Vary v_{GS} over the range −10V to 0 V.

Plot a graph similar to Fig 1.2 and from this calculate two or three values for g_m at different negative values of v_{GS}.

Compare these values with those obtained for the MOSFET in Activity 1.3.

Problems on FET amplifiers

1 Explain what is meant by transconductance. How is the output current of a MOSFET converted to an output voltage in a common-source amplifier?

2 Why is it important to keep signal amplitudes small in FET amplifiers?

3 Why is it usually preferable to use capacitor coupling at the input and output of an amplifier, rather than use wired connections?

4 Explain how a common-drain amplifier is used to match a signal source with high output resistance to an amplifier with low input resistance.

5 What is the advantage of the high input resistance of FET amplifiers? Give two practical examples of this.

6 Why do we usually try to make the quiescent output voltage of an amplifier equal to half the value of the supply voltage?

7 Explain what the –3 dB point means.

8 Outline the practical steps in testing the frequency response of an amplifier. Why are the results usually plotted on logarithmic scales?

9 Explain why the amplitude of the output of an amplifier usually falls off at low and high frequencies.

10 What is the difference in the working of an enhancement FET and a depletion FET?

11 What are the main similarities and differences in the performance of MOSFET and JFET amplifiers?

12 Explain the action of a by-pass capacitor.

Multiple choice questions

1 A MOSFET is described as a unipolar transistor because:

A current flows through it in only one direction.
B it may explode when connected the wrong way round.
C it has only one type of charge carrier.
D the channel is made of n-type material.

2 A follower amplifier is often used as a buffer between:

A a high resistance output and a low resistance input.
B a low resistance output and a high resistance input.
C a high resistance output and a high resistance input.
D a low resistance output and a low resistance input.

3 A follower amplifier has a voltage gain of:

A approximately 100.
B exactly 1.
C slightly less than 1.
D slightly more than 1.

4 The total parallel resistance of resistors of 56 kΩ and 120 kΩ is:

A 38.2 kΩ.
C 64 kΩ.
B 176 kΩ.
D 6720 kΩ.

5 The advantage of using a capacitor to couple an amplifier to a signal source is that:

A it stabilises the gain.
B the amplifier is not affected by the quiescent output voltage of the source.
C the capacitor acts as a highpass filter.
D the capacitor filters out noise from the signal.

6 The operating gate voltage of an n-channel JFET is:

A always positive.
B always negative.
C lower than –5 V
D less than 0.5 V.

2 BJT amplifiers

Summary

There are three basic ways in which a BJT (bipolar junction transistor) may be used as an amplifier: common-emitter amplifier, common-collector amplifier and common-base amplifier. These each have very different attributes and applications. We look at the circuits of these three amplifiers and consider their input and output characteristics and their frequency responses. We compare BJT amplifiers with their FET equivalents and note the advantages and disadvantages of each kind.

Common-emitter amplifier

The BJT amplifies current. In the simple common-emitter amplifier of Fig. 2.1, based on an npn BJT, a small current i_B flows through R1 into the base of Q1. Because of transistor action, this causes a much larger current i_C to flow in at the collector of Q1. The currents combine and flow out of Q1 at the emitter.

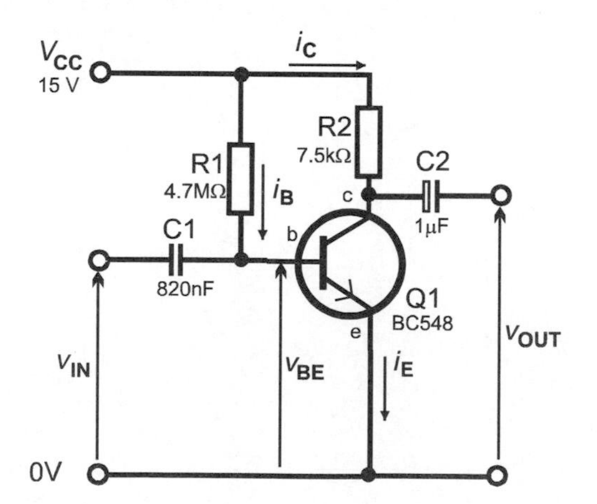

Figure 2.1 *In a common-emitter BJT amplifier the emitter terminal is common to both the input and output circuits. Compare with Fig. 1.1.*

The emitter current i_E is:

$$i_E = i_B + i_C$$

Current gain

Fig. 2.2 shows the relationship between i_B and i_C. It is known as the *forward transfer characteristic*. This graph is plotted by using a test circuit like Fig. 2.3. Compare this with the circuit in Fig. 1.3, as used for testing FETs. This one has two current meters, because both the input and output of a BJT are currents. Note that the meter for measuring i_B is a microammeter, while the meter for i_C is a milliammeter. This is because i_C is always much larger than i_B.

Test your knowledge 2.1

What is the h_{fe} of a BJT if i_C increases by 2.06 mA when i_B is increased by 5 µA?

Example

We can measure how much larger by plotting a graph of i_C against i_B, as in Fig. 2.2. This shows that, if i_B is increased by 10 µA, i_C increases by 3.84 mA. The ratio between the two (or the

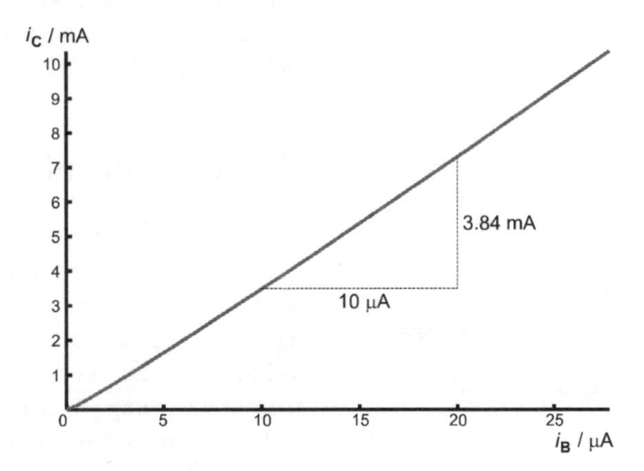

Figure 2.2. *The graph of output current against input current of a BJT is almost a straight line. Compare this with Fig. 1.2, where the gradient of the curve increases with increasing input voltage.*

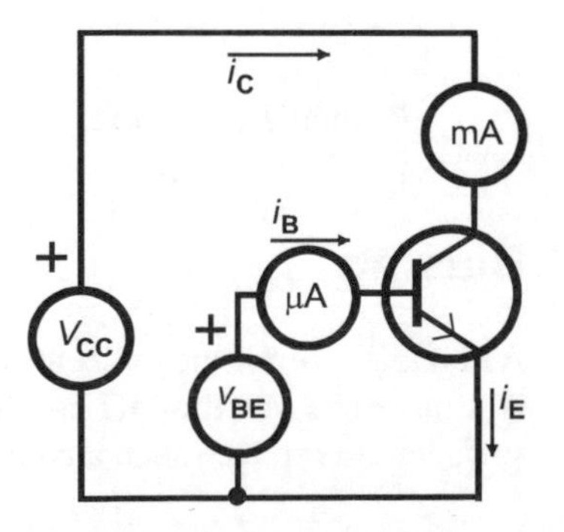

Figure 2.3 *This is the kind of circuit used for measuring the forward transfer characteristic of a BJT.*

Activity 2.1 Forward transfer characteristic

Set up the circuit of Fig. 2.3. V_{CC} is a fixed voltage source of 15 V DC, v_{BE} is a variable voltage source. Adjust the v_{BE} to produce a range of base currents, from zero to 25 μA in steps of 5 μA.

Measure the corresponding values of i_C.

Plot i_C against i_B, as in Fig. 2.2 and calculate the small signal current gain h_{fe}.

slope of the curve in Fig. 2.2) is known as the *small signal current gain*, symbol h_{fe}. In this example:

$$h_{fe} = 3.84/10\ [\text{mA}/\mu\text{A}] = 384.$$

The gain for the BC548 is often listed as 400 (but see below). Note that the graph is almost a perfect straight line. We say that the relationship between i_B and i_C is *linear*.

Voltage output

Quiescent: when no signal is present.

CE: common-emitter.

In the amplifier of Fig. 2.1, the function of R2 is to convert the current i_C into a voltage, according to Ohm's Law. The values given in Fig. 2.1 have been calculated to give a quiescent collector current i_C of 1 mA. This is a typical value for a CE amplifier. R2 is chosen so that the voltage drop across it is equal to half the supply voltage. This allows the output voltage to rise and fall freely on either side of this half-way voltage without distortion.

Example

Current = 1 mA, required voltage drop = 7.5 V, so:

$$R_1 = 7.5/0.001 = 7500$$

R_1 should be 7.5 kΩ.

Biasing

A suitable size for the collector current in a common-emitter amplifier is 1 mA. The value of R1 is often chosen to provide a base current sufficient to produce such a collector current.

Example

Current gain = 380, and i_C = 1 mA, so:

$$i_B = 1/380\ [\text{mA}] = 2.63\ \mu\text{A}$$

v_{BE} is usually between 0.6 V and 0.7 V for a silicon transistor.

Assuming that the base-emitter voltage drop v_{BE} is 0.6 V, the voltage drop across R1 is 15 – 0.6 = 14.4 V. The resistance of R1 is:

$$R_1 = 14.4/2.63\ [\text{V}/\mu\text{A}] = 5.48\ \text{M}\Omega$$

We could use a 5.6 MΩ resistor. But to allow for the gain of the transistor to be a little less than 380, make i_B a little larger by making R1 a little less. We decide on 4.7 MΩ.

This illustrates one of the problems with such a simple circuit. The correct value of R1 depends on the gain of the transistor. But transistors of the same type vary widely in their gain. Any given BC548 may have h_{fe} between 110 and 800. When building this circuit it is almost essential to match R1 to the particular transistor you are using.

Input resistance

From Fig. 2.1 it looks as if this circuit has a very high input resistance, the resistance of R1. But, unlike the gate of an FET, the base of a BJT is *not* insulated from the body of the transistor. Current i_B flows into the base, through the emitter layer and out by the emitter terminal. Along this path it encounters resistance, typically about 25 Ω. This is the emitter resistance r_e, which can be thought of as a resistor inside the transistor (Fig. 2.4). In the way it affects current entering the base, this resistance *appears* to be h_{fe} times its actual value.

Test your knowledge 2.2

If the transistor in the CE amplifier of Fig. 2.1 has h_{fe} = 120, what is the input resistance?

Example

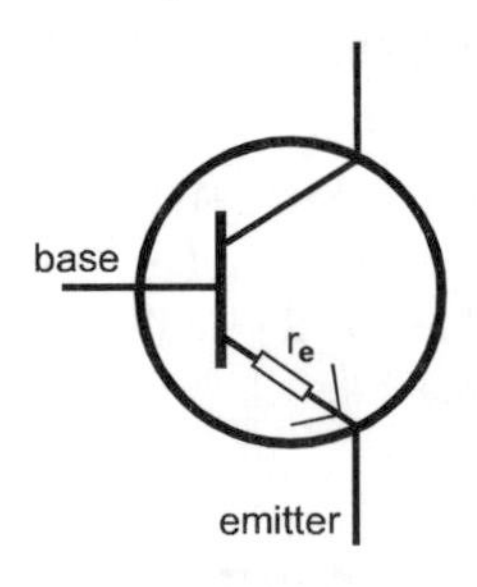

Figure 2.4 *When current flows from the base of a BJT to the emitter it is subject to the emitter resistance.*

If h_{fe} is 400, r_e appears to be 400 × 25 = 10 kΩ. In comparison with this, the value of R1 can be ignored. The amplifier has a low input resistance of only 10 kΩ.

Voltage gain

In this discussion, v_{in} means 'a small change in v_{IN}', and the same with other symbols.

In Fig. 2.5, the signal passes more-or-less unchanged across C1:

$$v_{IN} = v_b$$

where v_B is the base voltage. This stays a constant 0.6 V higher than v_E, the emitter voltage. We can write:

$$v_{in} = v_b = v_e$$

Applying Ohm's Law, we also know that:

$$i_e = v_e / R_4 \qquad \text{and} \qquad v_{out} = -i_c R_3$$

The negative sign indicates that we are dealing with a voltage *drop* across R_3.

From these equations we find that:

$$v_{out} = -i_c R_3 = -i_e R_3 = -v_e R_3/R_4 = -v_{in} R_3 / R_4$$

From this we obtain:

$$\text{voltage gain} = v_{out}/v_{in} = -R_3 / R_4$$

The result shows that the voltage gain depends only on the values of the collector and emitter resistors.

Frequency response

Test your knowledge 2.3

Confirm that the value of C1 in Fig. 2.1 produces a cut-off point close to 20 Hz.

As with the MOSFET amplifiers, C1 and R1 form a highpass filter. The value of C1 in Fig. 2.1 is chosen to pass all signals above 20 Hz. The values of R2 and C2 are chosen to pass the same frequencies on the output side.

Voltage gain

We will work through a rough calculation to find the voltage gain of the amplifier.

Example

If the input voltage v_{IN} is increased by a small amount, say 10 mV, this increases the base current i_B. If the input resistance is 10 kΩ, the increased current is:

$$i_b = 10/10\ [\text{mV/k}\Omega] = 1\ \mu\text{A}$$

Given that h_{fe} is 400, the increased current through R3 is:

$$i_c = 400 \times i_b = 400\ \mu\text{A}$$

This produces an increased voltage drop across R3:

$$v_{out} = -400 \times 7.5\ [\mu\text{A} \times \text{k}\Omega] = -3\ \text{V}$$

The voltage gain of this amplifier is:

$$A_v = -3/10\ [\text{V/mV}] = -300$$

There are several approximations in this calculation but the result agrees well with a practical test, which showed a voltage gain of –250.

Improving stability

The main disadvantage of the amplifier of Fig. 2.1 is its lack of stability. Its performance depends on the value of h_{fe}, which varies from transistor to transistor. Also, h_{fe} is affected by temperature. The amplifier may be operating in a particularly hot or cold environment. Even in a room at a comfortable temperature, the temperature of the amplifier increases after it has been run for a while.

The circuit is improved if we connect the positive end of R1 to the collector of Q1. The positive end of R1 is now held at *half* the supply voltage, so we halve its resistance to obtain the same i_B as before. This does not affect the performance of the amplifier but has an effect on its stability.

To see how this works, suppose that for some reason h_{fe} is higher than was allowed for in the design. Perhaps the transistor has a higher than average h_{fe} or perhaps h_{fe} has been raised by temperature. If h_{fe} is higher than usual, i_C is *larger* than usual. If i_C is larger, the voltage drop across R2 is greater. This lowers the voltage at the collector and so lowers the voltage at the positive end of R1. This reduces i_B, which in turn *reduces* i_C. In summary, a change which increases i_C is countered by a reduction in i_C. The result is that i_C tends to remain unaffected by variations in h_{fe}. The gain of the amplifier is stable.

Further improvement in stability is obtained by biasing the base with two resistors, as in Fig. 2.5. These act as a potential divider to provide a fixed quiescent voltage, independent of h_{fe}. At the same time, we introduce a additional improvement, the *emitter resistor* R4. This is not to be confused with the emitter *resistance*, shown in Fig. 2.4.

Test your knowledge 2.4

Confirm that the values of R_1 and R_2 in Fig. 2.5 produce the required base voltage.

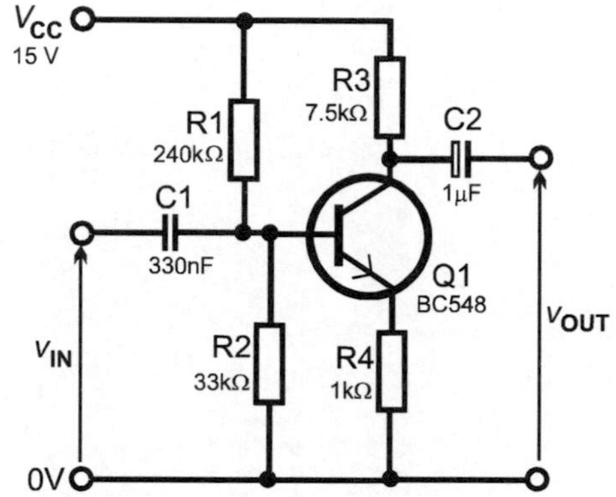

Figure 2.5 *Biasing the base with a voltage divider helps stabilise the common-emitter amplifier.*

The emitter resistance and emitter resistor are in series but the emitter resistance is usually about 25 Ω and may be ignored when the emitter resistor has a value of 1 kΩ or more. If i_C is 1 mA, a convenient value for R4 is 1 kΩ, which produces a voltage drop of 1 V in the quiescent state. Now that the emitter voltage is 1 V, the base voltage must be 0.6 V more than this, to maintain the required v_{BE} of 0.6 V. The values of R1 and R2 are calculated so as to produce 1.6 V at the base.

Input resistance

The input resistance is the total of three resistances in parallel: R_1, R_2, and R_4. As with emitter resistance, a current flowing into the base is affected by h_{fe} times the resistance between the base and the 0V line. Now that there is a 1 kΩ resistor there, we can ignore r_e. The resistance seen from the base is h_{fe} times 1 kΩ, that is, 400 kΩ. The total of these three resistances in parallel is 27 kΩ. This gives the amplifier a moderate input resistance, but one which is far less than that of FET amplifiers.

Test your knowledge 2.5

For the circuit of Fig. 2.5, find a value for C1 which will filter out frequencies below 1 MHz.

Given the new value of the input resistance, we can calculate a new value for C1.

Voltage gain

As explained in the box on p. 24, the voltage gain of this amplifier depends only on the values of the collector and emitter resistors:

$$\text{voltage gain} = -R_3 \,/\, R_4$$

Voltage gain does not depend on h_{fe}, so *any* transistor, either of the same type or of a different type, will produce the same voltage gain. Voltage gain is also virtually independent of temperature. The amplifier has high stability at the expense of low gain.

Test your knowledge 2.6

Use the voltage gain equation to find the gain if we omit R4.

Example

In Fig. 2.5 the expected voltage gain is:

$$-7.5/1\ [\text{k}\Omega/\text{k}\Omega] = -7.5$$

A test on this circuit showed an actual voltage gain of –7.25.

Frequency response

With the capacitor values given in Fig. 2.5, the frequency response of the circuit shows the lower cut-off point to be at 20 Hz, as calculated. In the higher frequencies, the cut-off point is approximately 11.7 MHz. This bandwidth of 11.7 MHz is satisfactory for many applications, but is not as high as the 100 MHz upper cut-off point found with the MOSFET circuit of Fig. 1.1. An upper cut-off point of 11.7 MHz may be essential for an amplifier in a radio receiver, but it is unnecessary for an audio amplifier. The human ear is not sensitive to frequencies higher than about 20 kHz. In Chapter 17, we show that there are disdvantages in amplifying frequencies outside the required range.

Bypass capacitor

Gain can be restored by wiring a high-value capacitor across R4, as in Fig. 2.6. This holds the emitter voltage more or less constant, so the calculations in the box on p. 24 do not apply. Without the capacitor, the voltage at the emitter rises and falls with the signal. This provides negative feedback. For example as i_B rises (tending to increase v_{BE}), i_C rises, and the emitter voltage rises. This tends to decrease v_{BE}, which decreases i_C and counters the rise in emitter voltage. Another way of looking at this is to say that the capacitor shunts the signal at the emitter through to the ground. For this reason C3 is called a *bypass capacitor*. With this capacitor in place, the voltage gain of the amplifier is about 280. The lower cut-off point is raised to 130 Hz, so bandwidth is slightly reduced.

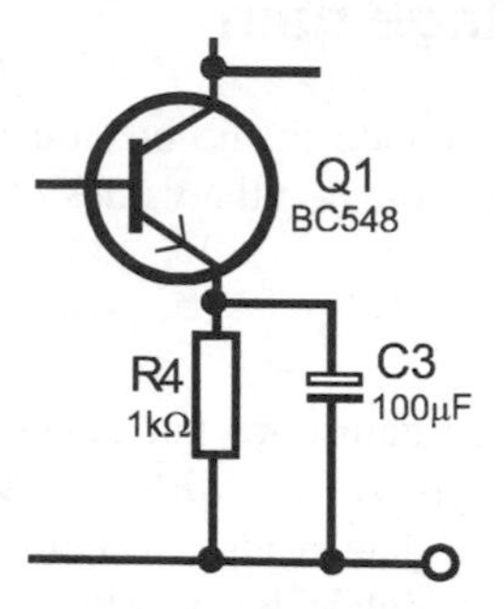

Figure 2.6 *The addition of a bypass capacitor holds the emitter voltage steady and maintains voltage gain.*

Common-emitter amplifiers have current flowing through the transistor even when no signal is being amplified. This type of amplifier is known as a Class A amplifier. Power is being wasted when there is no signal which makes Class A amplifiers unsuitable for high-power amplification. We discuss power amplifiers in Chapter 8.

Activity 2.2 BJT c-e amplifier

Using Fig. 2.5 as a guide, build a single-stage BJT common-emitter amplifier. Connect a signal generator to its input and an oscilloscope to its output. Using a 100 mV sinusoidal signal, find the voltage gain of the amplifier at 1 kHz.

Vary the values of R3 and R4 to confirm that these determine the voltage gain of the amplifier. Try this with several different transistors of known h_{fe} (see Activity 2.1) to show that voltage gain is independent of h_{fe}.

Investigate the effect of the bypass capacitor.

Try varying the frequency of the input signal in steps over the range 10 Hz to 100 MHz. Keep the input amplitude constant at 100 mV. Measure the output amplitude at each frequency. Plot a graph of output amplitude against frequency. It is best to plot the frequency on a logarithmic scale, as in Fig. 1.8.

Activity 2.3 BJT transconductance

Using the circuit of Fig. 2.3, adjust v_{BE} until i_c = 1 mA. Calculate g_m. Make further measurements to show that g_m depends on i_c, but is not dependent on the type of BJT. Compare value obtained with those found in Activity 1.6.

Transconductance

For a BJT, this is defined as:

$$g_m = i_C / v_{BE}$$

It can be shown that, if the collector current is 1 mA, then:

$$g_m \approx 40\ i_C$$

when i_C is rated in milliamperes. Transconductance does not depend on the type of transistor or on h_{fe}. It depends only on the collector current.

A result of this is that, as i_C increases and decreases with the signal, there are corresponding changes in g_m. Amplification varies with i_C, causing distortion.

Common-collector amplifier

The common-collector amplifier shown in Fig. 2.7 has an emitter resistor but the collector is connected directly to the positive rail. The base is biased by two resistors. Assuming that the quiescent emitter current is 1 mA, the voltage across the emitter resistor R3 is 7.5 V, bringing the output to exactly half-way between the supply rails. To provide for a v_{BE} of 0.6 V, we need to hold the base at 8.1 V. The values of R1 and R2 are calculated so as to provide this.

Voltage gain

We calculate the voltage gain of the amplifier by considering what happens when the input voltage v_{IN} changes by a small amount v_{in}. A change in v_{IN} is carried across C1 to the base of Q1.

The change in base voltage is given by:

$$v_b = v_{in}$$

As the base voltage changes, the drop (v_{BE}) across the base-emitter junction remains constant. The emitter voltage v_E changes by the same amount. This change is transferred across C2 to the output terminal. For all the small changes in voltage we can say that:

$$v_{out} = v_e = v_b = v_{in}$$

$$\Rightarrow \quad \text{voltage gain} = v_{out}/v_{in} = 1$$

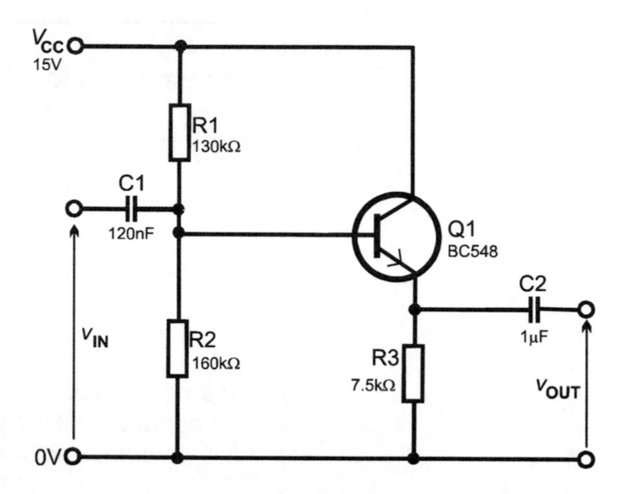

Figure 2.7 *The voltage output of a common-collector amplifier is always approximately 0.6 V lower than its input.*

The amplifier is a non-inverting amplifier with unity voltage gain. The output exactly follows all *changes* in the input, but is 0.6 V lower because of the drop across the base-emitter junction. If the signal is large enough, this drop can be ignored. Because the output follows the input, we refer to this kind of amplifier as a *voltage follower*. It is also known as an *emitter follower*.

The input resistance is high, being equal to the total resistance R_1, R_2, and $h_{fe} \times R_3$. With the values shown, the input resistance is 70 kΩ. The output resistance is the value of R3, which in this case is 7.5 kΩ. The output resistance in only about one tenth of the input resistance, making the amplifier suitable as a buffer (see Fig. 1.11).

Test your knowledge 2.5

Which MOSFET amplifier is the equivalent of this common-collector amplifier?

Current and power gain

Because input current is small (due to the high input resistance) and output current is moderately large, this amplifier has high current gain and high power gain.

Frequency response

The frequency response is very good. In the case of the amplifier of Fig. 2.7, the value of C1 is chosen to produce a highpass filter with lower cut-off point at 20Hz. The upper cut-off point is at 128 GHz. The reason for this high cut-off point is that a transistor with its collector connected directly to the positive rail is not so subject to the effects of capacitance as one which is connected through a resistor.

Common-base amplifier

In the amplifier in Fig. 2.8, the signal is applied to the emitter and the output is taken from the collector. The base is held at a fixed voltage by the voltage divider consisting of R1 and R2. A large-value bypass capacitor C3 is used to hold the base voltage steady.

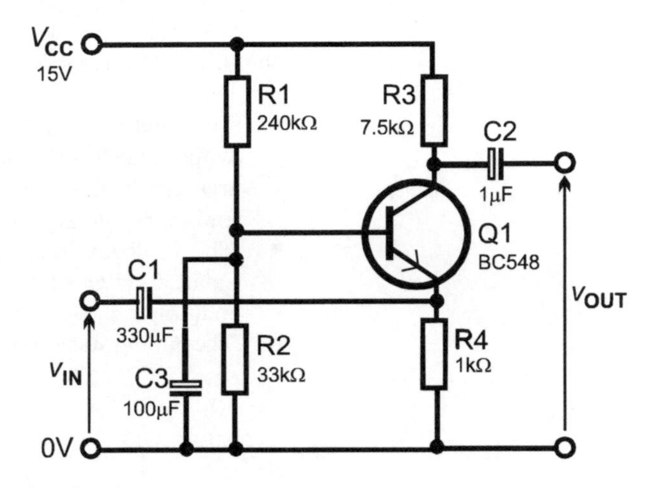

Figure 2.8 *The common-base amplifier has low input resistance, which is generally a disadvantage, but it has good high-frequency response.*

The signal passes through C1 to the emitter terminal. When v_{IN} increases, the emitter voltage increases, so reducing v_{BE}. This reduces i_B which in turn causes i_C to decrease. The voltage drop across R3 decreases, raising the output voltage. This is a non-inverting amplifier with a voltage gain about the same as that of a common-emitter amplifier. Its current gain is less than 1.

Its main disadvantage for most applications is that it has very low input resistance and an output resistance usually of a few kilohms. Its response is good at high frequencies because the effects of capacitance in the transistor are much reduced in this circuit. This makes it suitable as a VHF and UHF radio-frequency amplifier. Another example of its use is the Hartley oscillator illustrated in Fig. 5.1.

Comparing BJTs with FETs

The more important differences between FET and BJT amplifiers are:

- FET amplifiers have much higher input resistance.
- FETs have lower transconductance, especially JFETs.
- Individual FETs vary more widely from each other in their parameters, because of difficulties in manufacturing them.

Feedback in amplifiers

The topic of feedback has been mentioned a few times already in Chapter 1 and in this chapter. Now we will look at it in more detail.

Fig. 2.9. is a block diagram of a typical amplifier. It accepts an input signal v_{IN} and produces an output signal v_{OUT}. A small *fraction* of the output signal is routed back and added to the input signal. This 'fed back' signal is known as *feedback*. In the diagram we divide the amplifier into three blocks:

- The amplification stage (the large triangle) which has a gain of A_0, without feedback. This is the voltage gain which is determined by transconductance, by h_{fe}, or by the ratio of the collector and emitter resistors.
- The feedback loop. This represents the fact that feedback is a small *fraction* of the output by including a multiplier stage, to multiply v_{OUT} by β, a factor which is less than 1.
- The adding stage where βv_{OUT} is added to v_{IN}.

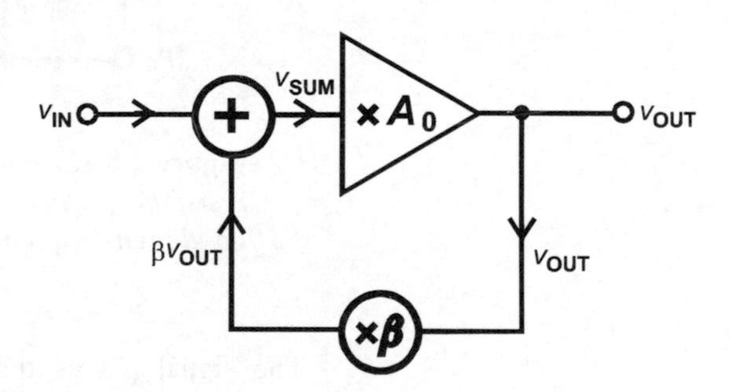

Figure 2.9 *This block diagram illustrates some of the features of feedback.*

In the amplifying stage:

$$v_{OUT} = A_0 \times v_{SUM}$$

A_0 is the open-loop gain of the amplifier, that it to say, the gain without feedback. In the feedback loop, the signal fed back to the adding stage is $\beta \times v_{OUT}$.

In the adding stage:

$$v_{SUM} = v_{IN} + \beta\, v_{OUT}$$

Substituting this value of v_{SUM} in the equation for the amplifying stage:

$$v_{OUT} = A_0(v_{IN} + \beta\, v_{OUT})$$

$$\Rightarrow \quad v_{OUT}(1 - \beta A_0) = A_0\, v_{IN}$$

$$\Rightarrow \quad \frac{v_{OUT}}{v_{IN}} = \frac{A_0}{1 - \beta A_0}$$

The expression on the left is the voltage gain A of the *whole* amplifier (with feedback) therefore:

$$A = \frac{A_0}{1 - \beta A_0}$$

If feedback is negative, either β or A_0 is negative (but not both). Therefore the denominator of the expression for A is always greater than 1, and A is less than A_0. Negative feedback reduces the gain of the amplifier, but usually brings an advantage such as stability.

Negative feedback

If an amplifier is an inverting amplifier (such as a common-emitter amplifier), then its gain A_0 is negative. The gain of the feedback loop (β) is positive and less than unity. The quantity fed back is therefore negative and is subtracted from the incoming signal.

If the amplifier is non-inverting (such as a common-base amplifier) we still feed back the output signal in a way that partly cancels the input signal. Since A_0 is positive, the circuit is such that β is negative. Again the quantity fed back is negative and is subtracted from the incoming signal.

Fig. 2.9 shows feedback taken from the output, but we can take feedback from other points in the amplifier, such as the emitter terminal of the transistor.

Problems on BJT amplifiers

1 Explain how a small increase in the voltage input of a common-emitter amplifier produces a much larger increase in the output voltage.

2 What is the difficulty with using a single resistor to bias a common-emitter amplifier?

3 Explain what is meant by the small signal current gain of a BJT.

4 Describe the action of the emitter resistor in stabilising the common-emitter amplifier.

5 Briefly describe two ways in which negative feedback helps to stabilise the common-emitter amplifier.

6 Describe the action and applications of a common-collector amplifier.

7 What are the disadvantages of the common-base amplifier? Why is it often used for radio frequencies?

8 Compare the common-emitter BJT amplifier with the common-source MOSFET amplifier.

9 Explain what is meant by negative feedback. Quote two examples of negative feedback in amplifiers.

Multiple choice questions

1 In a common-collector amplifier the output is taken from the:

A collector.
B base.
C positive supply.
D emitter.

2 A transistor has an h_{fe} of 80. When the base current is increased by 15 μA, the collector current increases by:

A 15 mA.
B 1.2 mA.
C 1 mA.
D 15 μA.

3 A common-emitter amplifier:

A has a voltage gain of 1.
B has high output resistance.
C has low input resistance.
D in an inverting amplifier.

4 The voltage gain of a common-emitter amplifier with an emitter resistor depends on:

A the value of the emitter resistor only.
B the value of the emitter resistance only.
C the values of the emitter and collector resistors.
D h_{fe}.

5 Of the three types of BJT amplifier, the one with the best high-frequency response is the:

A common-base amplifier
B common-drain amplifier.
C common-collector amplifier.
D common-emitter amplifier.

6 Compared with FET amplifiers, BJT amplifiers have:

A higher input resistance.
B smaller voltage gain.
C higher transconductance.
D higher output resistance.

3 Transistor switches

Summary

The best way to understand how transistors are used for switching is to study practical examples of switching circuits. We look at five examples of these and, while doing so, describe a number of different electronic sensors. We also show how to use a Schmitt trigger to make the switching action more positive.

Transistor switches are nearly always based on the common-emitter connection in the case of BJTs, and on the common-source connection with FETs. The controlling input goes to the base or gate, which generally has a directly wired connection to the controlling circuit. The device to be switched is placed in the collector or drain circuit. The switched device usually requires a current of several milliamperes, possibly several amperes. The current needed for operating the transistor switch is a lot smaller, often only a few microamperes. This makes it possible to control high-current devices from sensors, logic gates and other circuits with low-current output. The main limitation is that only devices working on direct current can be switched, but not devices powered by alternating current. These points are illustrated by the examples below.

Switching a lamp

This circuit (Fig. 3.1) uses a MOSFET transistor for switching a lamp. We might use this to control a low-voltage porch lamp which is switched on automatically at dusk. The sensor is a *photodiode*, which could be either of the visible light type or the type especially sensitive to infra-red radiation. It is connected so that it is reverse-biased. Only leakage current passes through it. The current varies according to the amount of light falling on the diode. In darkness, the current is only about 8 nA, but in average room lighting it rises to 3 μA or more.

The current is converted into a voltage by resistor R1. In light conditions the voltage drop across the resistor is several volts. The voltage at

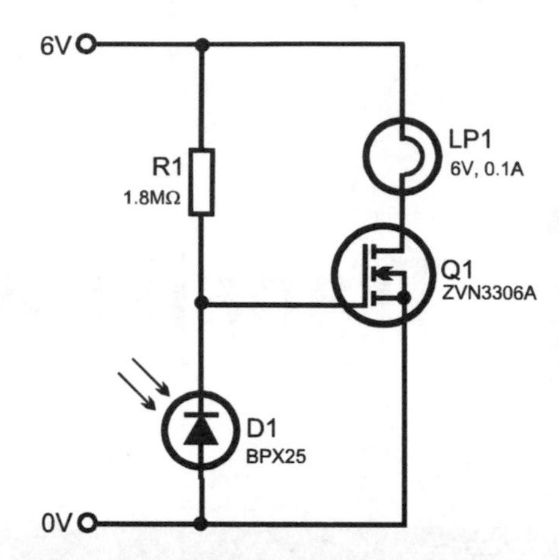

Figure 3.1 *The MOSFET switches on the filament lamp when the light intensity detected by the photodiode falls below a given level.*

Test your knowledge 3.1

In what connection is the transistor in Fig. 3.1?

the point between R1 and the diode is low, below the threshold of the transistor. The transistor is off and the lamp is not lit.

At dusk, light level falls, the leakage current falls, and the voltage drop across R1 falls too. This makes the voltage at the gate of the transistor rise well above its threshold, which is in the region of 2.4 V. The transistor turns fully on. We say that it is *saturated*. When the transistor is 'on', its effective resistance is only 5 Ω. Current flows through the lamp and transistor, lighting the lamp.

This circuit makes use of the high input resistance of MOSFETs. The sensor network (D1 and R1) has high output resistance. The high value of R1 means that it is unable to supply more than a few microamperes to drive the transistor. But, since the transistor has high input resistance, it requires only a few nanoamperes. If we were to use a BJT in this circuit we might find that the sensor network was unable to provide enough base current to turn it on.

Test your knowledge 3.2

A BUZ73L MOSFET has an 'on' resistance of 0.4 Ω and is rated at 7 A. If the supply voltage is 9V, what is the minimum resistance of a device that it can safely switch?

The lamp specified for this circuit has a current rating of 100 mA. The transistor has a rating of 270 mA so it is not in danger of burning out. A useful improvement would be to reduce R1 (say, to 820 kΩ) and wire a variable resistor (say 1 MΩ) in series with it. This would allow the sensitivity of the circuit to be adjusted. The circuit may be made to operate in the reverse way by exchanging the resistor and photodiode.

Light-sensitive alarm

The circuit of Fig. 3.2 uses a different type of light sensor, a *light-dependent resistor* (or LDR). This is made from a semiconductor material such as cadmium sulphide. The resistance of this substance varies according the the amount of light falling on it. We could have used this as a sensor in the previous circuit, but with a resistor of different value. The resistance of an LDR ranges between 100 Ω in bright light to about 1 MΩ in darkness.

The sensor R2 is connected in series with R1 to form a voltage divider.

The voltage at the point between the resistors is low (about 10 mV) in very bright light, but rises to about 5 V in darkness. With LDRs other than the ORP12, voltages will be different and it may be necessary to change the value of R1 or replace it with a variable resistor.

The function of the circuit is to detect an intruder passing between the sensor and a local source of light, such as a street light. When the intruder's shadow falls on the sensor, its resistance increases, raising the voltage at the junction of the resistors. Increased current flows through R3 to the base of Q1. This turns Q1 on and the siren sounds.

Test your knowledge 3.3

What happens to the resistance of an LDR when the amount of light falling on it increases?

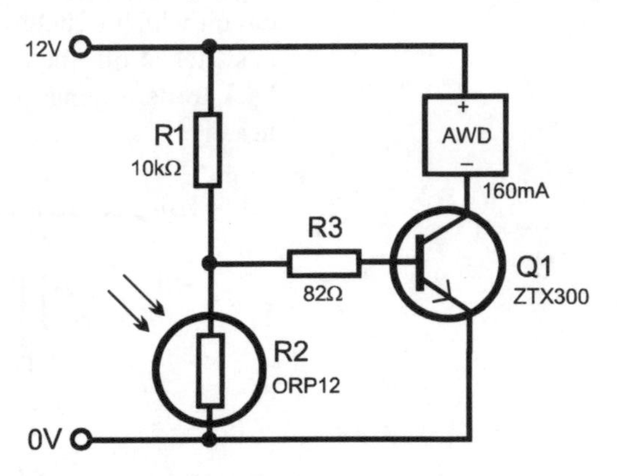

Figure 3.2 *The sensor in this circuit is a light-dependent resistor. An audible warning device sounds when light intensity falls below a given level.*

This circuit uses a BJT; almost any type will do provided it is able to carry the current required by the audible warning device, which in this example is a piezo-electric siren. A typical ultra-loud (105 dB) warbling siren suitable for use in a security system takes about 150 mA when run on 12 V. The ZTX300 transistor is rated at 500mA.

This circuit is able to make use of a BJT because the sensor network is able to deliver a current of the size required to turn on Q1. It is also possible to use a MOSFET.

Heater switch

Fig. 3.3 is a circuit for maintaining a steady temperature in a room, a greenhouse or an incubator. The sensor is a *thermistor,* a semiconductor device. Its resistance decreases with increasing temperature, which is why it is described as a negative temperature coefficient (ntc) device. Thermistors are not ideal for temperature *measuring* circuits, because the relationship between temperature and resistance is far from linear. This is no disadvantage if only one temperature is to be set, as in this example. Thermistors have the advantage that they can be made very small, so they can be used to measure temperatures in small inaccessible places. They also quickly come to the same temperature as their surroundings or to that of an object with which they are in contact. This makes them good for circuits in which rapid reading of temperature is important. These advantages offset the disadvantage of non-linearity in many applications.

Test your knowledge 3.4

An ntc thermistor has a resistance of 10 kΩ at 25°C. What can we say about its resistance at 20°C?

The switched device in this example is a *relay*. This consists of a coil wound on an iron core. When current passes through the coil the core becomes magnetised and attracts a pivoted armature. The armature is pulled into contract with the end of the core and, in doing so, presses two spring contacts together. Closing the contacts completes a second circuit which switches on the heater. When the current through the coil is switched off, the core is no longer magnetised, the armature springs back to its original position, and the contacts separate, turning off the heater.

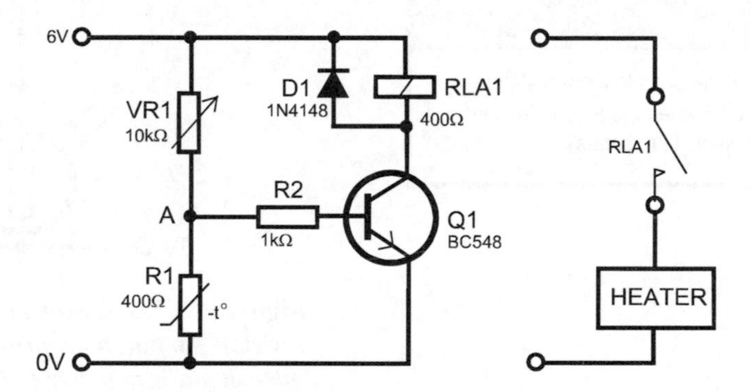

Figure 3.3 *A relay is necessary to switch a heater which runs from the 230 V AC mains.*

Test your knowledge 3.5

What is the minimum contact rating for a relay required to switch a 1000 W, 230V floodlamp?

Most relays have contacts capable of switching high voltages and currents, so they may be used for switching mains-powered devices running on alternating current. Current ratings of various types of relay run from 1 A up to 40 A. Various arrangements of switch contacts are available, including simple change-over contacts and relays with 3 or more independent sets of contacts. These can be wired so that some devices can be turned off at the same time that others are turned on. Relays are made in an assortment of sizes, including some only a little larger than a transistor, suitable for mounting on printed circuit boards.

WARNING

The circuit in Fig. 3.3 illustrates the switching of a device powered from the mains. **This circuit should not be built or tested except by persons who have previous experience of building and handling mains-powered circuits.**

Thermistors are available in several resistance ratings, the one chosen for this circuit having a resistance of 400 Ω at 25°C. VR1 is adjusted so that the voltage at point A is close to 0.6 V when the temperature is close to that at which the heater is to be turned on. The transistor is *just* on the point of switching on. As temperature falls further, the resistance of the thermistor R_1 increases, raising the voltage at A, and increasing i_B. This switches on the transistor and current passes through the relay coil. The relay contacts close, switching on the heater. Note that the heater circuit is entirely separate from the control circuit. This is why many types of relay can be used for switching mains-voltage alternating current. After a while, as the heater warms the thermistor, R_1 decreases, the voltage at point A falls and the transistor is turned off. The relay coil is de-energised and the heater is turned off.

This circuit works by negative feedback between the heater and the thermistor. The action of the circuit, including the extent to which the temperature rises and falls, depends on the relative positions of the heater and thermistor and the distance between them.

Emf: electromotive force: A potential difference produced by con=version of another form of energy into electrical energy. Other forms include chemical (cell), mechanical (dynamo, friction), and magnetic (collapsing field, car ignition system).

The diode D1 is an important component in this circuit. When the transistor switches off, current ceases to flow through the coil and the magnetic field in the coil collapses. This causes an emf to be induced in the coil, acting to oppose the collapse of the field. This emf may result in hundreds of volts being developed across the transistor, producing high currents which may permanently damage or destroy it. If the diode is present, it protects the transistor by conducting currents away to the power line. The diode is useful in any circuit which rapidly switches current through an inductive element.

Schmitt trigger

The resistance of the thermistor (Fig. 3.3) changes relatively slowly, so that the circuit spends several minutes in an intermediate state in which the transistor is not fully switched on. The magnet is not fully energised and the armature tends to vibrate rapidly, switching the heater on and off several times per second. This leads to sparking at the contacts, which may become fused together so that the relay is permanently 'on'. The circuit of Fig. 3.4 avoids this, giving a sharply trigger switching action. This *Schmitt trigger* consists of two transistors. Q1 is switched by the sensor network. Q2 is switched by Q1 and in turn switches the relay.

The thermistor and variable resistor are arranged so that the voltage at point A falls as temperature falls. This is the opposite action to that in Fig. 3.3. This is necessary because we have *two* inverting transistors in this circuit. The resistance of R2 must be greater than that of the relay coil, so that more current flows through R6 when Q2 is on, and less flows when Q1 is on.

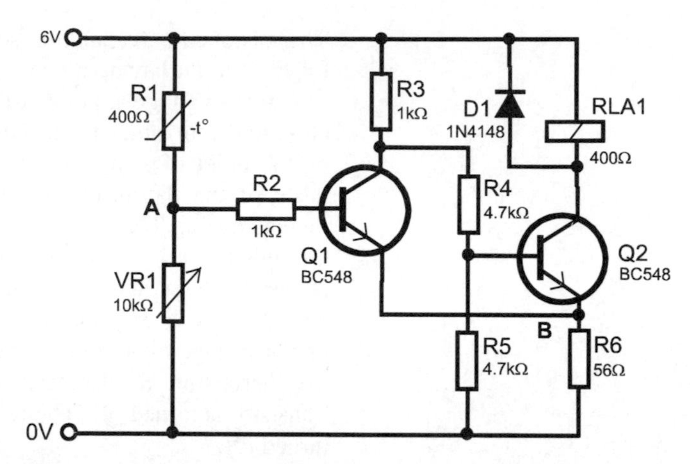

Figure 3.4 *The Schmitt trigger circuit has a 'snap' action which is suitable for switching heaters and many other devices.*

There are two ways in which the trigger circuit improves on the action of the simpler circuit of Fig. 3.3:

- It has a sharp 'snap' action. As temperature falls and Q1 *begins* to turn off, Q2 *begins* to turn on. The increasing current through R6 raises the voltage at the Q1 emitter. As the Q1 base voltage continues to fall its emitter voltage starts to rise, rapidly reducing v_{BE} and turning Q1 off very rapidly. Q1 turns off faster and Q2 turns on faster. The circuit spends much less time in its intermediate state.
- The 'turn-on' temperature is lower than the 'turn-off' temperature. We call this *hysteresis*. If the heater is off, the current through R6 and the voltage across it are relatively small. Measurements on the circuit of Fig. 3.4 found the voltage at B to be 0.4 V. This means that as soon as the voltage at A falls to 1 V (0.4 + 0.6 due to v_{BE}) the circuit starts to change state, turning the heater on. But this increases the current through R6 and the voltage at B rises to 0.9 V. Now the voltage at A has to rise to 1.5 V (0.9 + 0.6 due to v_{BE}) before Q1 can be turned on again and the circuit returns to its original state. In terms of heating, the heater comes on when the room has cooled to a set temperature, say 15°C. The heater stays on, perhaps for 10–15 minutes, until the room has warmed to a higher temperature, say 17°C, then goes off. This is a much better action that having the heater continually switching on and off several times a minute.

Schmitt trigger circuits are often used to improve switching action. The circuit of Fig. 3.1 could use a trigger to avoid the lamp bring switched on and off with every slight up and down variation of light level at dusk. The Schmitt trigger can be built from other inverting circuit units such as operational amplifiers and logic gates.

Logic control

The input of the circuit in Fig. 3.5 taken from the output of a logic gate. This would not be able to supply enough current to light the lamp, but can raise the gate voltage sufficiently to turn the transistor on. When the logic output is low (0 V) the voltage at the gate of Q1 is below the threshold (2.5 V) and the transistor is fully off. There is no drain current and the lamp is unlit. When the logic output goes high (say, +5V for a TTL output) this puts the gate of Q1 well above the threshold. The transistor is turned fully on and the maximum drain current flows, turning the lamp on. It is saturated and is working well beyond the range of Fig 1.2.

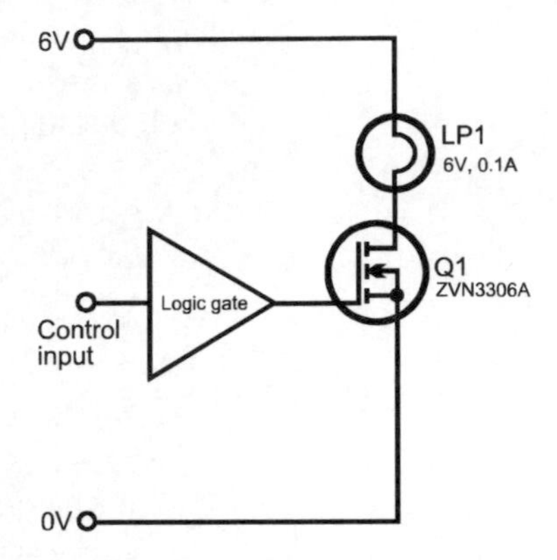

Figure 3.5 *This is how a lamp may be put under the control of a logic circuit, such as a computer or microcontroller.*

The power supply for the lamp may be at a higher voltage than that used to supply the logic circuit. If the logic is operating on 5V, the lamp or other device could be powered from, say, a 24V DC supply, provided that the current rating (0.5 A) and power rating (1 W) of the transistor are not exceeded.

This circuit could also be used to switch other devices, including a *solenoid.* A solenoid is a coil wound on a former, with an iron core or *armature* which slides easily in and out of the coil. Usually the armature is held by a light spring so it rests only partly inside the coil when there is no current flowing. When current flows through the coil, the magnetic field strongly attracts the armature, which is drawn forcefully into the coil. There is some kind of mechanism linked to the armature. For example, the armature may be attached to bolt, so that energising the coil causes a door to be bolted. In this way the door may be bolted by computer control. As with the relay circuit of Fig. 3.3, a solenoid switching circuit requires a diode to protect the switching transistor against high induced voltage.

An important point when operating switching transistors is that they should be switched *quickly* between the on and off states. When the transistor is off there is no current through the transistor and no heat is generated there. When the transistor is on, it has very low resistance and so only a small voltage is developed across it. The heating effect is small. But, if switching is slow, the transistor may be left for too long in an intermediate state, in which the voltage across the transistor is relatively high. A relatively large amount of power is being dissipated, which may overheat and possibly damage the transistor.

Activity 3.1 Switching circuits

On a breadboard or on stripboard, build the switching circuits illustrated in this chapter and test their action. There are a number of simple ways in which these circuits can be altered to change their action:

Modify the circuit of Fig. 3.1 so that the lamp is turned on when light level increases.

Add a variable resistor to the circuit of Fig. 3.2 to make it possible to set the light level at which the switch operates.

Try increasing and decreasing the resistance of R6 in Fig. 3.4 to see what effect this has on the hysteresis. Also look at the effect of changing the value of R3.

In Fig. 3.5, build a flip-flop based on a pair of CMOS NAND gates (Fig. 12.1) and use this to drive the MOSFET.

Activity 3.2 New switching circuits

Use Figs. 3.1 to 3.5 as a guide when designing and building switching circuits with the following functions:

A photodiode and a BJT to switch a lamp on when light level falls below a preset value.

A light dependent resistor and a MOSFET to switch an LED on by detecting the headlamps of an approaching car at night.

A circuit to sound a siren as a garden frost warning.

A circuit to switch on an electric fan (a small battery-driven model) when the temperature falls below a given level. This circuit is better if it has hysteresis.

Any other switching function that you think is useful.

Problems on transistor switches

1 Describe the action of a MOSFET switch used with a photodiode as sensor for switching on a filament lamp.

2 Design a fire-alarm circuit for switching a warning lamp on when the temperature rises above a given level.

3 Design a "Dad's home!" circuit for switching on a small piezo-electric sounder when the sensor detects approaching car headlamps at night. Calculate suitable component values.

4 Explain why it is essential to include a protective diode in a circuit that switches an inductive device.

5 Describe the operation of a simple change-over relay, illustrating your answer with diagrams.

6 Draw a diagram of a Schmitt trigger circuit based on BJTs and explain how it works.

7 Design a Schmitt trigger circuit based on MOSFETs.

8 Design a circuit for operating a magnetic door release automatically when a person approaches the doorway. The release operates on a 12 V supply and requires 1.5 A to release it.

9 Compare the features of BJTs and MOSFETs when used as switches.

Multiple choice questions

1 When used as a switch, a BJT is connected in:

A the common-emitter connection.
B the common-collector connection.
C the common-base connection.
D none of the above.

2 Which of these features of a MOSFET is not of importance in a transistor switch?

A High input resistance.
B Low 'on' resistance.
C Medium transconductance.
D Low input capacitance.

3 A light-dependent resistor is usually made from:

A silicon oxide.
B p-type semiconductor.
C cadmium sulphide.
D carbon.

4 A thermistor is not ideal for a temperature measuring circuit because:

A its resistance is too high.
B it responds too slowly to changes of temperature.
C its response is not linear.
D it has a negative temperature coefficient.

5 The maximum current that a MOSFET can switch when operating on a 9 V supply and with an on resistance of 2 Ω :

A depends on the device being switched.

B depends on its maximum current rating.

C is 4.5 A.

D is 222 mA.

6 The voltage level at which a Schmitt trigger switches on is different from that at which it switches off. We call this:

A negative feedback.

B current gain.

C hysteresis.

D positive feedback.

4 More BJT amplifiers

Summary

Variations and additions to the basic BJT amplifiers from Chapter 2 provide a range of different amplifier types with special features and characteristics. These comprise the Darlington pair, the long-tailed pair and tuned amplifiers.

Darlington pair

A Darlington pair consists of two BJTs connected as in Fig. 4.1. The emitter current of Q1 becomes the base current of Q2. The current gain of the pair is equal to the product of the current gains of the individual transistors.

Example

If h_{fe} = 100 for each transistor in a Darlington pair, the gain of the pair is 100 × 100 = 10 000.

Test your knowledge 4.1

A Darlington pair is made up of a BC548 with h_{fe} = 400 and a TIP31A (a power transistor) with h_{fe} = 25. What is the gain of the pair?

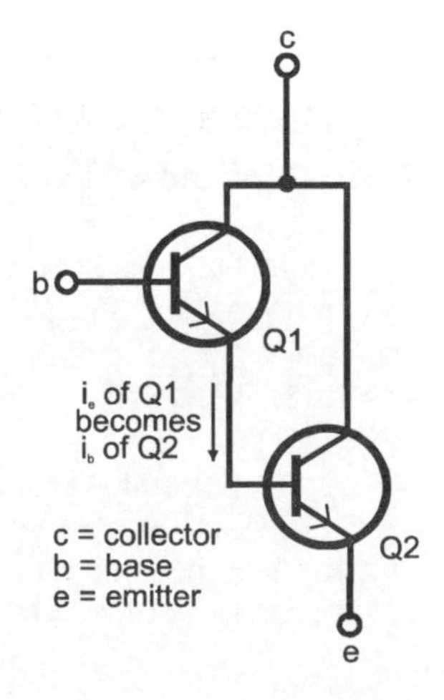

Figure 4.1 *A Darlington pair has very high current gain.*

Wiring two individual transistors together may make up a pair. The two transistors may also be manufactured as a unit on a single silicon chip. This has three terminals, collector, base and emitter, as labelled in Fig. 4.1. The chip is enclosed in a normal three-wired package looking just like an ordinary transistor. When the pair is made from individual transistors, they need not be of the same type. Very often Q1 is a low-voltage low-current type while Q2 is a power transistor.

A Darlington may replace the single transistor in any of the BJT amplifiers of Chapter 2. This gives increased sensitivity to small base currents and usually to increased gain. Darlingtons may also be used as switching transistors, as in Chapter 3. They require only a very small current to trigger them.

The Darlington pair is also used as a switching transistor. In Fig. 4.2 two metal plates about 1 cm × 2 cm are mounted side by side with a 0.5 mm gap between them. When a finger touches the plates so as to bridge the gap, a minute current flows to the base of Q1. This current is only a few tens of microamperes, but the pair of transistors has a current gain of about 160 000. The resulting collector current through Q2 is sufficient to make the solid-state buzzer sound.

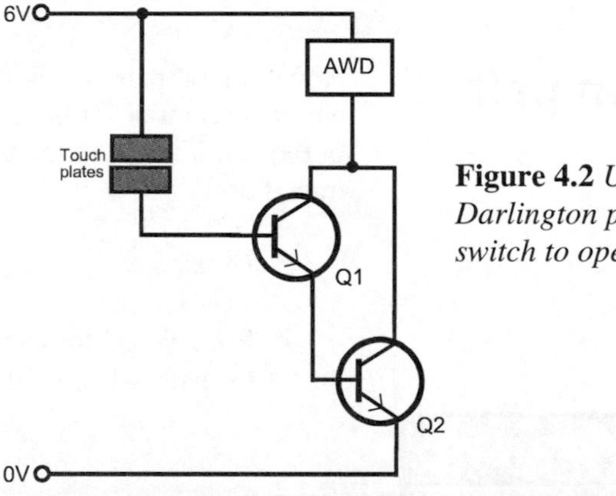

Figure 4.2 *Using a Darlington pair in a touch-switch to operate a buzzer.*

Activity 4.1 Darlington pair

Set up a Darlington pair as a switch controlling an audible warning device (Fig. 4.2). A pair of BC548 transistors are suitable. Alternatively, use a Darlington pair transistor such as an MPSA13.

When the circuit is working, insert a microammeter and a milliammetre into it to measure the current through the finger and the current through the AWD. What is the current gain of your circuit?

Differential amplifier

This differential amplifier (Fig. 4.3) is also known as a *long-tailed pair*. A differential amplifier has two inputs v_{IN+} and v_{IN-}. The purpose of the amplifier is to amplify the voltage *difference* between its inputs. When it does this, we say that it is operating in the *differential-mode*.

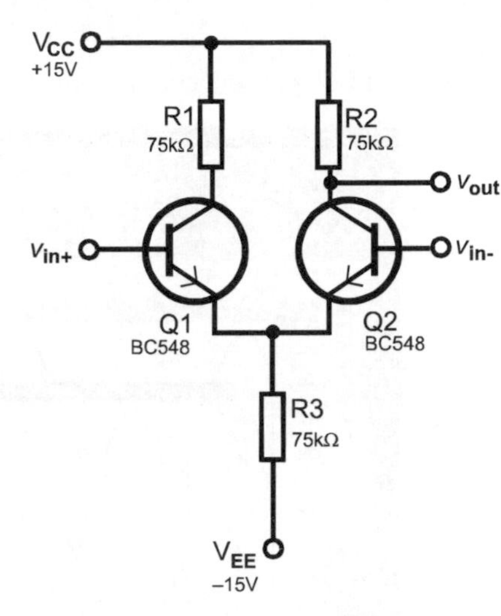

Figure 4.3 *A differential amplifier has two inputs and one or two outputs. The second output (not shown here) is taken from the collector of Q1.*

To demonstrate this, two sinusoid signals that are 90° out of phase are fed to the two input terminals (Fig. 4.4). The output signal is taken from the collector of Q2. At any instant:

$$v_{OUT} = (v_{IN+} - v_{IN-}) \times \text{differential-mode voltage gain}$$

Measurements on the graph show that the differential voltage gain is approximately 120.

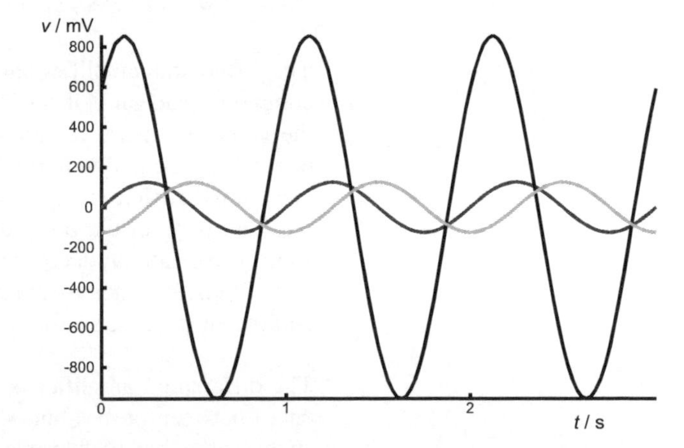

Figure 4.4 *The amplifier of Fig. 4.3 in differential mode. The output (black curve) is proportional to the difference between the non-inverting input (dark grey curve) and the inverting input (light grey curve). The inputs have 5 mV amplitude and are plotted on a 25 times scale.*

If we feed the *same* signal to both inputs, we are operating the amplifier in *common-mode*. We obtain the result shown in Fig. 4.5. The amplitude of the output signal is only half that of the input signal and it is inverted:

$$v_{OUT} = v_{IN} \times \text{common-mode gain}$$

In Fig. 4.5, the common-mode gain is –0.5.

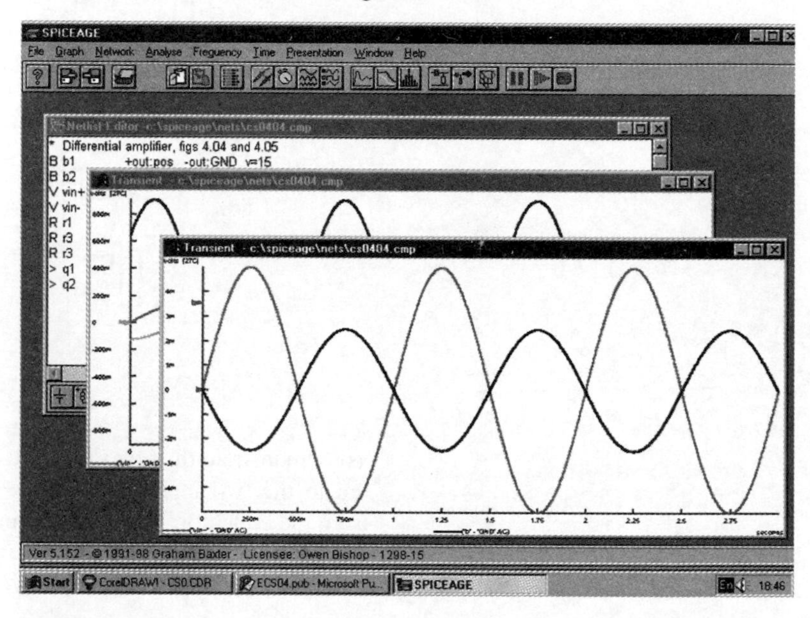

Figure 4.5 *A simulation of the amplifier of Fig. 4.3 in common mode. The output (black curve) has smaller amplitude than the input (grey curve) and is inverted, demonstrating a gain of –0.5.*

The differential amplifier has high differential-mode gain but very low common-mode gain. It is suited to measuring voltage differences when the voltages themselves are subject to much larger changes affecting both voltages equally. In medical applications we may need to measure very small differences of voltage between two probes attached to different parts of the body surface. At the same time both probes are subject to relatively large voltage changes resulting from electromagnetic fields from nearby mains supply cables and from other electrical equipment in the same room.

The differential amplifier is able to amplify the small voltage differences between probes but is much less affected by the larger voltage swings affecting both probes. It is able to *reject* the common-mode signals. We express its ability to do this by calculating the *common-mode rejection ratio*:

$$\text{CMRR} = \frac{\text{diffl. mode voltage gain}}{\text{common mode voltage gain}}$$

Example

We ignore negative signs when calculating the CMRR:

- The CMRR is 144/0.5 = 288.
- The CMRR can also be expressed in decibels:

$$\text{CMRR} = 20 \times \log_{10}(144/0.5) = 49.2 \text{ dB}$$

Increasing CMRR

It can be shown that the differential-mode voltage gain is *approximately* equal to $10i_T R_C$, where i_T is the current in the tail resistor (R_3) and R_C is the resistance of the collector resistors (R_1 and R_2). The current in the tail resistor has a constant value. In the circuit of Fig. 4.2, for example, it is 192 μA. It can also be shown that the common-mode voltage gain is equal to $-R_C/2R_T$.

Example

For the circuit of Fig. 4.2 the differential-mode voltage gain is:

$$10 \times 192 \times 75 \; [\mu\text{A} \times \text{k}\Omega] = 144$$

This agrees approximately with the value obtained from the graph. The common-mode voltage gain is:

$$-75/(2 \times 75) = -0.5$$

Combining these two formulae we obtain:

$$\text{CMRR} = 10i_T R_C, \times 2R_T/R_C = 20i_T R_T$$

This result shows that we can increase the CMRR by increasing i_T or R_T or both. But, according to Ohm's law, $i_T R_T$ is equal to the voltage across the tail resistor. The voltage at the emitters of Q1 and Q2 is close to 0V, and the voltage at the other end of R_3 is V_{EE}.

Example

The CMRR can be calculated as:

$$20 \times V_{EE} = 20 \times 15 = 300$$

There are approximations in the formula but the result agrees well with the CMRR of 288 previously calculated.

Test your knowledge 4.2

Calculate the new CMRR if V_{EE} is increased to –24V?

Looking at the formula it appears that we can increase the CMRR by operating the circuit with an increased negative supply voltage. This is practicable only for a limited range of supply voltages.

Constant current sink

Another approach to increasing CMRR is to increase i_T, without having to increase V_{EE}. Using a sub-circuit known as a constant current sink can do this. Fig. 4.6 shows a sink based on a BJT. The base of Q3 is held at a constant voltage by the Zener diode. With the values given in Fig. 4.6, i_T is 336 μA (compared with 198 μA in the original circuit). The effect is to reduce the common-mode gain to –0.4 and to increase the differential-mode gain to 211. In total, the CMRR is increased to 528, or 54.4 dB.

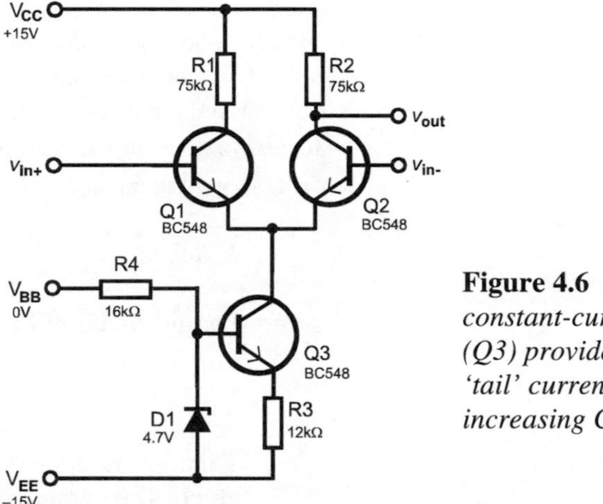

Figure 4.6 *The constant-current sink (Q3) provides larger 'tail' current, so increasing CMRR.*

Constant current circuits

Figure 4.6 contains a BJT with its base voltage stabilised by a Zener diode. Because this gives v_{BE} a fixed value, a constant collector current passes through the transistor. The current is independent of the collector voltage, provided that the transistor does not become saturated.

A JFET is often used to produce constant current, as in Fig. 4.7. On the left, the gate is at the same voltage as the drain, in which case a constant current I_{DSS} flows through the transistor for a wide range of drain voltages. The value of I_{DSS} may be found in the data sheets and is generally a few tens of milliamperes. A JFET may be programmed to produce constant currents less than I_{DSS} by wiring a resistor between source and gate, as on the right.

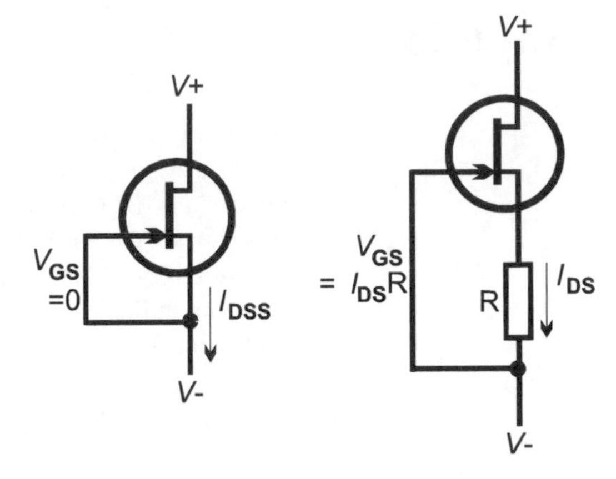

Figure 4.7 *Two JFET constant current sources or sinks.*

Activity 4.2 Differential amplifier

Build the differential amplifier (Fig. 4.3) on a breadboard or simulate it on a computer.

First measure the *differential-mode voltage gain.* Connect a voltage source to each of the inputs and a voltmeter to the output. Draw a table with 5 columns. Head the 3 columns v_{IN+}, v_{IN-} and v_{OUT}. Set the two voltage sources to two *different* voltages within the range ±100 mV. Record the two voltages (v_{IN+} and v_{IN-}). Measure and record v_{OUT}. Repeat this for 10 pairs of input voltages. For each pair calculate ($v_{IN+} - v_{IN-}$) and write this in the fourth column. For each pair calculate the differential-mode voltage gain and write this in the fifth column. Calculate the average value of this for the 10 pairs.

Next measure the *common-mode voltage gain.* Connect the two inputs together and connect a single voltage source to each of them. Draw a table with three columns and in the first two columns record v_{IN} and v_{OUT} as you obtain 10 sets of readings with the input in the ±100 mV. For each set of readings, calculate the common-mode voltage gain and write this in the third column. Calculate the average value for this for the 10 sets.

Finally calculate the *common-mode rejection ratio.* Using the two gains found in (1) and (2), calculate the common-mode rejection ratio. Express this in decibels.

Tuned amplifier

Tuned amplifiers are used at radio frequencies. Their essential feature is a tuned network, usually consisting of a capacitor in parallel with an inductor. There are several ways in which the tuned network may be built in to the amplifier, but one of the commonest ways is to use it to replace the resistor across which the voltage output signal is generated. In a BJT common-emitter amplifier it replaces the collector resistor. In an FET common-source amplifier, it replaces the drain resistor. Fig. 4.8 is the basic common-emitter amplifier of Fig. 2.5 with its collector resistor R_3 replaced by a tuned capacitor-inductor network. The transistor has been replaced by a high-frequency type.

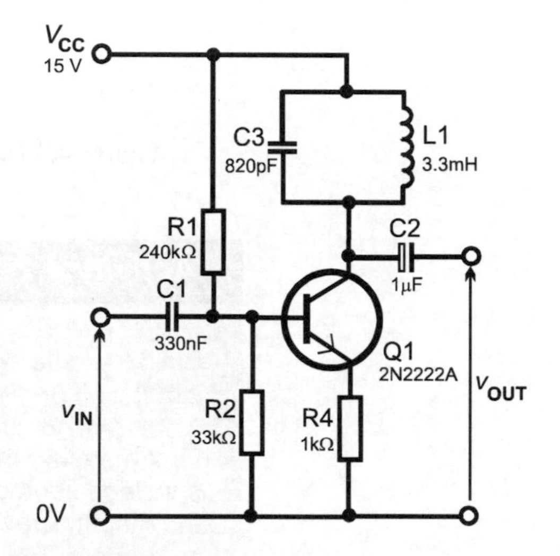

Figure 4.8 *A tuned amplifier based on a simple common-emitter amplifier.*

The resonant frequency of the network is given by:

$$f_0 = \frac{1}{2\pi\sqrt{LC}}$$

> **Test your knowledge 4.3**
>
> What are the reactances of C_3 and L_1 in Fig. 4.8 when the frequency is (a) 96 kHz, (b) 96.75 kHz and (c) 98 kHz?

Example

The resonant frequency of the network in Fig. 4.8 is:

$$f_0 = \frac{1}{2\times\pi\times\sqrt{3.3\times10^{-3}\times820\times10^{-12}}} = 96.75\text{kHz}$$

In Fig. 4.9 the amplifier is fed with a 10 mV signal in a range of frequencies from 95 kHz to 100 kHz. Looking at the curve nearest the front of the diagram, it can be seen that the output amplitude peaks sharply at just under 97 kHz, as predicted by the calculation above. The reason for this peak is the fact that, at the resonant frequency, the

Reactances

The reactance of a capacitor is:

$$X_C = 1/2\pi fC$$

The reactance of an inductor is:

$$X_L = 2\pi fL$$

As frequency increases, X_C decreases, but X_L increases. At one particular frequency they are equal. This is the frequency at which resonance occurs. If we put:

$$1/2\pi fC = 2\pi fL$$

and solve for f, we obtain the equation for resonant frequency given on the opposite page.

Test your knowledge 4.4

What is the resonant frequency of an LC parallel network in which C = 100 pF and L = 250 mH?

capacitor-inductor network behaves as a pure resistance of high value. This is equivalent to replacing the network with a resistor equivalent to the one that was present in Fig. 2.5. As the collector current flows through the network, a voltage is developed across it, and this is the output of the amplifier. At lower or higher frequencies the effective resistance of the network is not as great, so the voltage developed across it is less.

Because the resonating network has a high resistance (in theory, an infinite resistance, but less than this in practice due to losses in the capacitor and inductor) the voltage developed across it is high. In Fig. 4.9, the amplitude of the output signal at the resonant frequency is 15 V. This is a voltage gain of 1 500.

The output of the circuit varies mainly with the properties of the inductor, including the resistance of its coil. The *quality factor* of the inductor is given by $Q = 2\pi fL/r$, where r is the resistance of the coil. If the inductor has a high Q (say, 400), the amplifier output is high but falls off sharply on either side of f_0. If Q is low (say, 50) the peak output amplitude is not as high, and falls off more slowly on either side of f_0. This can be seen in Fig. 4.9 where the successive curves from front to back show the effect of decreasing Q from 412 down to 50. Although high gain may be desirable in some applications, a radio receiver usually has to amplify sidebands with frequencies spaced a little to either side of the carrier frequency. A broader bandwidth may be preferred at the expense of reduced gain. This can be obtained by choosing an inductor with a suitable Q and often by wiring a low-value resistor in series with the inductor.

Changing the value of the emitter resistor, R3, as explained in Chapter 2, alters the voltage gain of the amplifier. This has an effect on frequency response. In Fig. 4.10 the frequency response is plotted for values of R_3 ranging from 1 Ω to 1 kΩ. Gain is highest when R_3 is 1Ω, with a broad bandwidth (13 kHz). At the other extreme, when R_3 is 1 kΩ, gain is only 37 and the bandwidth is narrowed to 2 kHz.

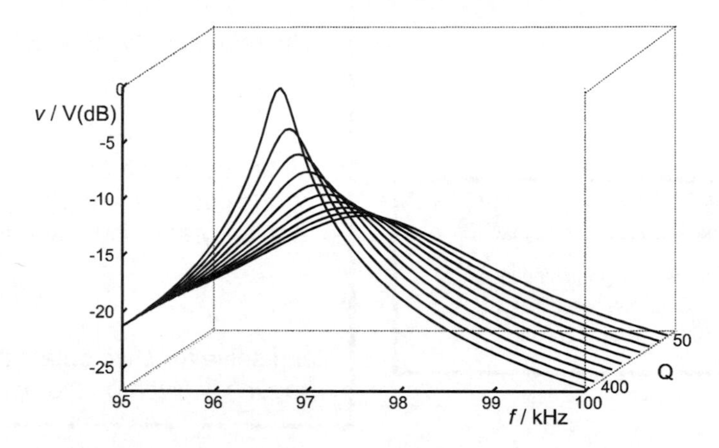

Figure 4.9 *The frequency response of the tuned amplifier varies according to the Q of the inductor.*

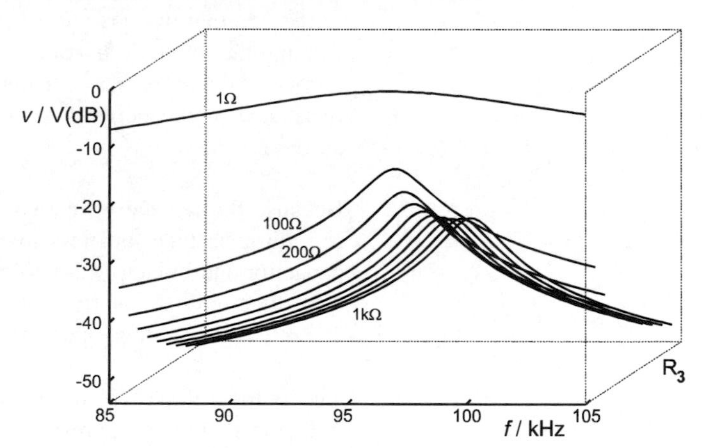

Figure 4.10 *Changing the amplifier gain by altering the emitter resistor has a marked effect on gain and bandwidth.*

The advantages of this amplifier for radio-frequency applications are:

- It has very high gain.
- The effects of capacitance, which would reduce gain in a normal amplifier, are eliminated here because the stray capacitances in the transistor become, in effect, part of the network capacitance and help to bring about resonance.

- A tuned amplifier is *selective*. It is sensitive only to signals within its bandwidth.
- It is easy to couple stages of a multi-stage tuned amplifier by winding a coil on the same former as the tuning inductor and picking up the signal for the next stage from that (inductive coupling).

Activity 4.3 Tuned amplifier

Build the common-emitter tuned amplifier shown in Fig. 4.8. Connect a signal generator to its input and an oscilloscope to its output. Set the signal amplitude to 10 mV. Vary the frequency over the range 10 kHz to 1 MHz and find the frequency (f_c) at which amplification is a maximum. Compare this with the resonant frequency of the loop formed by C3 and L1. Calculate Q for the inductor. Plot the frequency response (see Fig. 4.9) for a range of frequencies on either side of the resonant frequency.

Repeat, using a different capacitor and inductor for C3 and L1.

You could also use a simulator for the investigations above.

Activity 4.4 Bandwidth

Use the tuned amplifier built for Activity 4.3 and find the bandwidth for 2 or 3 different values of R3.

Find the resonant frequency (f_c) as above, and note the voltage amplitude of the output signal at that frequency.

Find the lower –3dB point. This is the frequency (lower than f_c) at which the amplitude is 0.7 times the amplitude at the resonant frequency.

Find the upper –3dB point. This is the frequency (higher than f_c) at which the amplitude is 0.7 times the amplitude at the resonant frequency.

The difference between the frequencies found in the previous two tests is the bandwidth.

You could also use a simulator for the investigations above.

Problems on BJT amplifiers

1 Describe a Darlington pair and state how to calculate its current gain.

2 What are the advantages of having a Darlington transistor as a single package?

3 Why may we want to build a Darlington transistor from two separate BJTs?

4 Design a rain-alarm circuit for washday, using a Darlington transistor.

5 Draw the circuit of a BJT differential amplifier and state the relationship between its inputs and output.

6 Suggest two applications for a differential amplifier.

7 Describe a BJT constant current sink and explain how it works.

8 Explain why a tuned amplifier operates over a narrow frequency range.

9 What are the advantages of a tuned amplifier for use at radio frequencies?

Multiple choice questions

1 If the differential mode gain of a differential amplifier is 50 and the common mode gain is –0.25, the CMRR is:

A 50.25. C 23 dB.
B 200. D 0.005.

2 If a differential amplifier with a simple tail resistor operates on a ±9 V supply, its CMRR is approximately:

A 180. C –0.5.
B 300. D 9.

3 The transistors of Darlington pair have current gains of 150 and 200. Their combined current gain is:

A 30 000. C 400.
B 350. D 50.

4 The effects of stray capacitance are eliminated in a tuned amplifier because:

A capacitance has no effect at high frequency.
B the capacitance acts as part of the resonant network.
C the capacitance is cancelled by the inductance.
D the transistor is connected to avoid capacitance effects.

5 The output of a tuned network shows a peak at the resonant frequency because:

A the capacitor and inductor have equal reactance.
B the resonant network has high resistance.
C the inductor has high Q.
D resonance generates high voltages.

5 Oscillators

Summary

An oscillator produces a periodic output signal. The most common signals are sinusoidal signals and pulsed signals. This chapter describes oscillators which produce sinusoids, examples being the Hartley oscillator based on a tuned circuit, and the phase shift oscillator. The precision of oscillators may be improved by including a quartz crystal in the circuit.

Hartley oscillator

In this oscillator (Fig. 5.1), the source of oscillation is a capacitor-inductor network of the kind used in tuned amplifiers (Chapter 4).

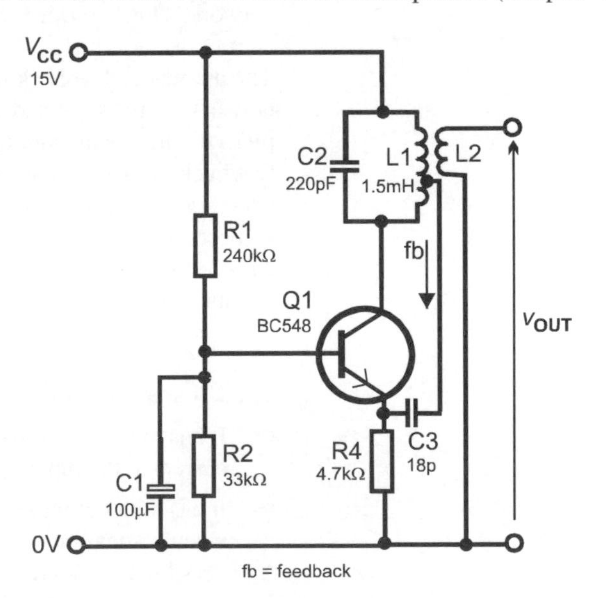

***Figure 5.1** The Hartley oscillator consists of a common-base amplifier with a resonant network in its collector circuit.*

The network oscillates at a rate determined by the values of the capacitance and the inductance:

$$f = \frac{1}{2\pi\sqrt{LC}}$$

Test your knowledge 5.1

Given a 1µ capacitor, calculate the value of the inductor in a 200 Hz Hartley oscillator.

Once the network is oscillating, it continues for a while but energy is lost in heating the conductors and also in the dielectric of the capacitor and the armature of the inductor. The network can be kept oscillating by supplying it with a relatively small pulses of energy in phase with its oscillations. This is similar to the way in which the pendulum of a clock is kept swinging by small amounts of energy transferred to it from the spring or weights through the escapement mechanism. In Fig. 5.1 the common-base amplifier based on Q1 supplies the energy. This is very similar to the CB amplifier shown in Fig. 2.8. R1 and R2 bias its base and the voltage there is stabilised by the bypass capacitor C1. The tuned network replaces the collector resistor.

The output of a CB amplifier is taken from its collector, so it is the collector current which supplies energy to the tuned network. But it is essential that this energy be supplied at exactly the right moments during the oscillating cycle. The inductor coil is tapped a short way along its length and a pulsing signal is fed back through capacitor C3 to the emitter of Q1. In a CB amplifier, the emitter is the input terminal. The CB amplifier is non-inverting so, whenever current in L1 is increasing, the increased voltage passed across C3 turns Q1 more fully on and causes increased collector current to flow through the network. This is *positive* feedback.

The amount of feedback must be adjusted to balance that lost from the network by resistive and other effects, including the loss to L2 which picks up the signal from L1 to provide the output from the oscillator. If feedback is insufficient, the oscillations will die away. If feedback is excessive, the transistor may become saturated. This causes the output to become distorted from its naturally sinusoidal shape. Feedback is adjusted by selecting the tapping point on L1 and also by choosing a suitable value for C3.

The CB amplifier is very suitable for use in this application because:

- It has low input resistance, but the feedback signal provides ample current at the input.
- It has high output resistance. The network has very high resistance when resonating, so we are feeding from a high resistance into a very high resistance.
- It operates well at high frequencies, which is an important feature for an oscillating circuit.

An oscillating network

At resonance a tuned network is in a state of oscillation. It oscillates between four states (Fig. 5.2). Starting at top left, the capacitor is fully charged and no current is flowing. Then the capacitor discharges though the inductor, generating a magnetic field in its coil. This builds up until the capacitor is fully discharged (top right). The current stops and the field collapses, but self-induction results in the generation of an emf in the inductor, tending to keep the current flowing. This current charges the capacitor in the opposite direction (bottom right), and the field in the inductor has decayed. The capacitor now discharges, generating a magnetic field but in the opposite direction (bottom left). As this field collapses, an emf is generated which re-charges the capacitor in the original direction.

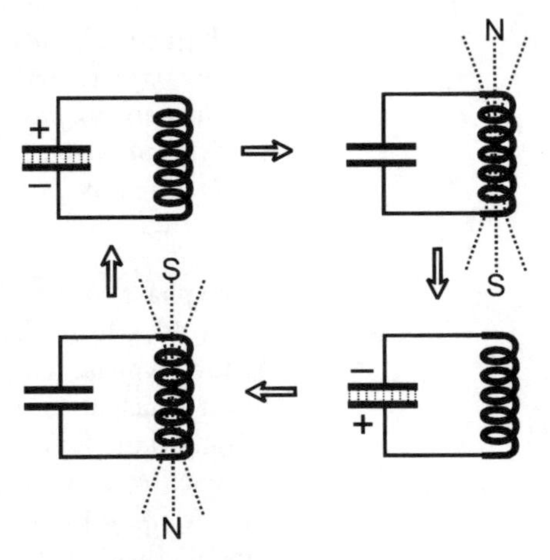

Figure 5.2 *A capacitor and inductor connected in parallel form a resonant network. The energy in the network is stored alternately as charge on the capacitor or as a magnetic field in the inductor. The network alternates between the two states at a rate depending on the capacitance and inductance.*

A disadvantage with this oscillator is that it needs a large-value inductor to operate at audio frequencies. The answer to *Test your knowledge 5.1* demonstrates this fact. Large self-inductance means large size and weight.

We have described a version of the Hartley oscillator in which a BJT is used to provide the required amplification. Other versions of the amplifier, operating on the same principle, may be built around a MOSFET or a JFET.

Phase shift oscillator

The phase shift oscillator (Fig. 5.3) works on a principle that is entirely different from that of the tuned network (Hartley) oscillator. It has no inductors, so the size and weight of large-value inductors do not limit the frequency range.

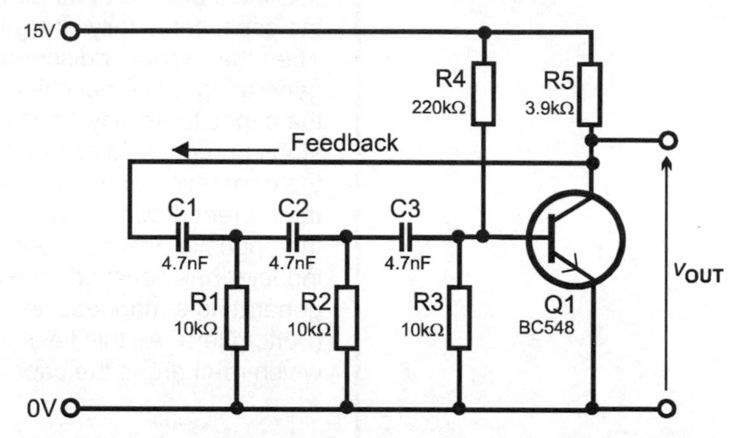

Figure 5.3 *The output from the collector of the transistor is 180° out of phase with its input at the base. The signal is fed back and the network makes it a further 180° out of phase when it gets back to the base. This makes it* in phase *(giving positive feed-back) at the base, causing the circuit to oscillate.*

Phase lead: if two signals have the same frequency, but are out of phase, the one which reaches its peaks earlier is said to have a *phase lead*.

The phase shift oscillator is based on a chain of three resistor-capacitor networks. It is a feature of each resistor-network that its output signal shows a phase lead on the input signal. In Fig. 5.4 we see what happens to a sinusoid as it passes along a chain of three such networks. The input to the first network is a 1.5 kHz sinusoid of amplitude 1 V. The capacitors and resistors have the same values as in Fig. 5.3, except that R3 is *equal* to the other resistors as there is no resistor such as R4 in parallel with it. At each stage the signal undergoes two changes:

- Phase lead increases, by about 50° each time (slightly more at the third stage).
- Amplitude falls.

As a result of this, the signal that finally leaves the chain has a phase lead of 156° and an amplitude of 42 mV. By experimenting with signals of other frequencies it is found that the phase lead and output amplitude depend on frequency. The important point is that the phase lead is exactly 180° at one particular frequency. At this frequency the output signal is exactly *out of phase with,* or *inverted,* with respect to the input signal.

In the circuit of Fig. 5.3 the transistor is connected as a common-emitter amplifier. This amplifies *and inverts* the signal coming from

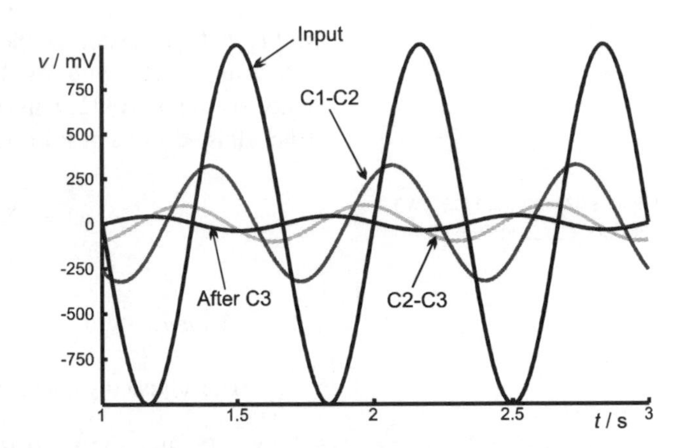

Figure 5.4 *This shows what happens in a phase-shift network like R1-R3 and C1-C3 of Fig. 5.3, when there is a 1.5 kHz sinewave generator feeding a 1 V signal into C1. Between C1 and C2 the signal lags by 60° and is reduced in amplitude (328 mV). Between C2 and C3 it lags by another 60° and the amplitude is 98 mV. On the other side of C3, which would normally be connected to the base of Q1, the signal lags another 60°, making a total phase lag of 180°. The amplitude is only 44 mV now, but Q1 brings this back to 1 V in the complete circuit.*

the network chain. The amplified *inverted* signal is fed back to the beginning of the chain. The amplification compensates for the loss of amplitude as the signal passes along the chain. The *double* inversion (in the chain *and* in the transistor) means that feedback is positive. As a result of this, the feedback maintains the oscillations indefinitely. But this only happens at the one particular frequency at which the total phase lead in the chain is exactly 180°.

It can be shown that the frequency of oscillation is:

$$f = \frac{1}{2\pi\sqrt{6}RC}$$

Working to 3 significant figures, this equation can be simplified to:

$$f = \frac{1}{15.4RC}$$

Test your knowledge 5.2

Calculate the frequency of a phase shift oscillator with 47 kΩ resistors and 47nF capacitors.

Example

With the values quoted in Fig. 5.4 the frequency is:

$$f = \frac{1}{15.4 \times 10 \times 10^3 \times 4.7 \times 10^{-9}} = 1382\,\text{Hz}$$

In Fig. 5.4 all resistors in the chain have the same value. But R3 is, in effect, in parallel with R1. For highest precision their parallel resistance should be 10kΩ. This means a slight increase in the value of R3. The adjusted value of R3 is calculated using the equation:

This equation is derived from the ' two parallel resistors' equation on p. 3.

$$R_3 = \frac{R_4 \times R_1}{R_4 - R_1}$$

Example

The adjusted value of R3 in this oscillator is:

$$R_3 = (220 \times 10)/(220 - 10) = 2200/230 = 10.48\ \text{k}\Omega$$

Using the E24 series, the closest value is still 10kΩ, so all three resistors can have the same value. If higher precision is required, we can use a high-precision 10.5 kΩ resistor or wire a 470Ω resistor in series with R3. But there is little point in doing this unless the capacitors too are high-precision types. In general, if R4 is very much greater than the network resistors, it is unnecessary to adjust R3.

Other versions of a phase shift oscillator are based on a MOSFET, a JFET or an operational amplifier.

Precision

The frequency of the phase shift oscillator depends upon the values of the capacitors and the resistors R1 to R4. If we want the frequency to be precise, we use high-precision components, such as 1% tolerance metal film resistors and precision polystyrene capacitors (also 1% tolerance). This degree of precision is not as easy to attain with the Hartley oscillator because inductors typically have a tolerance of 10%. One solution is to use a variable capacitor, but it is a tedious matter to set the frequency precisely to the required value.

One solution to the problem is to include a quartz crystal in the tuning circuit to force the circuit to oscillate at the natural frequency of the crystal. Crystals are available cheaply in a wide range of frequencies from 100 kHz up to many tens of megahertz. Most have a precision of ±30 ppm or better. This allows us to build high-precision oscillators at low cost.

Fig. 5.5 shows how a crystal may be included in a *Colpitts oscillator*. A Colpitts oscillator has a capacitor-inductor network similar to that of the Hartley oscillator of Fig. 5.1. Instead of taking the feedback from a tapping of the inductor, we replace the capacitor by two capacitors in series and take the feedback from the connection between them. This has been done in Fig. 5.5 and, in addition, the inductor has been replaced by a quartz crystal. An extra capacitor C4 has been added to the network to balance its action.

Vibrating quartz crystals

A quartz crystal consists of atoms of silicon and oxygen arranged in a regular three-dimensional array, which we call a crystal *lattice*. The atoms are held in their positions by electrical forces of attractions and repulsion between them. The lattice is not rigid; think of the atoms as being held together by a framework consisting of coiled springs. If we apply a force to the lattice, it is squashed out of shape and the distances between the atoms are slightly altered. This unbalances the electrical fields in the lattice with the result that one surface of the crystal becomes more positive than the opposite surface. In other words, a voltage difference develops across the crystal. This is the basis of the action of the crystal microphone. Sound waves make the crystal vibrate, distorting it and causing a varying voltage to appear across it.

The crystal also operates in the reverse way. If we apply a varying voltage across a crystal, this alters the electrical fields in the lattice. Some atoms to move closer together and others to move further apart. The result is that the crystal changes shape. This is the basis of the piezo-electric sounder, which emits sounds when it is subject to an alternating audio-frequency voltage.

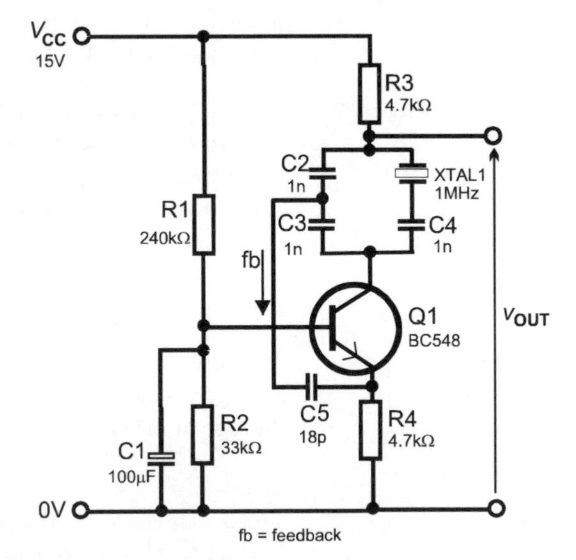

Figure 5.5 *A Colpitts oscillator has positive feedback from a capacitor network, which may include a quartz crystal to fix the oscillating frequency with high precision.*

> **Precision of quartz crystals**
>
> Like any 'springy' structure, the crystal has a natural frequency of vibration. This depends on its size and the direction in which the crystal is cut from the block of quartz. Crystals are cut to oscillate at one particular frequency, and this frequency can be very accurately set. For example a watch crystal is cut to have a natural frequency of 32.768 kHz with a precision of ±0.0015%.

Electronically a crystal *behaves* just like a network including some capacitance, some resistance, and some inductance, so we still have an 'inductor' in the circuit but in a different form. The difference in its action is that the crystal has a precision of ±15 ppm (± 0.000 015%) compared with the 10% of a typical wirewound inductor. Because there is no inductor as such, it is not possible to use a second coil (L2, Fig. 5.1) for output. Instead we place a resistor R3 in series with the network and take the voltage generated across this as the output of the oscillator.

When power is applied, the network begins to oscillate. The feedback path runs from between C1 and C2, through C5 to the emitter of Q1. As in the Hartley oscillator, this is *positive feedback*. At one moment the crystal changes its shape as it absorbs electrical energy from the network. At the next moment it returns to its original shape, and transfers electrical energy back to the network. Only a small amount of energy is required from the network to keep the crystal oscillating.

Q: quality factor, as explained on p. 53.

A very important feature of the crystal is that it has a very high Q. This means that it oscillates very strongly at its fixed frequency and exchanges maximum energy with the network. At frequencies close on either side of this it hardly oscillates at all. For this reason, the exact values of the capacitors in the network are not important. The crystal dominates the network and forces it to oscillate at its own frequency. Adding a crystal to the network gives high precision to the oscillator.

The crystal is very stable too, being little affected by temperature changes in the region of 25°C. Its frequency also varies very little with the age of the crystal.

Positive feedback

The discussion of feedback in Chapter 2 (Fig. 2.9) described *negative* feedback. This is generally an advantage in a circuit such as an amplifier or in any other circuit in which amplification plays a part. It makes the amplifier more stable. The amplifier is less affected by changes in temperature and in the variability between components.

Negative feedback reduces the gain of the amplifier, which may be a disadvantage, but it usually ensures that the gain is constant under all conditions.

A is gain with feedback.
A_0 is gain without feedback.
β is fraction fed back.

The discussion of feedback showed that:

$$A = \frac{A_0}{1 - \beta A_0}$$

With *positive* feedback, both A_0 and β are positive, or both are negative. In the case of the Hartley oscillator, we use a common-base amplifier, which is non-inverting. Thus A_0 is positive. Also part of the signal is fed back from the inductor to the emitter of the transistor so as to be in phase with the output of the transistor. So β is positive. In the phase shift oscillator we have a common-emitter amplifier, which is inverting, so A_0 is negative. The total phase shift is 180°, so β is negative too. A negative quantity multiplied by a negative quantity gives a positive quantity. In *both* amplifiers, βA_0 is positive.

The value of βA_0 varies with frequency. At one particular frequency:

$$\beta A_0 = 1$$

When this happens, the denominator of the expression for A becomes zero, and the value of A becomes infinite. This means that we get an output from the circuit without any input. The circuit oscillates at that one frequency.

Activity 5.1 Oscillators

Build one of the oscillators shown in Fig. 5.1 or Fig. 5.2, and connect an oscilloscope to its output. Observe the waveform and measure its frequency.

In Fig. 5.1 the output of the circuit may be taken from the collector of Q1, so coil L2 is not required. Investigate the effect of increasing or decreasing the feedback, either by adjusting the tapping point on L1 or by altering the value of C3. Also alter the frequency by changing C2 and/or L1.

In Fig. 5.2 alter the oscillating frequency by changing resistors R1 to R3 and/or capacitors C1 to C3.

Problems on oscillators

1 What is meant by a 'tuned network'? Explain how this can be used as part of an oscillator.

2 What is positive feedback? Quote two examples.

3 Why is a common-base amplifier suitable for use in an oscillator of the Hartley type?

4 Describe how a phase shift oscillator works.

5 Explain how a quartz crystal is used to increase the precision of an oscillator.

6 Describe the operation of a Colpitts oscillator.

Multiple choice questions

1 A Hartley oscillator is not suitable for audio frequencies because:

A its gain is too low.
B its gain is too high.
C it needs large inductors.
D its output is very distorted.

2 At each stage in the capacitor-resistor chain of a phase shift oscillator, there is:

A an increase of amplitude.
B a phase lead.
C positive feedback.
D a phase lag.

3 A Hartley oscillator has a tuned network consisting of a 22 nF capacitor and a 150 mH inductor. The frequency of the oscillator is:

A 19.7 MHz.
B 48.2 MHz.
C 2.77 kHz.
D 8.70 kHz.

4 A Hartley oscillator is based on a:

A common-collector amplifier.
B BJT amplifier.
C MOSFET amplifier.
D common-base amplifier.

5 The tolerance of a typical inductor is:

A 5%.
B 10%.
C 15 ppm.
D 1%.

6 The precision of a typical quartz timing crystal is:

A 1%.
B 10%.
C 200 ppm.
D 30 ppm.

7 A crystal-controlled oscillator, runs at the frequency of the crystal because the crystal has high:

A *Q*.
B precision.
C frequency.
D capacitance.

6 Operational amplifiers

Summary

An operational amplifier is a differential amplifier with distinctive features, which are described at the beginning of this chapter. This is followed by descriptions of the kinds of circuit in which operational amplifiers are commonly employed: voltage comparator, inverting amplifier, non-inverting amplifier, voltage follower, adder, and difference amplifier.

Operational amplifiers are precision, high-gain, differential amplifiers. They were originally designed to perform mathematical *operations* in computers, but nowadays this function has been taken over by digital microprocessors. Operational amplifiers (or 'op amps' as they are usually called) are still widely used in many other applications. An op amp can be built from individual transistors and resistors but practically all op amps are manufactured as integrated circuits. Dozens of different types of op amp are available with various combinations of characteristics.

Terminals

All op amps have at least five terminals (Fig. 6.1):

- Positive and negative supply. Most op amps run on a *dual supply* (or *split supply*). Typical supply voltages are ±9 V, ±15 V or ±18 V. Some op amps run on low voltages such as ±1 V or ±2V. Certain op amps that are capable of accepting input voltages close to the supply rails can also run on a single supply, such as 2V, up to 36V.
- Inverting and non-inverting inputs to the first stage of the amplifier, which is a differential amplifier (see Chapter 4). In this book we refer to these inputs as the (–) and (+) inputs.
- Output terminal.

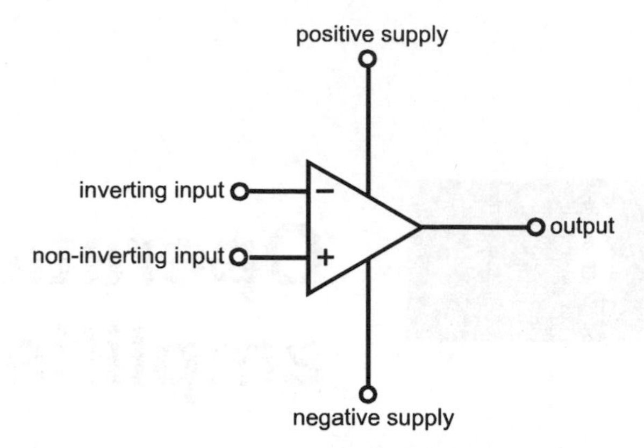

Figure 6.1 *All operational amplifiers have these five terminals. Some may have additional terminals. The power supply terminals are often omitted from circuit diagrams to simplify the layout.*

Amplifiers may also have two or three other terminals, including the *offset null* terminals.

Packages

Most op amps are available as a single amplifier in an 8-pin integrated circuit package, and practically all have the standard pinout shown in Fig. 6.2. Pins 1 and 5 are used for other functions, depending on the type. Many op amps are also available with 2 or 4 amplifiers in a single 14-pin integrated circuit package. They share power supply pins and usually do not have terminals for special functions.

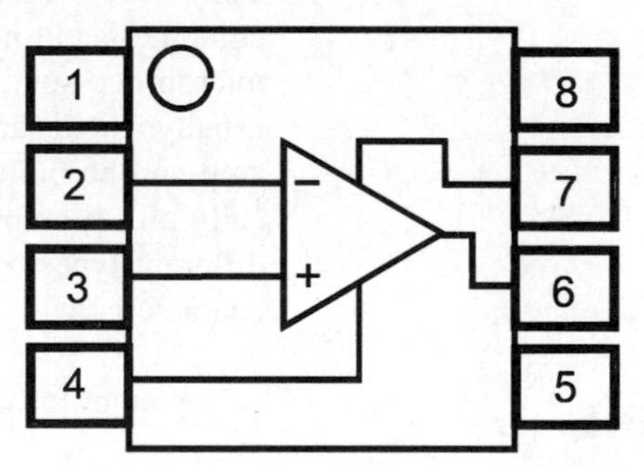

Figure 6.2 *Op amp integrated circuits have these pin connections. A circular dimple indicates which is pin 1.*

Output polarity

Assuming that the op amp is working on a dual supply (the mid-rail voltage is 0 V), output is positive when the (+) input voltage exceeds the (–) input voltage. It is negative when the (–) input voltage exceeds the (+) input voltage. It is zero when inputs are equal.

Ideal op amp

An ideal op amp has the following features:

- Infinite voltage gain.
- Gain is independent of frequency.
- Infinitely high input resistance.
- Zero output resistance.
- Zero input voltage offset.
- Output can swing positive or negative to the same voltages as the supply rails.
- Output swings instantly to the correct value.

Practical op amps

All op amps fall short of these ideals. A more practical list is:

- Very high voltage gain. The gain without feedback (known as the *open-loop gain*) is of the order of 200 000.
- Gain falls with frequency. It is constant up to about 10 kHz then falls until it reaches 1 at the transition frequency, f_T. Typically, f_T is 1 MHz, but is much higher in some op amps.
- High input resistance. This is usually at least 2 MΩ, often much more.
- Low output resistance. Typically 75 Ω.
- Input voltage offset is a few millivolts.
- The output voltage swings to within a few volts of the supply voltages (typically ±13 V for an amplifier run on ±15 V).
- Output takes a finite time to reach its correct value and may take additional time to settle to a steady value.

Input voltage offset

Because an op amp is a differential amplifier, its output should be 0 V when its inputs are at equal voltages. In practical amplifiers, the output is 0 V when the inputs *differ* by a small amount known as the *input offset voltage.* Many op amps have a pair of *offset null* terminals (Fig. 6.3), by which the input offset voltage may be nulled, that is, made equal to zero. The same voltage is applied to both inputs and the variable resistor is adjusted until the output is 0 V.

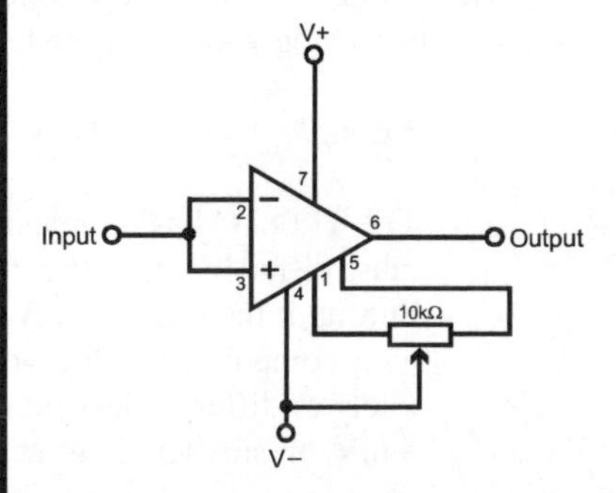

Figure 6.3 *Offset null adjustment requires a variable resistor with its wiper connected to the negative supply or (with some op amps) to 0 V. Some op amps require a 100 kΩ variable resistor.*

Slew rate

The maximum rate at which the output voltage can swing is called the *slew rate*. It is usually expressed in volts per microsecond. Slew rates range from a fraction of a volt to several hundred volts per microsecond, but a typical value is 10 V/μs. Op amps with the highest slew rates are the best ones for use at high frequencies.

Voltage comparator

This is the only application in which we make use of the op amp's very high voltage gain. The object of the circuit (Fig. 6.4) is not to *measure* the difference between the two input voltages but simply to indicate which is the higher voltage. Output swings negative if v_1 is more positive than v_2. Output swings positive if v_2 is more positive than v_1.

Output is zero or close to zero if the inputs differ by only a small amount, but usually they differ sufficiently to make the output swing as far as it can go in one direction or the other.

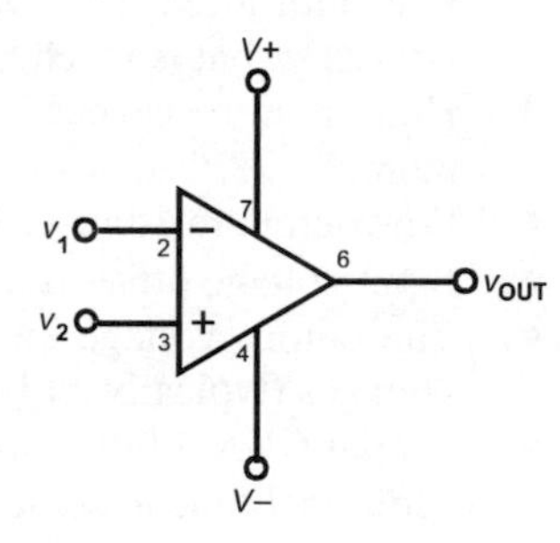

Figure 6.4 *Demonstrating an op amp wired as a voltage comparator.*

Example

A TL081C op amp has an open-loop gain of 200 000. If the amplifier is operating on a power supply of ±15 V, its output can swing to within 1.5V of either rail. Its swing is ±13.5 V. To produce a swing of this amount, the inputs must differ by 13.5/200 000 = 67.5 μV. As long as the differences are 67.5 μV or more, the amplifier saturates and its output swings to ±13.5 V.

Data sheets often quote the open-loop gain in decibels. A gain of 200 000 is equivalent to 106 dB.

In this application it is essential to choose an op amp with a small input offset voltage. If this is not done, input offset could make the output swing the wrong way when inputs are close.

Example

The TL081C has a typical input offset voltage of 5 mV. Inputs must differ by at least 5 mV to obtain a reliable output swing. Putting it the other way round, this op amp is unsuitable for use as a comparator if the voltages that we want to compare are likely to differ by less than 5 mV. If differences are likely to be 5 mV or smaller, it is essential to use an op amp with small offset. Bipolar op amps are best in this respect and a suitable one is the OP27, which has an input offset voltage of only 30 µV.

Activity 6.1 Voltage comparator

Demonstrate the action of an op amp as a voltage comparator using the circuit of Fig. 6.5. The meters must be able to read both positive and negative voltages. Either use digital multimeters with automatic polarity, or centre-reading moving-coil meters. Keep the (+) input constant (adjust VR1 to bring ME1 to +1V, for example) and then set adjust VR2 to set the (+) input at a series of values ranging from –6V to +6V.

Record the readings of the three meters in a table. Repeat with a different voltage on ME1, or hold the (–) input constant and vary the (+) input. Write a summary interpreting your results. What is the reading on ME3 when the input voltages are equal? If ME3 does not show a reading of 0V, can you explain why? Does the output voltage ever swing fully to +6V or –6 V? How far does it swing in each direction? Repeat this investigation for another type of op amp.

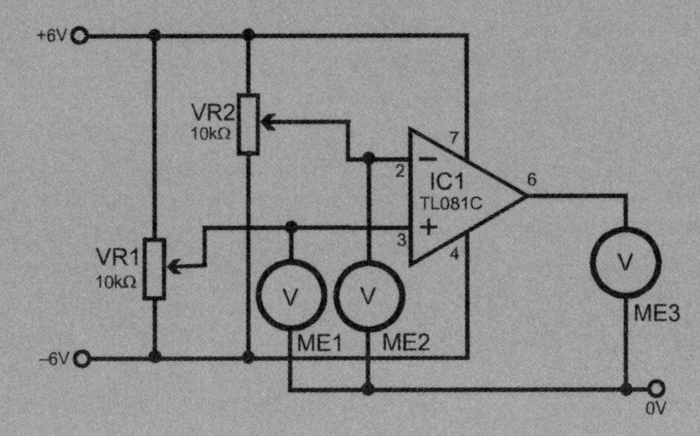

Figure 6.5 *Measuring the input offset voltage of an op amp.*

Test your knowledge 6.1

Which way does the output of an op amp swing when:

1 The (+) input is at +2.2 V and the (–) input is at +1.5 V?
2 The (+) input is at –4.6V and the (–) input is at + 0.5 V?

Inverting amplifier

The inverting amplifier (Fig. 6.6) has part of the output signal fed back to the inverting input. This is *negative feedback*, and the circuit has three interesting properties as a result of this.

Equal input voltages

The (+) input has high input resistance. Only a minute current flows into it and so only a minute current flows through R_B. As a result, there is almost no voltage drop across R_B and the (+) input is practically at 0 V.

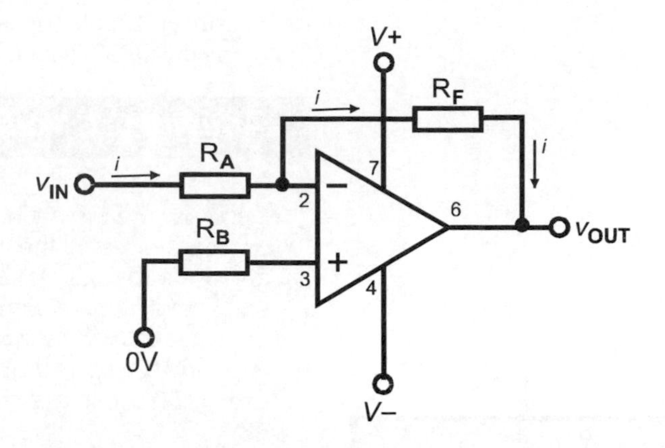

Figure 6.6 *In an inverting amplifier circuit the input goes to the (–) input terminal and there is negative feedback. Suitable values forActivity 6.3 are* R_A = *10 kΩ,* R_F = *100 kΩ and* R_B = *9.1 kΩ.*

If v_{IN} is made positive v_{OUT} becomes negative. There is a fall in voltage across the resistor chain R_A and R_F. At the v_{IN} end it is positive and at the v_{OUT} end it is negative. Somewhere along the chain the voltage is zero. As v_{OUT} falls, it pulls down the voltage at the (–) input until this is the point at which the voltage is zero.

If v_{OUT} falls any further, the (–) voltage becomes *less* than the (+) voltage and v_{OUT} begins to rise. If the voltage is exactly zero (taking this to be an ideal amplifier in which we can ignore input offset voltage), v_{OUT} neither rises or falls. Feedback holds it stable. The important thing is that the (–) input has been brought to the *same voltage* as the (+) input.

The same argument applies if v_{IN} is made negative. It also applies if the (+) input is held at any voltage above or below zero. This allows us to state a rule:

An inverting amplifier comes to a stable state in which the two inputs are at equal voltage.

This is a helpful rule to remember when working out how some of the op amp circuits operate.

Gain set by resistors

Fig. 6.7 shows the voltages when the circuit has reached the stable state referred to above. Because of the high input impedance of the (–) input, no current flows into it (ideally). All the current flowing along

R_A flows on along R_F. Call this current i.

Because the (–) input is at 0 V, the voltage drop across R_A is:

$$v_{IN} = iR_A$$

Similarly, the voltage drop across R_F is:

$$v_{OUT} = iR_F$$

Combining these two equations, we obtain:

$$\text{Voltage gain, } A_v = \frac{v_{OUT}}{v_{IN}} = \frac{iR_F}{iR_A} = \frac{R_F}{R_A}$$

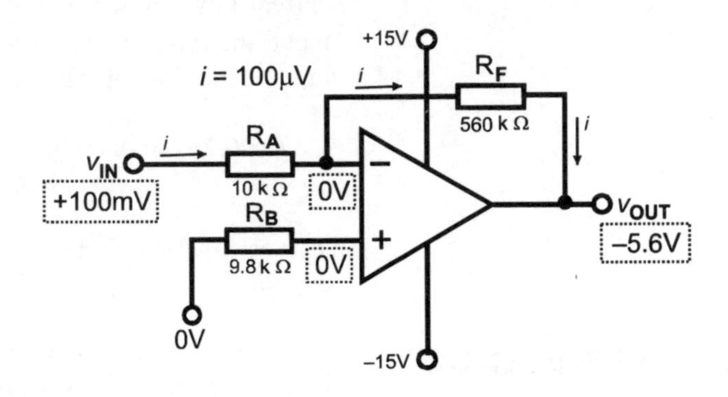

Figure 6.7 *The output voltage stabilises when the voltages at the (+) and (–) inputs are equal.*

This leads to another rule for op amps:

The voltage gain of the circuit is set only by the values of the two resistors.

Test your knowledge 6.2

An inverting amplifier has R_A = 2.2kΩ and R_F = 820kΩ. Calculate its closed-loop gain.

Example

An inverting op amp amplifier circuit has an input resistor R_A of 10 kΩ, and a feedback resistor R_F of 560 kΩ. Its voltage gain is:

$$A_V = \frac{R_F}{R_A} = \frac{560000}{10000} = 56$$

Although the op amp may have an open-loop gain of 200 000, the amplifier *circuit* has a gain of only 56. This is known as the *closed-loop* gain, and must always be substantially less than the open-loop gain of the op amp. The open-loop gain is subject to tolerance errors resulting from differences arising during manufacture. But the closed-loop gain depends only on the precise values of the resistors. If 1% tolerance resistors are used the gain is precise to 1%.

The *closed-loop* gain obtained from a given combination of R_A and R_F can never be as great as the *open-loop* gain of the op amp at any given frequency (see box).

Virtual earth

If the (+) input is at 0 V, the circuit stabilises with (–) at 0 V too. The input current flows through R_A toward the (–) input and an *equal* current flows on through R_F. Although no current flows into the (–) input, the action is the same as if there is a direct path from the (–) input to the 0 V rail. The rule is:

The (–) input of an op amp connected as an inverting amplifier acts as a virtual earth.

This is a helpful fact to remember when analysing certain op amp circuits.

Input resistance

As a consequence of the virtual earth, a signal applied to the input in Fig. 6.6 has only to pass through R_A to reach a point at 0 V. The input resistance of the amplifier in this example is only 10 kΩ. Although the input resistance *of the op amp* may be at least 2 MΩ and possibly as high as 10^{12} Ω, the input resistance *of the amplifier circuit* is nearly always considerably less. If resistor values are being chosen for high gain, R_F is made large and R_A is made small. As a result, the input resistance of a high-gain inverting amplifier is nearly always a few tens of kilohms. For higher input resistance we use an additional op amp as a voltage follower, as explained later in this chapter.

FETs have the gate insulated by a layer of silicon oxide or a reverse-biased depletion layer.

In Fig. 6.7 we show zero voltage at both ends of R_B. Ideally no current flows through it. In that case, omitting R_B altogether should make no difference to the operation of the circuit. If some loss of precision is acceptable, R_B can be left out of this circuit and the other circuits in which it occurs. This saves costs, makes circuit-board layout simpler, and reduces assembly time.

Test your knowledge 6.3

What is the input resistance of the amplifier described in Test your knowledge 6.2?

The conclusion reached in the previous paragraph relies on the 'ideal' assumption that no current flows into the inputs. In practice a current known as the *input bias current* flows into the inputs. In bipolar op amps, which have BJT input transistors, this is the base current that is necessary to make the transistor operate, and is around 100 nA.

Effect of frequency on open-loop gain

At DC and low frequencies the *open-loop* gain of an op amp is the value quoted in the data tables, often 200 000. The gain falls off at higher frequencies (Fig. 6.8). The op amp is said to operate at 'full power' at all frequencies from 0 Hz (DC) up to the frequency at which the power output is half its DC power. This is the frequency at which output is 3 dB down on the maximum. From 0 Hz up to this frequency is the *full-power bandwidth.* Beyond this frequency the gain falls at a steady rate such that the product of the open-loop gain and the frequency is constant .

Example:

If the gain is 20 at 100 kHz, it is 10 at 200 kHz, 5 at 400 kHz, and so on to a gain of 1 at 2 MHz. In every case:

Gain-bandwidth product = gain $\times$ frequency = 2 MHz

The frequency at which gain becomes 1 is known as the *transition frequency*, and is numerically equal to the gain-bandwidth product.

Op amp type	Full-power bandwidth	Gain-bandwidth product
741	10 kHz	1 MHz
TL081CP	100 kHz	3 MHz
EL2044CN	5.2 MHz	60 MHz

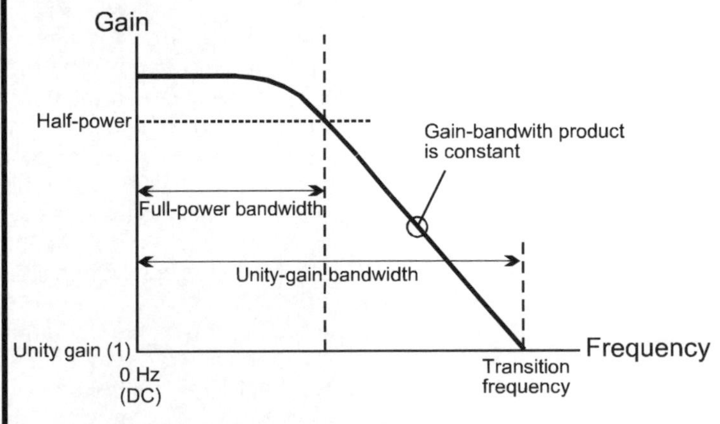

Figure 6.8 *The gain of an op amp falls off steadily at frequencies beyond the half-power bandwidth.*

Op amps with FET inputs do not require base current but there is leakage current of a few picoamperes. The leakage current can generally be ignored but the base current of bipolar can not. Although it is small, it flows through a high input resistance (around 2 MΩ) which leads to a voltage of a hundred or more millivolts at the inputs. The input bias currents of the two inputs differ due to manufacturing differences in the internal circuitry, so the voltages differ, leading to an additional input voltage offset in addition to the one already described.

Calculating R_B

In practical circuits it is possible to minimise the voltage offset by providing paths to the 0V line through the resistors. If we make these paths to 0 V equal in resistance, the voltages developed across them are equal and the voltage offset is zero. If v_{IN} is zero in Fig. 6.7, then v_{OUT} is also zero. The small base current flows through R_B to the (+) input, so there is a small voltage drop across it and the (+) input is at a small negative voltage. Bias current for the (–) input comes along R_A and R_F from two points both at 0 V. If R_B has the same resistance as R_A and R_F in parallel, the voltage drop is the same for both inputs. The voltage of the (–) input is equal to that at the (+) input. This minimises the effect of input bias currents.

Example

In Fig. 6.7 R_B must equal 10 kΩ and 560 kΩ in parallel:

$$R_B = \frac{10 \times 560}{10 + 560} = \frac{5600}{570} = 9.8$$

The calculation is in kilohms, so the value of R_B is 9.8kΩ.

Test your knowledge 6.4

Calculate a suitable resistance for R_B in the amplifier described in *Test your knowledge 6.2.*

Activity 6.2 Input voltage offset

Fig. 6.9 is a circuit for measuring the input voltage offset of an op amp. R3 and R4 forma potential divider holding the (–) input close to 0 V. It will not be exactly 0 V as the resistors are not likely to be exactly equal, but is will be near enough for the purpose of this measurement. R_1, R_2, and VR1 form a second potential divider to adjust the voltage at the (+) input about 100 mV on either side of zero. VR1 should preferably be a multi-turn variable resistor to allow the voltage to be adjusted precisely. One with 18 or 22 turns is suitable.

Adjust VR1 until the output of the amplifier rests at 0 V. Then use a multimeter to measure the voltages at the inputs. Subtract one from the other to obtain the input offset voltage.

The TL081C may be expected to have an offset of several millivolts, but it varies from one ic to another. Repeat the experiment with other FET-input op amps, and also with some bipolar-input op amps, such as the popular 741 and the more recent OP-177GP, which has an offset so small that you may not be able to measure it.

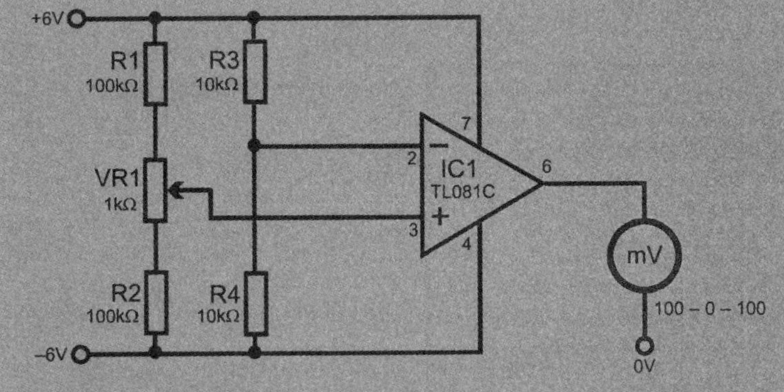

Figure 6.9 *A circuit for measuring input offset voltage.*

Activity 6.3 Transition frequency

The inverting amplifier circuit of Fig. 6.6 is fed with a sinusoid from a signal generator. Adjust the amplitude to 100 mV and set the frequency to 100 Hz. Observe the output on an oscilloscope.

Proceed by running the circuit at frequencies from 100 Hz to 10 MHz, in steps of ten times. At each frequency, measure and record the output amplitude. The first frequency of importance is that at which the output amplitude is 3 dB down on the input. For an input of 100 mV, and a gain reckoned at 56 times, the –3 dB point is 100 × 56 × 0.7071 = 3.96 V. This is the edge of the *full-power bandwidth* of the amplifier. Continue until you have found the *transition frequency*, at which the gain of the amplifier is 1 (amplitude = 100 mV).

Non-inverting amplifier

Potential divider = voltage divider

This amplifier uses negative feedback taken from a potential divider connected between the output and the 0 V rail (Fig. 6.10). The input signal is fed to the non-inverting input. The voltage v_F at the junction of R_A and R_F is given by the usual equation for a potential divider:

$$v_F = v_{OUT} \times \frac{R_A}{R_A + R_F}$$

This is the voltage at the inverting input. Assuming that the current flowing through R_B is so small that it can be ignored, the voltage at the non-inverting input equals v_{IN}. Applying the same reasoning as for the inverting amplifier, we find that the circuit becomes stable when the input voltages are equal:

$$v_{IN} = v_F = v_{OUT} \times \frac{R_A}{R_A + R_F}$$

Rearranging the equation:

$$A_V = \frac{v_{OUT}}{v_{IN}} = \frac{R_A + R_F}{R_A}$$

The rule we derive from this is the same as for the inverting amplifier:

The voltage gain of the circuit is set only by the values of the two resistors.

Test your knowledge 6.5

A non-inverting amplifier based on a TL081C op amp has R_A = 33 kΩ and R_F = 270kΩ.

1 What is its voltage gain?

2 What is its input resistance?

3 What is a suitable value for R_B?

Example

For comparison, we take resistors of the same values as in the inverting amplifier. Working in kilohms:

$$A_V = \frac{10 + 560}{10} = 57$$

The voltage gain is 57 times. This is very slightly greater than that of the inverting amplifier.

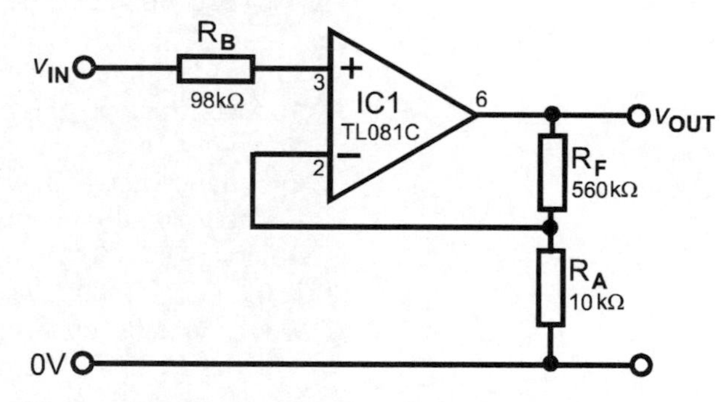

Figure 6.10 *The power rails and their connections to the op amp are omitted from this diagram of a non-inverting amplifier and in all op amp circuits after this one.*

For the same reasons as in the inverting amplifier, the value of R_B is equal to the value of R_A and R_F in parallel. If the op amp has FET inputs, the current flowing through R_B is so small that R_B can be omitted.

The input resistance of this amplifier is the input resistance of the op amp itself, which is very high. With typical bipolar op amps it is 2 MΩ, and with FET op amps it is up to 10^{12} Ω, or 1 teraohm. A non-inverting amplifier is often used when high input resistance is important.

Voltage follower

A particular instance of the non-inverting amplifier is illustrated in Fig. 6.11. There is no feedback resistor R_F and R_A is omitted. The full v_{OUT} is fed back to the inverting input. As before, the circuit settles so that the inputs are at equal voltage. The non-inverting input is at v_{IN}, so the inverting input must be at v_{IN} too. But the inverting input is at v_{OUT}, being connected directly to the output. As a result of this:

$$v_{OUT} = v_{IN}$$

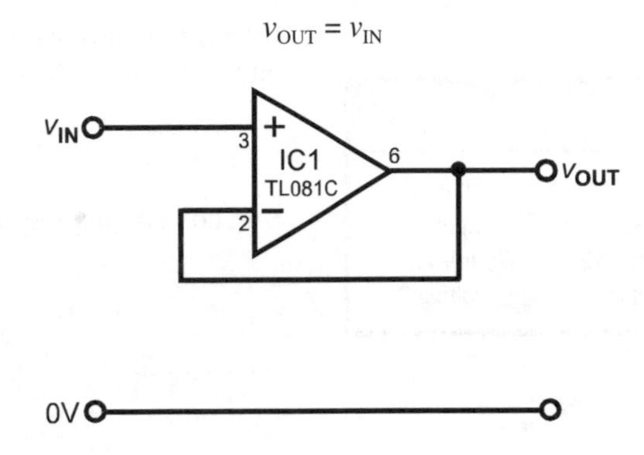

Figure 6.11 *A voltage follower is useful as a buffer between circuits.*

Because the output voltage equals the input voltage, this is called a *unity-gain* amplifier. It is also known as a *voltage follower*. It has the same applications as the emitter follower and source follower transistor amplifiers, to act as a buffer between a circuit with high output resistance and a circuit with low input resistance. The op amp is well suited as a buffer because the input resistance can be up to 1 TΩ, and its output resistance is only 75 Ω. The op amp follower is preferred to the transistor followers because its output follows input exactly, apart from a small error due to input offset voltage. In contrast, the transistor followers always have a much larger offset (0.6 V in the case of the emitter follower) and their gain is not *exactly* 1 (usually nearer to 0.9 in the source follower).

Adder

The amplifier in Fig. 6.12 is an extension of the inverting amplifier and depends on the virtual earth at the inverting input of the op amp. The version shown has four inputs, but there may be any practicable number of inputs, each with its input resistor. In the figure, all resistors have the same value and a suitable value is 10 kΩ.

When voltages (the same or different) are applied to the inputs, currents flow through each input resistor to the (–) input of the op amp. Because of the virtual earth each input resistor has its input voltage at one end and 0 V at the other. The currents through each input resistor are independent of each other and their values are determined by

Ohm's Law. The currents through the resistors are:

$$i_1 = \frac{v_1}{R_1} \qquad i_2 = \frac{v_2}{R_2}$$

$$i_3 = \frac{v_3}{R_3} \qquad i_4 = \frac{v_4}{R_4}$$

In fact, the currents join, and R_F carries the *sum* of the currents to the output. If all resistors are equal, and of resistance *R*, we can write:

$$i = i_1 + i_2 + i_3 + i_4 = \frac{1}{R}(v_1 + v_2 + v_3 + v_4)$$

But, considering the current through R_F, Ohm's Law tells us that:

$$i = -\frac{v_{OUT}}{R}$$

Rearranging:

$$v_{OUT} = -iR = -(v_1 + v_2 + v_3 + v_4)$$

The output voltage is the negative of the sum of the input voltages. A circuit such as this can be used as a mixer for audio signals. If the input resistors are variable, the signals can be mixed in varying proportions.

This leads to the concept of *weighted inputs*, a practical example of which is described in Chapter 14.

Test your knowledge 6.6

The resistors of a 3-input op amp adder are all 22 kΩ. The input voltages are 542 mV, 375 mV and 1.5 V. What is the output voltage?

Weighting: signals are passed through resistors of different values to multiply them by different factors before adding them.

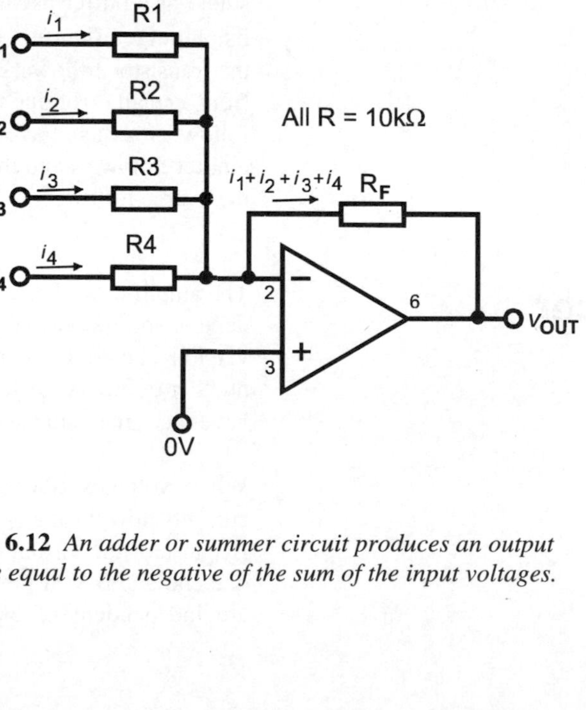

Figure 6.12 *An adder or summer circuit produces an output voltage equal to the negative of the sum of the input voltages.*

Difference amplifier

One form of the difference amplifier or *subtracter* is shown in Fig. 6.13. All its resistors are equal in value. Consider the resistor chain R3/R4 . The (+) input of the op amp has high input resistance, so no current flows out of the chain through this. One end of the chain is at 0 V and the other end is at v_2. There is a voltage drop of v_2 along the chain and, because R3 and R4 are equal, the voltage at the (+) input is $v_2/2$.

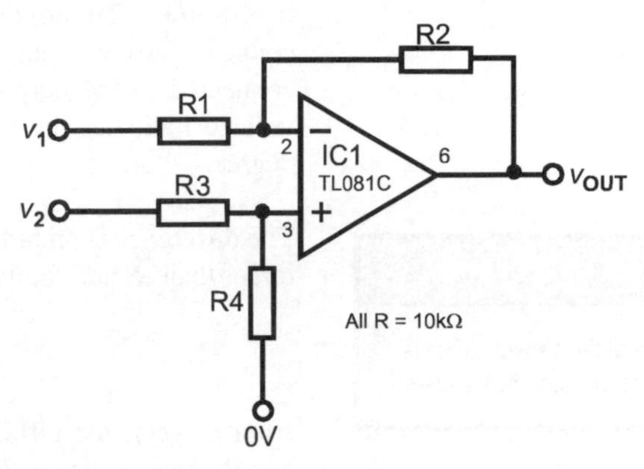

Figure 6.13 *A subtracter circuit produces an output equal to the difference between its inputs.*

The voltage at the (–) input is $v_2/2$. For the same reasons as for the other amplifiers, the circuit comes to a stable state when the voltages at the (+) and (–) inputs are equal.

Consider the resistor chain R1/R2. This has a voltage v_1 at one end, v_{OUT} at the other end and $v_2/2$ in the centre between the two resistors. No current leaves the chain to flow into the (–) input. Because the resistors are equal in value there are equal voltage drops across them. The voltage drops are:

$$v_1 - v_2/2$$

and

$$v_2/2 - v_{OUT}$$

These are equal:

$$v_1 - v_2/2 = v_2/2 - v_{OUT}$$

Rearranging gives:

$$v_{OUT} = v_2 - v_1$$

The output voltage equals the difference between the input voltages.

This circuit has applications when a constant voltage has to be subtracted from a varying voltage. Examples are found when the output from a sensing circuit needs such a correction so as to give a more useful reading. The LM335Z temperature sensor gives an output that is proportional to temperature measured on the kelvin scale. At 0°C, which is 273 K, the output is 2.73 V. The output increases by 10 mV for each degree rise in temperature. For example at 10°C, which is 283 K, the output is 2.83 V. If we subtract 2.73 V from the output, we obtain voltages that are directly proportional to the Celsius temperature. To do this, the v_1 input of the subtracter circuit is connected to a voltage reference set at 2.73 V. The v_2 input is connected to the output of the sensor circuit. The subtracter output voltage indicates the Celsius temperature on a scale of 10 mV per degree.

A degree *difference* of temperature is the same on both Celsius and kelvin scales.

The difference is amplified if $R_1 = R_3$ and $R_2 = R_4$, with R_3 and R_4 being larger than R_1 and R_3. Then:

$$v_{OUT} = (v_2 - v_1) \times \frac{R_2}{R_1}$$

Test your knowledge 6.7

If the output of the sensor circuit is 550 mV, what is the temperature?

Alternatively, the difference may be reduced in scale by making R_2 and R_4 smaller than R_1 and R_3.

A disadvantage of this circuit is that the input resistances of the two input terminals are unequal. When input voltages are equal, the (–) input of the amp is a virtual earth so the input resistance at the v_1 input is R_1. But the input resistance at the v_2 input is R_3 and R_4 in series. If all resistors are equal, the input resistance of the v_2 terminal is double that of the v_1 terminal. If the resistances of R_2 and R_4 are large because the voltage difference is to be amplified, the input resistance of the v_2 terminal is much greater than that of the v_1 terminal.

The effect of v_1 having a smaller input resistance is to pull down all voltages applied to that terminal, making v_{OUT} smaller than it should be. In some applications, both input signals include certain amounts of mains hum or other interference picked up equally on the input cables. A differential amplifier has common-mode rejection (p. 48) so this normally does not matter. But if the inputs have unequal input resistances, the interference is stronger on the input with the higher input resistance, and is not rejected.

Activity 6.4 More op amp circuits

Set up some of the op amp circuits shown in Figs. 6.10 to 6.13, either on a breadboard or by using a circuit simulator. Provide suitable input from voltage sources or signal generators and measure the output using a testmeter or oscilloscope, as appropriate. Confirm that the circuits function in the way described in this chapter.

Problems on operational amplifiers

1 Describe the appearance of a typical operational amplifier packaged as an 8-pin integrated circuit.

2 How would you measure the open-loop gain of an op amp? What result would you expect to get?

3 Describe what is meant by slew rate. Why may we sometimes need to employ an op amp with high slew rate?

4 List the terminals present on all op amps and state their function.

5 Describe how the open-loop gain of an op amp varies over the frequency range from 0 Hz (DC) to 100 MHz.

6 State two rules that are useful when designing op amp circuits.

7. Design a circuit for an op amp inverting amplifier with a closed-loop gain of –200, assuming an ideal op amp.

8. Design a circuit for an op amp non-inverting amplifier with a closed-loop gain of 101, assuming an ideal op amp.

9. Explain why op amp inverting amplifiers have a virtual earth.

10. What is input offset voltage, what is its effect, and how is it nulled?

11. Explain the meaning of (a) gain-bandwidth product, and (b) transition frequency.

12 Describe how an op amp voltage follower works and give an example of its applications.

13 The output of a sensor has high output resistance. Design a circuit (which may have more than one op amp in it) to take the output from the sensor, subtract 1.5 V from it and display the result on a multi-meter.

14 Design a circuit to mix three audio signals, two at equal amplitude and the third signal at half its original amplitude.

Multiple choice questions

1 The input resistance of a bipolar op amp is typically:

A 75 Ω.
B 75 kΩ.
C 200 kΩ.
D 2 MΩ.

2 The output resistance of an op amp is typically:

A 75 Ω.
B 75 kΩ.
C 200 kΩ.
D 2 MΩ.

3 The open-loop gain of an op amp falls to 1 at:

A 1 MHz.
B the transition frequency.
C 10 kHz.
D 75 kHz.

4 The input resistor of an inverting op amp is 12 kΩ and the feedback resistor is 180 kΩ. The input voltage is –0.6 V. The output voltage is:

A 1.5 V.
B 9 V.
C –9.5 V.
D –16 V.

5 In the amplifier described in Question 4, the value of the resistor between the (+) input and 0V should be:

A 192 kΩ.
B 11.25 kΩ.
C 180 kΩ.
D 2 MΩ.

6 The input resistor of a non-inverting op amp is 39 kΩ and the feedback resistor is 1.2 MΩ. The output voltage is –0.5 V. The input voltage is:

A –15.7 mV.
B –31.8 mV.
C 16.3 mV.
D –16.3 mV.

7 An op amp has an open-loop gain of 10 at 250 kHz. Its gain-bandwidth product is:

A 250 kHz.
B 25 kHz.
C 1 MHz.
D 2.5 MHz.

8 At 400 kHz the open-loop gain of the op amp described in Question 7 is:

A 160.
B 1.
C 6.25.
D none of these.

7 Applications of op amps

Summary

Operational amplifiers are such a useful and generally inexpensive building block that they appear in a wide range of applications. These include constant current circuits, ramp generating circuits, triggering circuits, sample and hold circuits, astables, precision rectifiers, and active filters.

Constant current circuit

The circuit of Fig. 7.1 is intended to make a constant current flow through R_L, whatever the value of R_L within limits. It does this by making a constant current flow through R_3. This current flows on through the transistor, then flows through R_L.

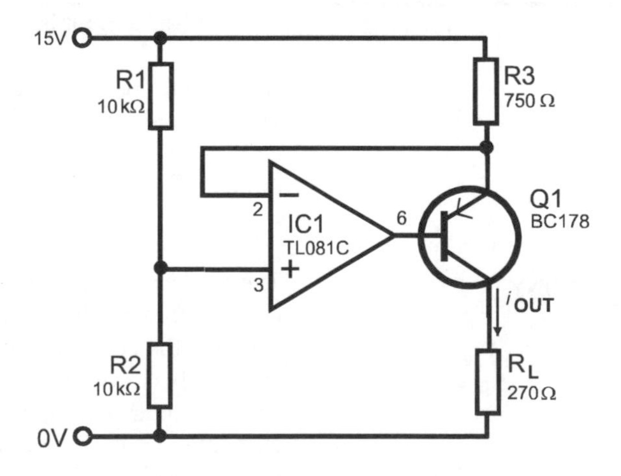

Figure 7.1 *In this constant current source, negative feedback holds the R3 current constant, and the same constant current flows to the load.*

The action of the circuit depends on the fact that negative feedback operates to bring both inputs of the op amp to the same voltage. The voltage at the (+) input is set to a fixed voltage, v_0 by the potential divider consisting of R1 and R2. Therefore the voltage at the (–) input settles at v_0. When this happens the voltage across R3 is $V_+ - v_0$.

The constant current is to be i_{OUT}. This generates a voltage across R3 and the circuit stabilises when this voltage is equal to $V_+ - v_0$:

Test your knowledge 7.1

In the circuit of Fig. 7.1, calculate the current through R_L if R_3 = 560 Ω.

$$R_3\, i_{OUT} = V_+ - v_D$$

$$\Rightarrow \quad i_{OUT} = \frac{V_+ - v_0}{R_3}$$

Example

In Fig. 7.1 R_1 and R_2 are equal, so v_0 = 7.5 V. The voltage drop across R3 is 15 – 7.5 = 7.5 V. Then i_{OUT} = 7.5/750 = 10 mA.

The equation above can be rearranged to make R_3 its subject:

$$R_3 = \frac{V_+ - v_0}{i_{OUT}}$$

Test your knowledge 7.2

In the circuit of Fig. 7.1, what value should R_3 have to provide a current of 7 mA?

This form of the equation is used to calculate what value R3 should have to generate a given current.

Example

If the current is to be 25 mA, then R_3 = 7.5/0.025 = 3 kΩ.

If we try to make i_{OUT} too large, or have too large a load resistor, the voltage across the load resistor may become so great that the base-emitter voltage of Q1 becomes less than 0.6 V. Then the transistor is turned off and no current flows.

Integrator

The best way to understand what the circuit in Fig. 7.2 does is to see how it behaves when a given input is applied to it. The (+) input is at 0 V, and the effect of negative feedback is to make the (–) input come to 0 V. It is a virtual earth, as described in Chapter 6. If there is a constant input v_{IN}, current flows through R to the virtual earth. The current is $i = v_{IN}/R$, because the virtual earth causes the voltage across R to be equal to v_{IN}. In fact, this current does not flow into the (–) input but flows toward the capacitor.

Putting this into equations, we can say that if current i flows for t seconds, the charge Q flowing toward plate A is (using the relationship

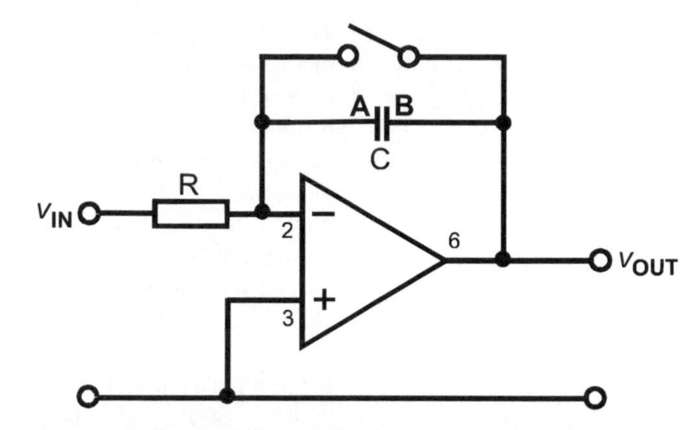

Figure 7.2 *In an integrator, the output falls as charge flows toward plate A of the capacitor.*

between *Q, i,* and *t* in the definition of the coulomb):

$$Q = it$$

If this charge were to accumulate on Plate A, it would produce a voltage *v* on the plate which is (by definition of the farad):

$$v = Q/C$$

Combining these two equations:

$$v = it/C$$

Substituting the value of *i*:

$$v = v_{IN} \times t/RC$$

This assumes that the charge accumulates and that the voltage increases. But negative feedback holds the input at 0 V. To make this happen, the output v_{OUT} falls, pulling down the voltage at plate B, which in turn pulls down the voltage at plate A, and holding it at 0 V. The fall in output exactly compensates for the charge that accumulates on plate A:

$$v_{OUT} = -v$$

$$\Rightarrow \qquad v_{OUT} = \frac{-1}{RC} \times v_{IN} t$$

If v_{IN} is constant, v_{OUT} equals v_{IN} multiplied by time *t*, multiplied by a scaling factor −1/RC. As time passes, v_{OUT} *decreases* at steady rate proportionate to 1/RC. The larger the values of R and C, the more slowly it decreases.

Test your knowledge 7.3

If v_{IN} is constant at 20 mV, R is 10 kΩ and C = 470 nF, what is v_{OUT} after 25 ms?

This discussion assumes that the capacitor is uncharged to begin with, that is, when t = 0. This is the purpose of the switch in Fig. 7.2, which is closed to discharge the capacitor at the beginning of each run. The result of *Test your knowledge 7.3* illustrates the fact that, with components of common values, the output voltage rapidly falls as far as it can in the negative direction and the op amp then becomes saturated. It also reminds us that quite a small input offset voltage can cause the circuit to swing negative in a second or so, even when v_{IN} is zero. Consequently it is important to use an op amp with small input offset voltage, and to use the offset null adjustment, as in Fig. 6.3.

If the input voltage is varying, we can work out what will happen by thinking of the time as being divided into many very short time intervals of length δt. We can consider that the voltage is constant during each of these, but changes to a new value at the start of the next interval. During each interval, charge accumulates according to the constant-voltage equation. The total charge, and also the fall in output voltage at any given time is the sum of the charges and voltage drops that have occurred during all previous intervals. For each interval, the voltage drop is $v_{IN}\delta t$ multiplied by $-1/RC$. For all the intervals between zero time and time t, we obtain the total voltage drop by calculating:

$$v_{OUT} = \frac{-1}{RC}\int_0^t v_{IN}\,dt$$

This is the *integral* of v_{IN} with respect to time, from time zero to time t, multiplied by the factor $-1/RC$. If we divide v_{OUT} by the factor $-1/RC$ we obtain the value of the integral, and this is the average value of v_{IN} during the period from 0 to t. In this way the circuit is a useful one for averaging a voltage over a period of time.

Ramp generator

The circuit has another important application, as a *ramp generator*. If we apply a constant positive or negative voltage to an integrator, v_{OUT} ramps down or up at a steady rate. Ramping voltages are often used in timing circuits. The output voltage is proportional to the time elapsed since the beginning of the ramp.

Sample and hold

If a voltage in a circuit is rapidly changing, it may sometimes be necessary to sample it and hold it, while its value is processed in some way. An example is the circuit in a digital audio recorder. The audio signal is a voltage that is rapidly changing. This signal is sampled at regular intervals, for example, 44100 times per second. The voltage *of the sample* then has to be held steady while it is being converted to its digital equivalent. Fig. 7.3 shows a simple way of doing this. A sample is taken by closing a switch, so that the capacitor is charged to the input signal voltage. A mechanical switch would not be able to operate fast

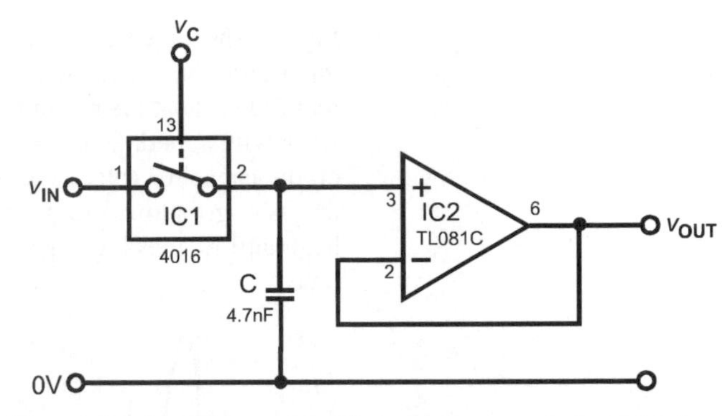

Figure 7.3 *A sample-and-hold circuit has a switch, which is opened at the instant a sample is to be taken.*

enough for sampling at high rates, so we use a switch based on CMOS transistors. The switch is closed by taking the control voltage v_C to logic low (0 V) and opened by taking it high.

Beginning with the switch closed, the voltage on the capacitor varies with v_{IN}. To take a sample, the control voltage is made high and the switch opens. The voltage on the capacitor remains at the level it had at the instant v_C went low. The voltage on the capacitor is sensed by IC2, which is a voltage follower. The output of this goes to a signal processing circuit such as an analogue-to-digital converter. An op amp with FET inputs is preferred in this application, because of its very high input resistance, which prevents charge leaking away from the capacitor. In addition, the capacitor is a low-leakage type.

CMOS switch

CMOS switches are also known as *transmission gates*. They are used when we want a switch that is under electronic (usually logic) control. The switch has a control input and is switched 'on' when this input is logic high. When it is 'on' the switch acts as a low-value resistor. Signals, either analogue or digital, can then pass through the switch in either direction. When the control input is low the resistance between the switch terminals is very high.

The switch shown in Fig. 7.3 is one of four identical switches in the 4016 ic. They are all single-pole single-throw switches but, by connecting their terminals in various ways, we can build changeover switches, double-pole switches, and other kinds. In the 4016 the low 'on' resistance is 300 Ω, but switches are available with 'on' resistances much less than this.

Fig. 7.4 shows typical behaviour of a sample-and-hold circuit. Here the input is a sine wave (grey curve). The control signal is plotted on a reduced scale at the bottom of the figure. In this example, it is a square wave with a mark-space ratio of 1:3. As long as v_C is high, the voltage on the capacitor (black curve) follows v_{IN}. Sampling occurs every 1.5 ms, as v_C goes low. From that instant the voltage across the capacitor is held until v_C goes high again. The voltage is processed during this hold time.

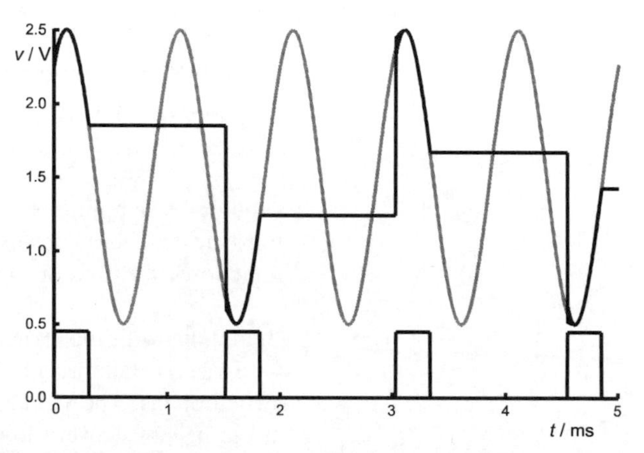

Figure 7.4 *A sample-and-hold circuit is sampling an input sine wave (grey) every 1.5 ms. The lower black curve shows the control signal v_C. The output (black) follows the input when v_C is high, but is held constant from the instant when v_C goes low.*

Inverting Schmitt trigger

This circuit (Fig. 7.5) has the same kind of action as the trigger circuit in Fig. 3.4. The voltage at the (+) input is set by the potential divider consisting of R_1 and R_2. With the values shown, the (+) input is at half the supply voltage. Note that the op amp uses a single supply in this application.

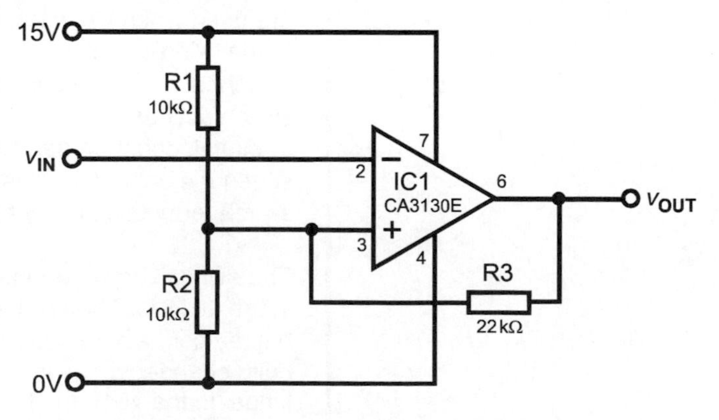

Figure 7.5 *An inverting Schmitt trigger circuit uses positive feedback to give it its 'snap' action.*

Without R3, this is just a comparator circuit like Fig. 6.4. Including R3 provides positive feedback which slightly raises or lowers the voltage at the (+) input. Starting with v_{IN} (light grey, Fig 7.6) at 0 V, the op amp is saturated, and its output (black) is high. Feedback through R3 pulls the (+) voltage (dark grey) slightly above the halfway value. This raised voltage is called the *upper threshold voltage* (Fig. 7.6, UTV). As a result of this, the input has to rise more than half way before the (–) input exceeds the (+) input and the output of the amplifier swings low.

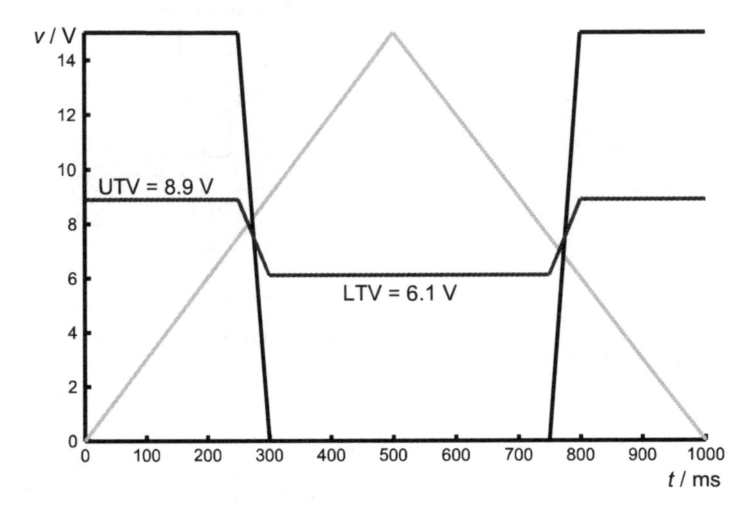

Figure 7.6 *As the input (light grey) to the trigger circuit ramps slowly up and down, the output (black) alters abruptly. Changes in the threshold are plotted in dark grey.*

This is called an *inverting* trigger because the output swings in the opposite direction to the change in input.

The triggering action is sharp because, as the (–) input *just* exceeds the UTV and the output *begins* to swing low, the falling output pulls the (+) input down. This *increases* the amount by which the (–) input exceeds the (+) input and the output falls even faster. The effect is cumulative, giving the 'snap' action characteristic of a Schmitt trigger. At the end of it, the (+) input has been pulled down to the *lower threshold voltage* (LTV).

Once triggering has occurred, v_{IN} has to fall *below the LTV* before the circuit can return to its original state. In the diagram, UTV = 8.9 V, and LTV = 6.1 V. The *hysteresis* of the trigger is 8.9 – 6.1 = 2.8 V. Feedback is raising or lowering the threshold voltage by 1.4 V. These amounts can be varied by varying the value of R3. If we want to alter the *average* threshold, we change one or both of R1 and R2.

Calculating thresholds

The calculation is best illustrated by a worked example. If we assume that the op amp swings its output fully to either power rail, R1 and R3 are in effect both connected to the positive rail when v_{IN} is low. They are in parallel (Fig. 7.7) and their effective resistance is 6.875 kΩ. We have a potential divider in which one resistor is 6.875 kΩ and the other is 10 kΩ.

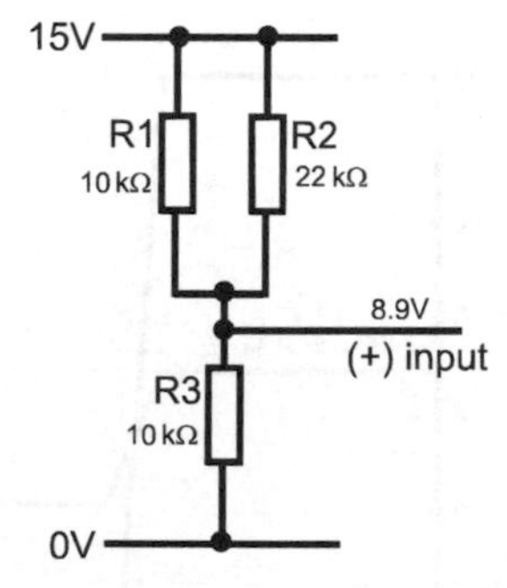

Figure 7.7 *When input is low, R1 and R3 are, in effect, connected in parallel to the positive supply rail.*

Hysteresis: difference between lower and upper thresholds.

Given that the positive supply is 15 V, the (+) input of the op amp is at:

$$\text{UTV} = 15 \times 10/16.875 = 8.89 \text{ V}$$

Similarly, when v_{IN} is high, the upper resistor in the potential divider is 10 kΩ, and the lower resistor is 10 kΩ in parallel with 22 kΩ. We calculate:

$$\text{LTV} = 15 \times 6.875/16.875 = 6.11 \text{ V}$$

The hysteresis is:

$$8.89 - 6.11 = 2.78 \text{ V}$$

These three values agree well with those obtained from measurements on Fig. 7.6.

Test your knowledge 7.4

Calculate the upper and lower thresholds in Fig. 7.5 if R3 is replaced with a 33 kΩ resistor. How much is the hysteresis?

Non-inverting Schmitt trigger

The inverting trigger may be made to operate in the reverse direction by holding the (–) input at a fixed voltage with a potential divider, and by feeding v_{IN} to the (+) input. The difficulty is that the positive feedback must still be applied to the (+) input. The effectiveness of positive feedback then depends on the output resistance of the source of v_{IN}. To eliminate this influence, we first feed v_{IN} to a voltage follower (Fig. 7.8). This isolates the source of v_{IN} from the effects of feedback swings. It also makes the trigger independent of any changes in the output resistance of the source.

The output resistance of the voltage follower is very low, typically

Designing inverting triggers

Follow this procedure to design an inverting Schmitt trigger, using the circuit shown in Fig. 7.5.

1 Decide on the lower threshold voltage (LTV) and the upper threshold voltage (UTV).

2 Find values of R1 and R2 to produce a voltage halfway between LTV and UTV.

3 Find what resistor R_P in place of R1 would produce the required UTV.

4 Find what resistor in parallel with R1 would give it the value R_P. This is the value of R3.

This gives the required UTV. Provided that R1 and R2 are fairly close in value, it gives a reasonable value for the LTV too. At this stage is it simplest to try out a few resistors in the circuit, either on the workbench or on a simulator.

Example

1 Given power supply of 15 V, find resistor values for thresholds UTV = 9 V, LTV = 7 V.

2 To produce 8 V (halfway). Take R_2 to be 10 kΩ. Then the total of R_1 and R_2 is $R_2 \times 15/8 = 18.75$ kΩ. $R_1 = 18.75 - 10 = 8.75$ kΩ.

3 To produce 9 V (UTV), a similar calculation to give the total of R_P and R_2 is $R_2 \times 15/9 = 16.67$ kΩ. $R_P = 16.67 - 10 = 6.67$ kΩ.

4 Use an equation based on the equation from p. 3:

$$R_3 = R_1R_P/(R_1 - R_P) = 8.75 \times 6.67/(8.75 - 6.67) = 28\text{k}\Omega$$

Result

$R_1 = 8.75$ kΩ, $R_2 = 10$ kΩ, $R_3 = 28$kΩ. It may be that standard E24 values are close enough: $R_1 = 8.2$ kΩ, $R_3 = 27$ kΩ.

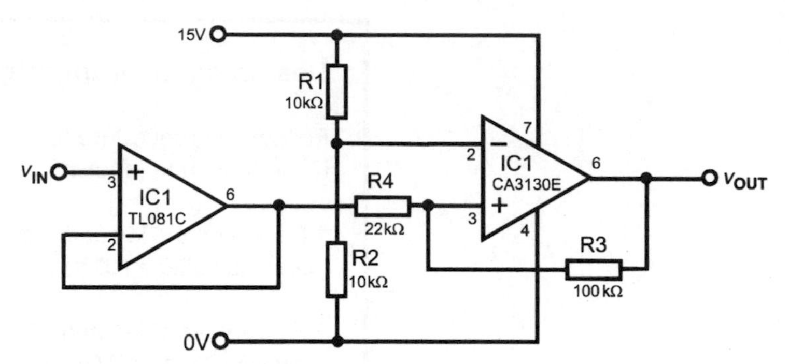

Figure 7.8 *The non-inverting Schmitt trigger needs a voltage follower as a buffer between the signal source and the trigger circuit.*

75 Ω, so a resistor R4 is needed to reduce the current to an amount comparable with the current through the feedback resistor. There is a voltage drop across R4, which means that the voltage at the (+) input is appreciably less than v_{IN} when v_{IN} is rising, and more than v_{IN} when v_{IN} is falling. This makes the action of the trigger different from that of the inverting trigger in which v_{IN} is fed directly to the (–) input of the op amp. The threshold points occur when the voltage at the (+) input is equal to the constant voltage (here 7.5 V) at the (–) input. In Fig. 7.9 we see the threshold points encircled and the values of v_{IN} marked as the upper and lower thresholds. The hysteresis is 9.04 – 5.95 = 3.09 V.

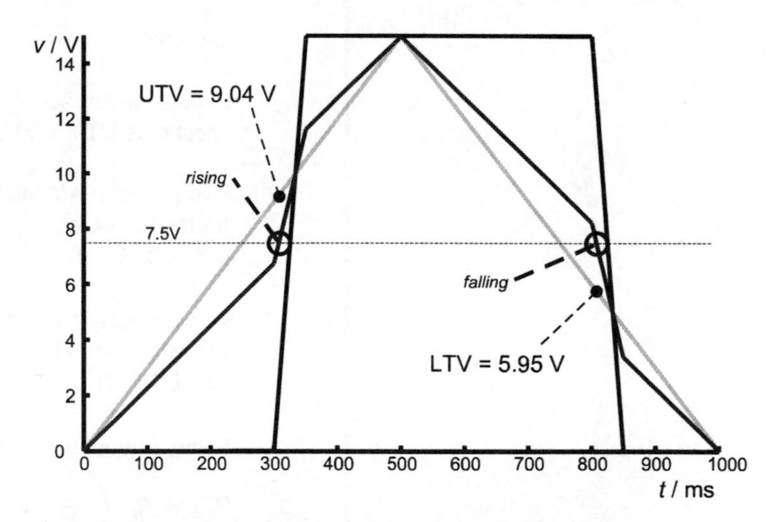

Figure 7.9 *In the non-inverting trigger the voltage at the (+) input (dark grey) ramps up and down, the ramp becoming much steeper as it crosses the ramping input voltage at the thresholds.*

The hysteresis shown in Figs. 7.6 and 7.9 is larger than normal for trigger circuits. Component values have been chosen to give large hysteresis, to make the graphs easier to read. In the non-inverting trigger, we can reduce hysteresis by increasing the value of R3.

Calculating thresholds

The calculation is best illustrated by a worked example. If we assume that the op amp swings its output fully to either power rail, the output is at 0 V while v_{IN} is rising from a low voltage (Fig. 7.10). When v_{IN} (or, more accurately, the output of the voltage follower) reaches the upper threshold, the voltage at the (+) input is 7.5 V. There is a voltage drop of 7.5 V across R_3. The voltage drop across R3 and R4 in series is:

$$v_{IN} = 7.5 \times 122/100 = 9.15 \text{ V}$$

This is the UTV and the value found from the graph of Fig. 7.8 is very close to this.

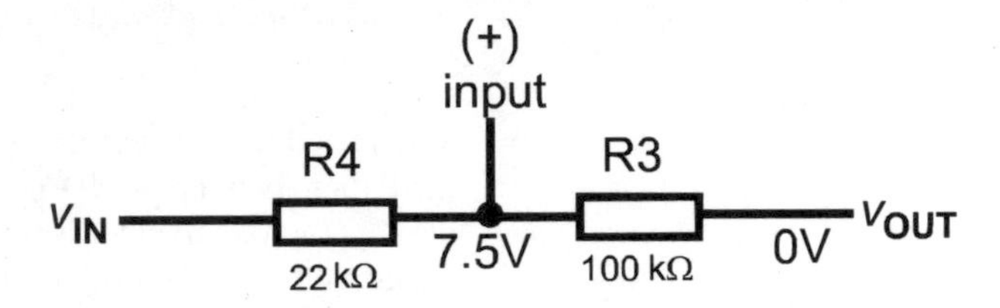

Figure 7.10 *For a non-inverting trigger, the input voltage at the threshold point is found by calculating the voltage drop along the R3 and R4 chain.*

Test your knowledge 7.5

Calculate the upper and lower thresholds in Fig. 7.8 if R3 is replaced with a 120 kΩ resistor.

Conversely, when v_{IN} is falling from a high value, v_{OUT} = 15 V. Again the voltage at the (+) input is 7.5 V, so:

$$v_{IN} = 15 - 9.15 = 5.85 \text{ V}$$

This is the LTV, as found from Fig. 7.9.

Astable circuit

One of the many ways in which an op amp can be used as an astable is shown in Fig. 7.11. The action of the circuit is plotted in Fig. 7.12. When the circuit is first switched on, the output (black) swings toward the negative rail and the capacitor charges in the negative direction through R3. The voltage at the (–) input (dark grey) falls below zero. The potential divider formed by R1 and R2 holds the (+) input (light grey) at half of v_{OUT}. If v_{OUT} is –7.5 V for example, the (+) input is at –3.75 V. The voltage is half of v_{OUT} because the resistors are of equal value, but other resistor ratios are often used.

After a period of time, depending on the values of R3 and C, the voltage across the capacitor reaches a negative value just less than that at the (+) input. This makes the (–) input less than the (+) input, causing v_{OUT} to swing toward the positive rail. The voltage at (+) is again half of v_{OUT}, but in the positive direction. Now the capacitor is

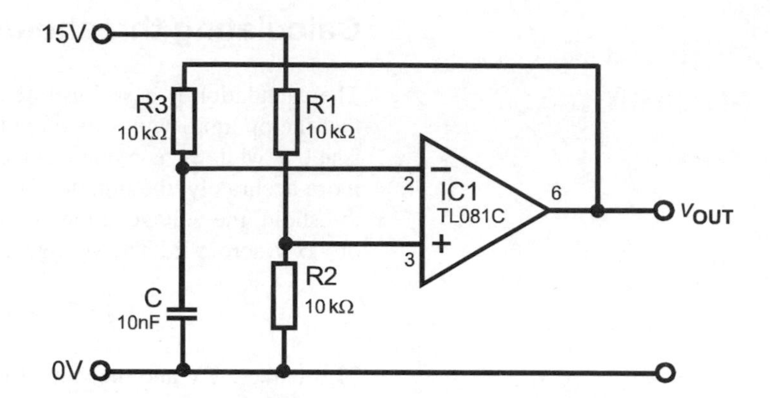

Figure 7.11 *This op amp astable is a comparator with negative feedback to charge and discharge the capacitor.*

connected through R3 to a positive voltage. It loses its negative charge and then begins to charge in the positive direction. Eventually the voltage across it (and at the (–) input) exceeds the voltage at the (+) input and v_{OUT} swings negative again. This process continues indefinitely, producing a square-wave output.

The period of the waveform is:

$$P = 2R_3C \log_e (1 + 2R_2/R_1)$$

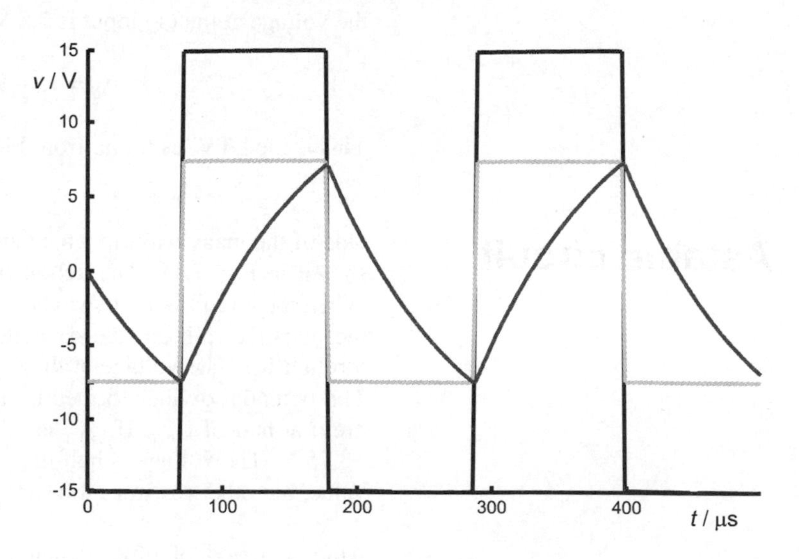

Figure 7.12 *Graphs of voltage levels in the astable show the output in black, the feedback in light grey, and the voltage across the capacitor in dark grey.*

Half-wave precision rectifier

A conventional rectifier circuit employs one or more diodes. As a result, the output is one or two diode drops (0.6 V or 1.2 V) below the input. This is acceptable for rectifiers in power supplies. But if we have an alternating signal, particularly one that is only a few hundred millivolts in amplitude, and want to rectify it, we need a rectifier, which eliminates the diode drop. Fig. 7.13 is a precision half-wave rectifier, and has an output unaffected by diode drop.

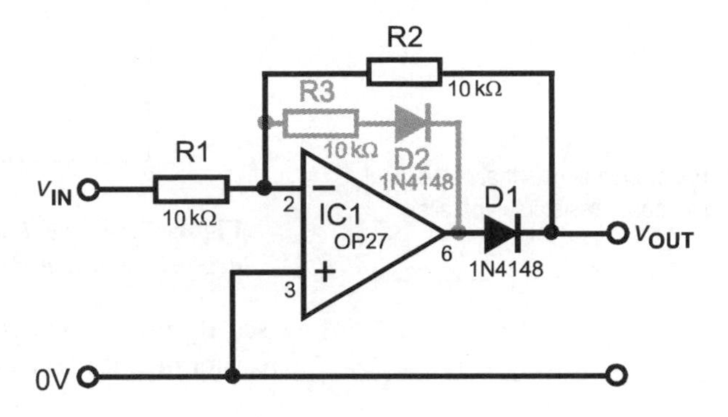

Half-wave rectifier: conducts only when the signal is of a given polarity, either positive or negative, but not both.

Figure 7.13 *A precision half-wave rectifier is an inverting amplifier, but with a diode in the feedback loop.*

The circuit behaves differently in the two half-cycles of the alternating signal:

In the positive half-cycles, the output of the op amp is negative. The diode is reverse-biased and no current flows. The output of the rectifier is zero (Fig. 7.14).

In the negative half-cycles, the output of the op amp is positive. The diode is forward biased and conducts. R_1 and R_2 are equal, so the open-loop gain is 1. There is a diode drop of 0.6 V across the diode, but this has no effect. At any instant, negative feedback operates to bring the (–) input to 0V, the same voltage as the (+) input. Since $R_1 = R_2$, the voltage drops across them are equal and $v_{OUT} = -v_{IN}$. The op amp output is always 0.6 V greater than v_{IN}, compensating for the diode drop. Diode drop varies with temperature and the amount of current flowing through the diode, but the action of the op amp automatically compensates for this.

For greater precision, a diode D2 and resistor R3 (shown in grey) may be added to provide feedback during the positive half-cycles. This prevents the amplifier from becoming over-loaded during the positive half-cycle and taking time to recover at the beginning of the negative half-cycle.

Another way of increasing precision is to connect the (+) input to 0 V through a resistor. This is to eliminate errors due to input bias current

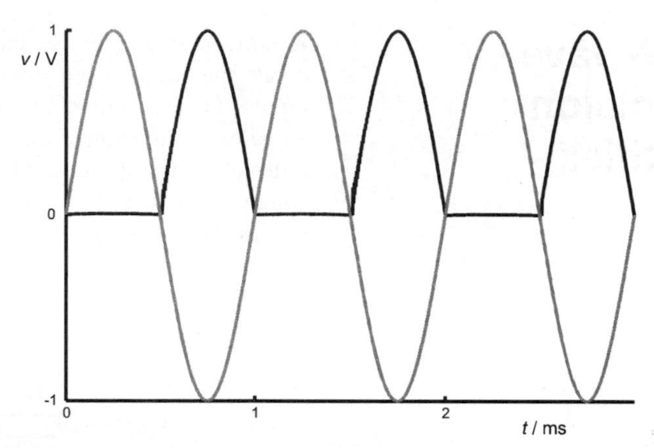

Figure 7.14 *The diode in the half-wave rectifier conducts on the negative-going half of the cycle.*

R_B **resistance** equals that of the input and feedback resistors in parallel.

(see R_B in Fig. 6.6). In this circuit R_1 and R_2 are both 10 kΩ, so a resistor of 5 kΩ is needed.

Full-wave precision rectifier

There are several designs for full-wave rectifiers. In Fig. 7.15 we use a half-wave rectifier (IC1) connected to a second op amp (IC2), as shown in Fig. 7.15. The circuit uses negative feedback in both the positive and negative half-cycles. For this reason R3 and D2 are drawn in black, as they are now an essential part of the circuit. The action of the circuit for each half-cycle is described separately:

In the positive half-cycle: IC1 is stable when both input terminals are at 0 V. D1 is reverse-biased so no current flows through it. The (+) input of IC2 is therefore at 0 V throughout the half-cycle. D2 is conducting so the output from IC1 is equal to $-v_{IN}$, as in the half-wave rectifier. This is fed to IC2, which acts as an inverting amplifier with a gain of 1. It simply re-inverts the signal:

$$v_{OUT} = v_{IN}$$

v_{OUT} is positive.

In the negative half-cycle: The inputs to IC1 are at 0V, as before. There is a resistor chain consisting of R3, R4 and R5 which connects the (–) terminal of IC1 to the output terminal of IC2. The total voltage drop across the chain is v_{OUT}. The three resistors have equal value so there is a drop of one-third of v_{OUT} across each. The voltage levels are as marked in Fig. 7.15. Because the (–) terminal of IC2 is at 2/3 v_{OUT}, the circuit stabilises with the (+) at this voltage too.

Currents flow through R1, R2 and R3 toward the virtual earth at the (–) terminal of IC1 and cancel out each other, for very little current can

Activity 7.1 Op amp applications

Work these activities on a breadboard or by using a circuit simulator.

Constant current source: Build the circuit of Fig. 7.1. Measure the current through the load (emitter) resistor, either by including a millammeter in the circuit or by measuring the voltage across the resistor and using Ohm's Law to calculate the current. Try smaller and larger values of the load to find out the range or resistance for which the current is constant.

Integrator: Build the circuit of Fig. 7.2, making $R = 1\ M\Omega$, $C = 10\ \mu F$, and $v_{IN} = 1$ V. Measure the rate at which v_{OUT} falls, in volts per second. Compare this with the rate calculated from theory, which is $-v_{IN}/RC$. Repeat for other values of *R* and *C*.

Sample and Hold: When assembling this circuit from Fig. 7.3, remember to connect all unused inputs of IC1 to 0 V or to the positive supply lines. Unused inputs are at pins 3–6 and 8–12. Connect the input of the circuit to a varying voltage source, such as a signal generator running at 0.5 Hz with an amplitude of 1 V. Connect the output to a voltmeter. Connect the control voltage input alternately to 0 V and to the positive supply line.

Schmitt trigger: Design and build an inverting Schmitt trigger using the circuit shown in Fig. 7.5. The box on p. 93 explains how to do the calculations.

Astable multivibrator: Using components of the values shown in Fig. 7.11, build an astable and measure its frequency (connect the output to an oscilloscope, measure its period and then calculate the frequency). Try different values for C and R3 to obtain different frequencies.

Precision rectifiers: Build the half-wave and full-wave rectifiers of Figs. 7.13 and 7.15. In each case connect the input to a signal generator running at 100 Hz, with 1 V amplitude. Connect the the input and output to a dual-trace oscilloscope. Examine and compare the input and output waveforms. If you have not previously done so, examine the signals from simple diode rectifiers, both half-wave and full-wave, for comparison.

actually flow into the terminal. As all three resistors are equal, the currents through them are proportional to the voltages across them:

$$v_{IN} + \tfrac{2}{3} v_{OUT} + \tfrac{1}{3} v_{OUT} = 0$$

$$\Rightarrow \quad v_{OUT} = -v_{IN}$$

As this is the negative half-cycle, v_{IN} is negative, so v_{OUT} is positive.

We have shown that v_{OUT} is positive in both half-cycles, so the alternating signal is rectified with positive polarity.

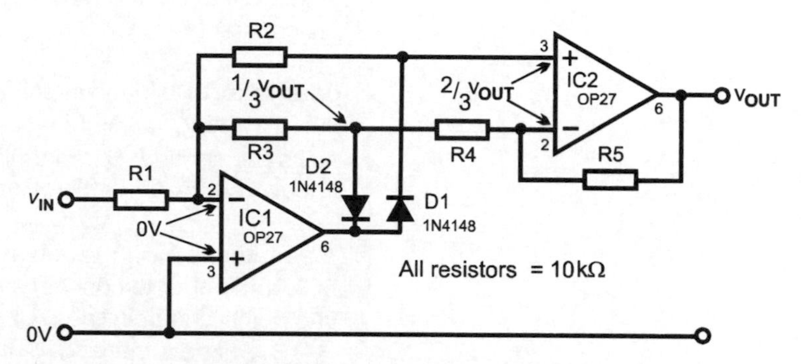

Figure 7.15 *A full-wave precision rectifier based on the half-wave rectifier of Fig. 7.13. The marked voltages refer to the negative half-cycle.*

Active filters

When a signal of mixed frequencies is passed through a filter circuit, some of the frequencies are reduced in amplitude. Some may be removed altogether. The simplest filter consists of a resistor and a capacitor (Fig. 1.5). As it is built from two passive components it is known as a *passive filter*. Other passive filters may consist of a network of several resistors and capacitors, or may be combinations of resistors and inductors, or capacitors and inductors. A passive filter must have at least one reactive component (a capacitor or an inductor) in order to make its action frequency-dependent.

Passive component: One that does not use an external power source. Examples are resistors, capacitors, inductors, and diodes.

Passive filters are often used but a large number individual filters must be joined together to effectively cut off unwanted frequencies. Such filters take up a lot of space on the circuit board and can be expensive to build. When filtering low frequencies the inductors are necessarily large, heavy and expensive. Another serious disadvantage of passive filters is that passive components can only *reduce* the amplitude of a signal. A signal that has been passed through a multi-stage passive filter is usually very weak. For these reasons we prefer active filters, based on active components, especially op amps. When these are used we can obtain much sharper cut-off with relatively few filtering stages, there is no need to use inductors, and the filtered signal may even show amplitude gain.

Active component: One that uses an external power source. Examples are transistors, operational amplifiers.

First-order active filter: Fig. 7.16 is one of the simplest active filters. It consists of a resistor-capacitor passive filter followed by a non-inverting amplifier to restore or possibly increase the amplitude of the signal. With the values shown, the voltage gain of the amplifier is 1.56. Fig. 7.17 displays the voltages in this filter when it is supplied with a 1V sinusoid at 1 kHz. The R-C network filters the input sinusoid (dark grey line). At the output of the network (light grey) the filtered signal has an amplitude of only 0.7 V. This is not the only change. The output reaches its peaks 125 μs behind the input. This is a delay of 1/8 of the period of the signal. In terms of angle, this is 360/8 = 45°. Because it is a delay, it has negative value. The output is *out of phase* with, or *lags* the input, by –45°. The black curve in Fig. 7.17 represents the output of the op amp.

The period of a 1 kHz signal is 1/1000 s, or 1 ms.

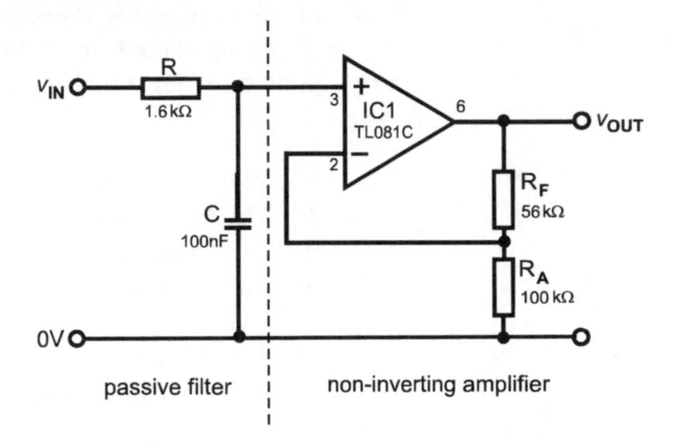

Figure 7.16 *In a first-order active filter, the signal from the RC filter network is amplified by a non-inverting op amp amplifier.*

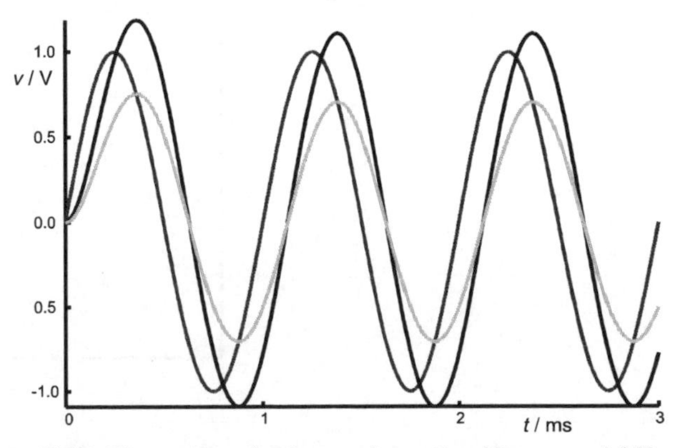

Figure 7.17 *The action of a first-order active filter on a 1 kHz sinusoid (dark grey). The RC filter network is reduces its amplitude and delays its phase (light grey). The amplifier then increases the amplitude (black).*

Example:

The gain of the amplifier in Fig. 7.16 is:

$$A_V = (100 + 56)/100 = 1.56$$

Gain of non-inverting amplifier = $(R_A + R_F)/R_A$

If the input signal has amplitude 0.7 V, the output amplitude is $0.7 \times 1.56 = 1.1$ V.

The amplifier does not change the phase any further, so the output lags the input by 45°.

We can study how the filter operates at a range of frequencies by connecting its input to a frequency generator and observing the output on an oscilloscope. Another way is to run a frequency response on a circuit simulator (Fig. 7.18). Both frequency and output amplitude are plotted on logarithmic scales. The amplitude scale is graduated in decibels.

The plot (black) shows that amplitude is the full 1.56 V up to about 300 Hz. From then on amplitude begins to fall off slightly, reaching 0.7 V (equivalent to –3dB, or half power) at 1 kHz. The bandwidth of the filter is 1 kHz.

The frequency of f_C, the –3 dB point or cut-off point, depends on the values of the resistor and capacitor.

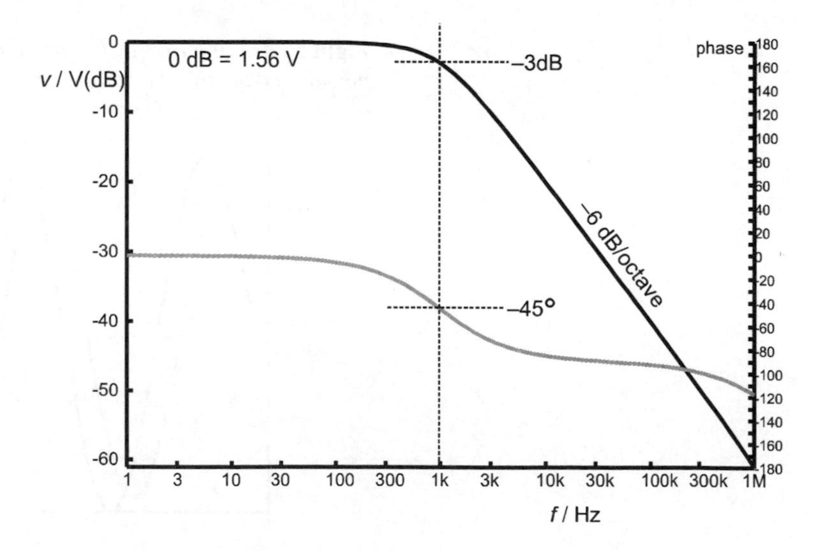

Figure 7.18 *The Bode plot of the frequency response of the first-order lowpass filter shows that it falls to its half-power level (amplitude 0.7 V) and has phase lag of 45° at the cut-off frequency of 1 kHz.*

Example

With the values given in Fig. 7.16 the cut-off point is:

$$f_C = \frac{1}{2\pi RC}$$

$$f_C = 1/(2 \times \pi \times 1.6 \times 10^3 \times 100 \times 10^{-9}) = 1 \text{ kHz.}$$

As frequency is increased beyond 1 kHz, the amplitude rolls off even faster and eventually the curve plunges downward as a straight line. If we read off frequency and amplitude at two points on this line we find that for every doubling of frequency, the amplitude falls by 6 dB. We say that the roll-off is –6 dB per octave.

Octave: an interval, over which frequency doubles or halves.

Another way of expressing this is to measure how much the amplitude falls for a ten-times increase in frequency, or decade. Fig. 7.18, the roll-off is 20 dB per decade.

These roll-off rates are characteristic of single-stage passive or active filters and do not depend on the values of the capacitor and resistor.

Fig. 7.18 also shows how the phase change of the output varies with frequency. The scale shows phase angle on the right margin of the figure. We can see that the phase change at 1 kHz is –45°, confirming what we found in Fig. 7.17. This is another characteristic of single-stage passive or active filters. The phase lag is less at lower frequencies. At 0 Hz (DC) the phase change is zero. It increases with higher frequencies.

Test your knowledge 7.6

A first-order lowpass filter has a 47kΩ resistor and a 220 pF capacitor. What is its cut-off frequency?

The filter of Fig. 7.18 is converted to a highpass filter by transposing the resistor and capacitor. When this is done, the frequency response has the appearance shown in Fig. 7.19. The amplitude curve rises up from 0 Hz at +6 dB per octave, or +20 dB per decade. The equation for calculating the cut-off point is the same for both lowpass and highpass filters, so this is 1 kHz as before. Beyond the cut-off point, amplitude rises to 0 dB above about 4 kHz.

Looking more closely at Fig. 7.19, we note that the amplitude starts to fall again from about 400 kHz upward. This is the not the effect of the RC filter but is due to the fall in gain of the op amp at high frequencies. The transition frequency of the TL081C is 3 MHz, so, if we were to plot Fig. 7.19 for higher frequencies, we should see a steep roll-off. In practice, if we extend the operation of this filter into higher frequencies it acts as a bandpass filter.

Second-order active filter: Adding an extra stage to the first-order lowpass filter allows feedback to be introduced. Fig. 7.20 has two RC

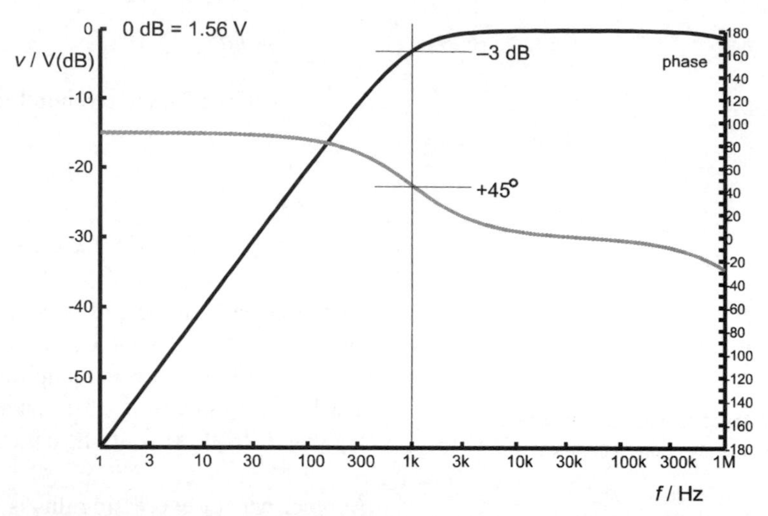

Figure 7.19 *Exchanging the capacitor and resistor produces a highpass filter with the same cut-off point as the lowpass filter, and a phase lead of +45°.*

Designing first-order active filters

1 Decide on a capacitor value.

2 Apply the equation: $R = 1/2\pi f_C C$.

Example

To design a lowpass filter with cut-off point at 40 kHz.

1 For stability, try the calculation for a 220 pF capacitor, so that a ceramic plate capacitor with zero tempco can be used.

2 Applying the equation, R = 18086 Ω

Use an 18 kΩ, 1% tolerance, metal film resistor. Assemble the filter as in Fig. 7.16. For a highpass filter, exchange the resistor and capacitor.

filters, one of which is connected to 0V, as in the first-order lowpass filter. The other is part of a feedback loop from the filter output. The feedback action makes the filter resonate at frequencies around the cut-off frequency. Resonance is not strong, but there is enough extra feedback around this frequency to push up the 'knee' of the frequency response curve (Fig. 7.21). The grey curve is a plot of the voltage across C1, and it can be seen that the strongest feedback occurs around 1 kHz. The response of the filter turns sharply down above 1 kHz. From then on, there is a steady and more rapid fall in response, the rate

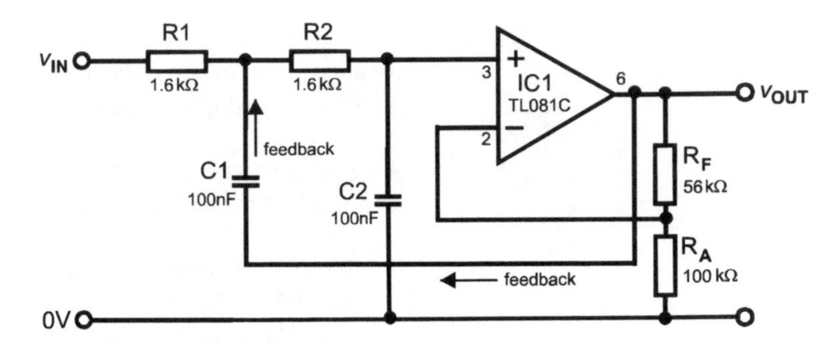

Figure 7.20 *The second-order active filter has positive feedback from the output to the capacitor of the first stage.*

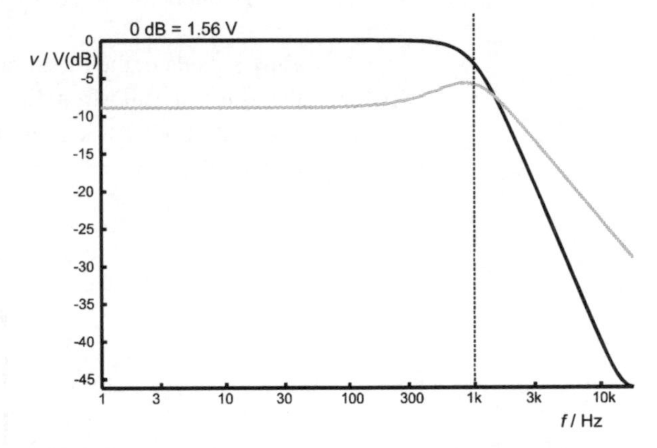

Figure 7.21 *The Bode plot of the response of the second-order filter (black) also shows the amplitude of the feedback signal (grey) which pushes up output at the knee of the curve. The doubled filter network produces a steep roll-off at high frequencies.*

depending on the number of stages. With two stages in the filter, the roll-off is −12 dB per octave or −40 dB per decade.

This sharpening of the 'knee' is accomplished in a passive filter by using an inductor-capacitor network to produce the resonance. Using an op amp instead provides the same effect but without the disadvantages of inductors.

A highpass version of the second order filter is built by substituting resistors for capacitors and capacitors for resistors. Higher-order filters with even steeper roll-off often consist of two or more second-order filters and possibly a first-order filter connected in series.

Bandpass filters: A bandpass filter may be constructed by following a lowpass filter with a highpass filter, or the other way about. The cut-off frequency of the lowpass filter is made *higher* than that of the highpass filter so that there is a region of overlap in which both filters pass frequencies at full strength. This area of overlap can be made narrow or broad by choosing suitable cut-off frequencies.

Another approach to bandpass filtering is the *multiple feedback filter* (Fig. 7.22). This has two loops providing *negative* feedback:

- R1 and C1 form a highpass filter; high-frequency signals passing through this are fed back negatively, partly cancelling out the high frequencies in the original signal.
- R2 and C2 form a lowpass filter; low-frequency signals passing through this are fed back negatively, partly cancelling out the low frequencies in the original signal.

Only signals in the medium range of frequency are able to pass through the filter at full strength. The frequency distribution graph shows a peak (at about 1.8 kHz in this instance) rolling off at –6 dB/octave on either side.

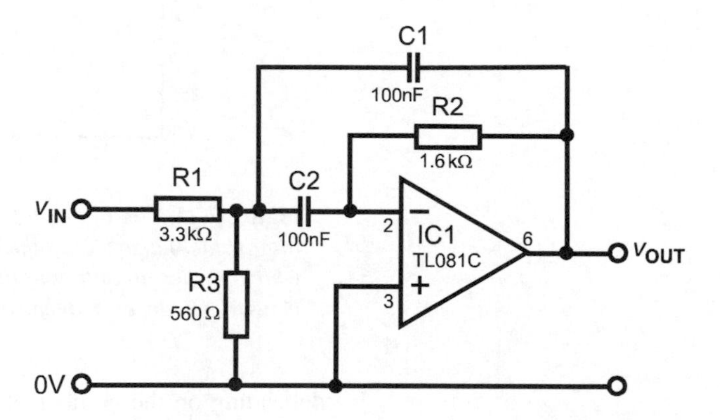

Figure 7.22 *A multiple feedback filter cancels out both high and low frequencies by feeding them back negatively, leaving the middle frequencies unaffected.*

Activity 7.2 Active filters

It is easier to work this activity on a circuit simulator, and this has the advantage that an analysis frequency response can be done very quickly. But you should build and test at least two filter circuits using a breadboard. The box on p. 104 explains the calculations you need for working out values of resistors and capacitors.

The active filter circuits that you can investigate include:

First-order (single stage) lowpass filter (Fig. 7.16).

First-order (single stage) highpass filter (Fig. 7.16 with R and C exchanged).

Second order lowpass filter (Fig. 7.20).

Second order highpass filter (Fig. 7.20 with Rs and Cs exchanged).

Bandpass filter consisting of a lowpass filter followed by a highpass filter (cutoff point for the lowpass filter must be *higher* than that for the highpass filter).

Multiple feedback bandpass filter (Fig. 7.22)

For each filter connect a signal generator to its input and a dual-trace oscilloscope to its input and output. Observe and explain what happens to the signal as it passes through the filter. Observe the effect of the filter on the amplitude of the signal and on the shape of the waveform, trying this with both sinusoidal and square waves, and at different frequencies.

For example, when testing a lowpass filter, use a signal with a frequency well below the cut-off point, try another signal at the cut-off point and a try a third signal well above the cut-off point. But remember that the transition frequency of the op amp itself may be about 1 MHz, so do not use frequencies higher than 100 kHz in these investigations.

Problems on op amp applications

1 Design an op amp constant current circuit to supply a current of 20 mA. What is the resistance of the maximum load that this circuit can take?

2 Explain how an op amp integrator works when the input voltage is constant.

3 How can an op amp integrator be used as a ramp generator?

4 Explain how an op amp integrator is used to measure the average value of the input voltage over a period of time.

5 Explain the working of an op amp sample and hold circuit. Sketch a graph showing a sinusoidal input, and its output, sampled at three different stages of the cycle.

6 Design an op amp inverting Schmitt trigger that has its upper threshold voltage at 8 V and its lower threshold at 6 V.

7 Explain the working of an op amp non-inverting Schmitt trigger circuit.

8 Why is a precision half-wave rectifier preferred to a one-diode rectifier in certain applications? Explain how it works.

9 List and discuss the advantages of op amp active filters, compared with passive filters.

10 Design a first-order lowpass filter with a cut-off frequency of 15 kHz.

11 Design a first-order highpass filter with a cut-off frequency of 20 Hz.

12 Explain the action of a second-order op amp lowpass filter and state why its performance is superior to that of a first-order filter.

13 Design a bandpass filter consisting of a lowpass and a highpass first-order filter in series. The specification is: low cut-off point = 2 kHz; upper cut-off point = 5 kHz; gain at the centre frequency = 1.5.

Multiple choice questions

1 An op amp used in an integrator circuit should have:

A low input offset voltage.
B high open-loop gain.
C low slew rate.
D high input resistance.

2 An op amp integrator receives a sinusoidal input, amplitude 50 mV. Assuming the integrator does not have time for its output to swing fully in either direction, its output after three complete cycles of the sinusoid is:

A 50 mV.
B 150 mV.
C 0 V.
D −25 mV.

3 The op amp used in a sample-and-hold circuit should have:

A high open-loop gain.
B high closed-loop gain.
C low slew rate.
D high input resistance

4 An inverting Schmitt trigger circuit changes state when:

A a rising input goes just above the upper threshold.
B a rising input goes just below the lower threshold.
C a falling input goes just below the upper threshold.
D the upper and lower thresholds are equal.

5 An op amp used in a Schmitt trigger should have:

A low input offset voltage.
B high input resistance.
C low output resistance.
D high common-mode rejection ratio.

6 The difference between the upper and lower thresholds of a Schmitt trigger is known as:

A feedback.
B input offset voltage.
C input bias current.
D hysteresis.

7 A passive filter can be built from:

A a capacitor, a resistor and an op amp.
B a capacitor and an inductor.
C resistors only.
D a resistor, and an op amp.

8 An op amp used in a highpass filter should have:

A a high gain-bandwidth product.
B low slew rate.
C low input bias current.
D high output resistance.

9 The cut-off frequency of a first-order active high pass filter is 5 kHz. It has a 1 nF capacitor. The nearest E12 value for the resistor is:

A 200 kΩ.
B 31.8 Ω.
C 33 kΩ.
D 220 kΩ.

10 A passive filter must have:

A a resistor.
B a diode.
C no more than two passive components.
D a least one reactive component.

11 At the cut-off point of a lowpass filter the phase change is:

A +90°.
B 0°.
C –45°.
D –90°.

12 The roll-off of a second-order highpass filter is:

A +6 dB per octave.
B +40 dB per decade.
C –12 dB per decade.
D +20 dB per octave.

8 Power amplifiers

Summary

After considering the relationships between voltage, current and power, as they apply to amplification, we look again at the common-collector and common-drain amplifiers, both of which are followers. These are examples of Class A amplifiers. Then we see how these are used as Class B current amplifiers in power amplifier circuits. Crossover distortion is a serious drawback of the simple Class B amplifiers but we show how this can be avoided. Power amplifiers inevitably generate relatively large amounts of heat, which can affect performance and may even destroy the semiconductors of the amplifier. Ways of minimising this risk are discussed. Finally, we look at a power amplifier available as an integrated circuit.

Voltage, current and power

A power amplifier is used to produce a major effect on the surroundings. Examples are power audio amplifiers producing sound at high volume, or motor-control circuits actuating the limbs of an industrial robot or aligning the dish of a radio telescope. Other examples include the circuits that produce the dramatic effects of disco lighting. The power of these devices is rated in tens or hundreds of watts, sometimes more. Even a pocket-sized tape player can have a 700 mW output, yet the signal its head picks up from the tape is far less powerful than that. The electrical signals that initiate any of these actions may be of extremely low power. For example, the power output of a microphone and many other types of sensor, or the control outputs from a microcomputer are usually rated at a few milliwatts, perhaps only a few microwatts. The aim is to amplify the power of the signals from these devices so that they can drive powerful speakers, motors or lamps.

The power at which a device is operating depends on only two quantities, the *voltage* across the device and the *current* flowing through it. The relationship is simple:

$$P = IV$$

The amplifiers that we have described in the early chapters of this book have mainly been voltage amplifiers. Some, such as those based on FETs, actually produce a current that is proportional to their input voltage but, even then, we usually convert this current to an output voltage by passing it through a resistor. Thus the early stages of amplification are usually voltage amplification. Since P is proportional to V, amplifying the *voltage* amplitude of a signal amplifies its *power* by the same amount.

Descriptions of voltage amplifiers often refer to the fact that the currents in the amplifiers are small. In many instances the collector current or drain current is only 1 mA. There is a good reason for this. As is explained in Chapter 17, large currents through a semiconductor generate noise. Noise is a random signal and this shows up as background hissing in an audio circuit. If the audio signal is weak, it may not be possible to pick it out against the noisy background. In other kinds of circuit it may become evident as unpredictable behaviour, making the circuit is unreliable.

When a signal has been through several stages of amplification it may have reached an amplitude of a few volts or perhaps a few tens of volts. But, for the reason given above, its current is rated in no more than a few hundred milliamps, perhaps less. Consequently, its power level is low. The next and final stage is to amplify the current.

Current amplifiers

The two current amplifiers most often used are the common-collector amplifier (Fig. 2.7) and the common-drain amplifier (Fig. 1.9). Both of these are voltage follower amplifiers. Their voltage gain just less than 1. They both have high current gain. In addition, they have low output resistance, a useful feature when driving high-power devices.

These amplifiers have one feature in common with all the amplifiers we have described in this book, so far. Passing a varying current through a resistor generates their output voltage. For example, in the common-emitter amplifier, the voltage-generating resistor is the collector resistor. In the common-collector amplifier, it is the emitter resistor. In the common-source amplifier, it is the drain resistor. The values of these resistors are chosen so that, when there is no signal, the output voltage sits halfway between the supply-line voltage. This gives the output voltage room to swing to the maximum extent in either direction without clipping or bottoming. An amplifier of this type is known as a *Class A* amplifier.

Test your knowledge 8.1

What are the alternative names for the CC and CD amplifiers?

Class A amplifiers

The main disadvantage of Class A amplifiers is that collector and emitter currents are flowing even when there is no signal. Power is being used but no sound or other form of output activity occurs. Such amplifiers are inefficient, since they waste 50% of the energy supplied to them. If an amplifier is to produce enough output power to drive a motor or a high-wattage speaker, we must design the circuit to avoid such waste.

Class B amplifiers

The *complementary push-pull* amplifier (Fig. 8.1) is called 'complementary' because it consists of two amplifiers, one based on an npn transistor and the other based on a pnp transistor. It is called 'push-pull' because one amplifier pushes the output in one direction and the other pulls it in the opposite direction. The amplifiers in Fig. 8.1 are both common-collector amplifiers. The circuit is intended as an audio amplifier, with the speaker taking the place of the emitter resistor of both amplifiers.

The amplifier runs on a spilt supply. When there is no signal the input is at 0 V. This means that both amplifiers are off, and no power is being wasted. This is a distinguishing feature of *Class B amplifiers*. When the input goes positive, the pnp amplifier remains off but the npn amplifier is turned on. The reverse happens when the signal goes negative. The resistance of the speaker coil is only a few ohms so a high current passes through it.

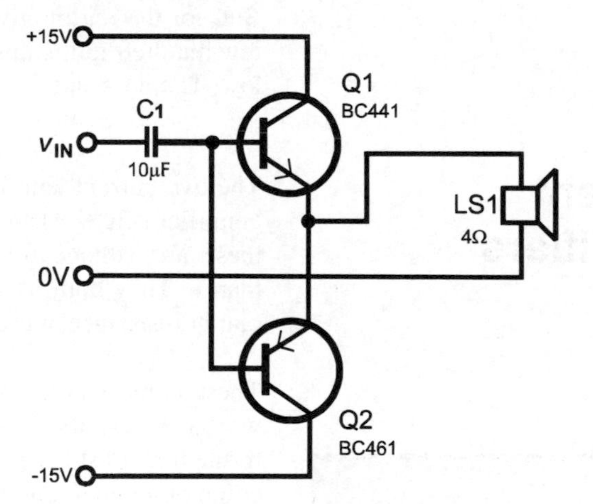

Figure 8.1 *In a Class B amplifier the transistors are biased so that they are off when there is no input signal.*

Example

A push-pull audio amplifier operates on a ±12 V supply. The resistance of the loudspeaker coil is 4Ω. What is the maximum current through the speaker?

$$\text{Current} = 12/4 = 3\ \text{A}$$

This result is only an approximation, since it ignores the base-emitter voltage drop of 0.6 V. Also the impedance of the coil varies with frequency and would be greater at high frequencies.

We can take the calculation further. If the current through the speaker is 3 A and the voltage across it is 12 V, the power dissipated is:

$$P = IV = 3 \times 12 = 36 \text{ W}$$

When the input to the amplifier goes negative, the npn transistor is switched off. Then the pnp transistor turns on and current flows through the speaker in the opposite direction.

The advantage of a Class B amplifier such as this, is that it consumes power only when there is a signal to be amplified. It is much more efficient than Class A. Its greatest drawback is illustrated in Fig. 8.2. An npn transistor does not conduct until the base-emitter voltage v_{BE} is greater than about 0.6 V. A pnp transistor does not conduct until the base is about 0.6 V below the emitter. For this reason, signals of amplitude less than 0.6 V have no effect on the amplifier. Output is zero. With signals greater than 0.6 V amplitude, there is a period of no output every time v_{IN} swings across from positive to negative or from negative to positive. This introduces considerable distortion, known as *crossover distortion.* It is proportionately greater in small signals. Fig. 8.2 also illustrates the fact that the voltage gain of each of the follower amplifiers is less than 1. It is approximately 0.8.

A Class B amplifier, similar to Fig. 8.1 can also be based on power MOSFETs. A complementary transistor pair is used, consisting of an n-channel MOSFET and a matching p-channel MOSFET. Usually the transistors are VMOS or HEXFET types. Their geometry is such that the conduction channel is very short but also very wide. It has very low

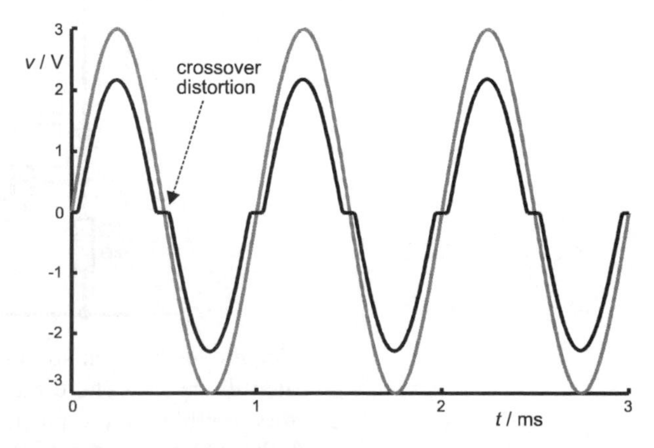

Figure 8.2 *The amplifier of Fig. 8.1 shows severe crossover distortion as the signal swings between positive and negative.*

resistance, allowing a high current to pass with minimum heat production. They have the usual MOSFET advantages of high input impedance and no thermal runaway.

Eliminating crossover distortion

A common technique for eliminating crossover distortion is to bias the transistors so that they are *just on the point of* conducting. With BJTs this means biasing them so that v_{BE} is equal to 0.6 V when no signal is present. The most convenient way to do this is to make use of the 0.6 V voltage drop across a forward-biased diode. In Fig. 8.3 current flows through the chain consisting of R1, D1, D2 and R2. There are voltage drops of 0.6 V across both transistors. These raise the base voltage of Q1 to 0.6 V above the value of v_{IN}. When there is no signal, v_{IN} is zero so the base of Q1 is at +0.6 V. It is ready to conduct as soon as v_{IN} increases above zero. Similarly, the base of Q2 is at −0.6 V and is ready to conduct as soon as v_{IN} falls below zero. In short, as soon as v_{IN} begins to increase or decrease, one or the other of the transistors is ready to conduct. There is very little crossover distortion, if any. This modification of the Class B amplifier is referred to as a *Class AB* amplifier.

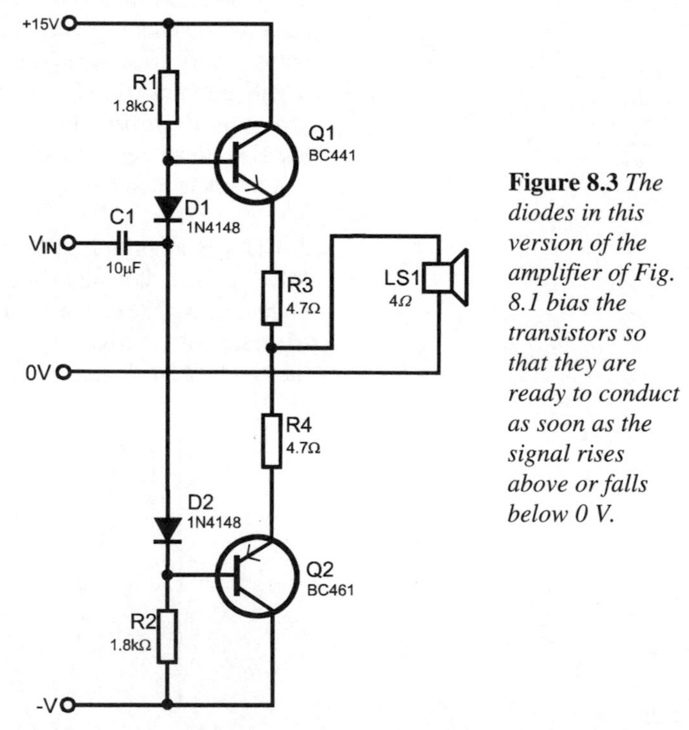

Figure 8.3 *The diodes in this version of the amplifier of Fig. 8.1 bias the transistors so that they are ready to conduct as soon as the signal rises above or falls below 0 V.*

The diodes have an additional advantage. Consider Q1 and its associated diode D1. The diode, like the base-emitter junction of Q1, is a forward biased pn junction. As temperature changes, the forward voltage drop (v_{BE}) across the base-emitter junction changes. But the diode compensates for this. Whatever the temperature, the voltage drops of D1 and Q1 are equal and Q1 remains biased ready to begin conducting with any increase of v_{IN}. The same applies to Q2 and D2. In this way the diodes improve the temperature stability of the amplifier.

Heating problems

Heat is generated whenever a current flows through a resistance. All components have resistance, so heat is generated in all components, though more in some than in others. Getting rid of this heat is a important aspect of practical circuit construction, especially in high-power circuits. If the temperature rises only a few tens of degrees, it alters the characteristics of a device and the performance of the circuit suffers. For example, as mentioned in the previous paragraph, temperature has a significant effect on the voltage drop across a pn junction. It may be possible to avoid this by using components with low tempcos or by including design features (such as the diodes) which compensate automatically for temperature changes. But when the power dissipation of a circuit is rated in watts, temperatures may quickly rise above 150°C, at which temperature semiconductor devices are completely destroyed.

Tempco: temperature coefficient.

The temperature that matters most is the temperature of the active regions of the device, the semiconductor junctions. This is where the maximum heat is generated and this is where the damage is done. The task is to remove this heat and transfer it to the air around the circuit. We must assume that the temperature of the environment, the ambient temperature t_A, is sufficiently low and that the air is able to circulate freely to carry the heat away.

Heat sinks

The usual way of increasing the transfer of excess heat is to attach a heat sink to the device. This is made of metal, usually aluminium, and is provided with fins to allow maximum contact with the air. The surface of the heat sink is usually anodised black, which increases heat loss by radiation by as much as ten times, compared with polished aluminium. Small heat sinks may clip on to the transistor but the larger ones are bolted on. They often have pegs for soldering the heat sink securely to the circuit board. The largest heat sinks of all are bolted to the circuit chassis or enclosure. Then one or more transistors are bolted to the heat sink. Components which are likely to become very hot in operation should be located near the edge of the circuit board, or even on a heat sink on the outside of the enclosure.

A heat sink must be mounted with its fins vertical to allow convection currents to circulate freely. Cooling is reduced by around 70% if the fins are horizontal. It is also helpful to mount circuit boards vertically if there are several of them. It is of course essential that the circuit enclosure should have adequate ventilation, aided by a fan in the case of a circuit in which a large amount of power is dissipated (example, a microcomputer).

Thermal resistance

Thermal resistance (R_q) is the ability of something to resist the transfer of heat, and is measured in degrees Celsius per watt (°C/W). If an object is being heated on one side by a source of heat rated at *P* watts (Fig. 8.4), the hot object is at *t*°C, and the ambient temperature is t_A°C, then the thermal resistance of that object is given by:

$$R_\theta = \frac{t - t_A}{P}$$

Examples

1 The metal tag of a power transistor operating at 2.2 W is at 90°C. It is bolted to a heat sink and the ambient temperature (inside the circuit enclosure) is 35°C. What is the thermal resistance of the heat sink?

For a relatively small object such as a heat sink, made from metal of good thermal conductivity and with free air circulation around it, we can assume that all parts of its surface, except that in contact with the transistor tag, is at ambient temperature.

Temperature difference = 90 – 35 = 55°C.

Thermal resistance = 55/2.2 = 25 °C/W

2 For the same transistor, the manufacturers give the thermal resistance between the tag and the actual junction on the chip as 0.9 °C/W. What is the junction temperature and is it dangerously high?

Test your knowledge 8.2

1 If the tag of a transistor is at 38°C when it is operating at 7.5 W and the ambient temperature is 20°C, what is the thermal resistance of the heat sink?

2 Given that the junction-to-tag thermal resistance is 0.8 °C/W, what is the temperature of the junction?

Test your knowledge 8.3

A transistor with junction-to-tag thermal resistance of 0.75 °C/W is mounted on a heatsink for which R_θ = 25°C/W. If t_A = 30°C, what is the maximum power at which the transistor can be safely run?

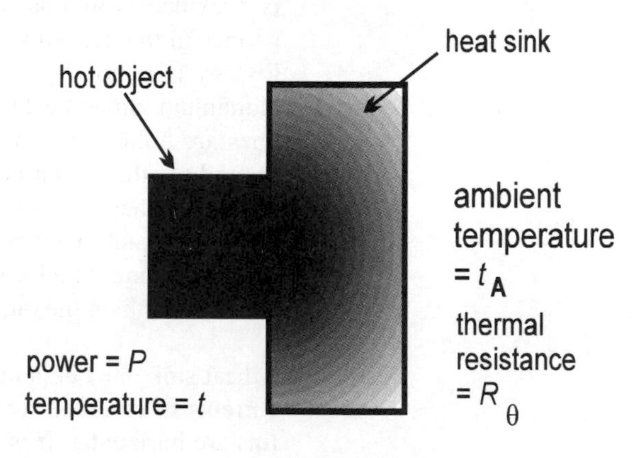

Figure 8.4 *A hot object, such as an operating transistor, is placed in contact with a material that conducts the heat away from it and passes it to the air around.*

Electrical and thermal resistance

The electrical resistance of a piece of conductor is defined as the *potential* difference across it when a given *current* is passing through it:

$$\text{Electrical resistance} = \frac{\text{potential difference}}{\text{current}}$$

Similarly, thermal resistance is defined as the *temperature* difference across it when heat dissipated with a given *power* is passing through it:

$$\text{Thermal resistance} = \frac{\text{temperature difference}}{\text{power dissipated}}$$

Rearranging the equation in the box:

$$\begin{aligned}\text{Temperature difference} &= \text{power} \times \text{thermal resistance} \\ &= 2.2 \times 0.9 \\ &= 1.98\ ^\circ\text{C}\end{aligned}$$

The junction is approximately 2 °C hotter than the tag, which is at 90 °C, so:

Junction temperature = 92 °C.

Temperatures up to 150°C are usually regarded as safe.

In these examples we have ignored the fact that parts of the transistor case and the tag are in direct contact with the air. Some heat is lost from these surfaces by convection, but usually the amount is so small that it can be disregarded.

The examples above take two thermal resistances into account:

- the resistance between the chip and the outside of the transistor case, usually a metal tag with power devices.
- the resistance between the area where the heat sink is bolted to the tag and the surfaces of the fins in contact with the air.

It is important for the tag to be in good thermal contact with the heat sink. Their surfaces must be as flat as possible. Often we coat the surfaces with silicone grease or special heat transfer compound to fill in the small gaps caused by slight unevenness of the surfaces. Greasing the surfaces can reduce the thermal resistance of the surface-to-surface interface by about a half.

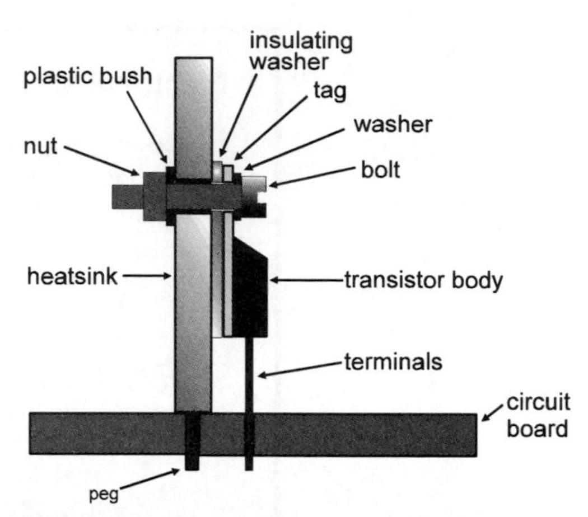

Figure 8.5 *Showing how a transistor in a TO220 or similar package is attached to a heat sink. The heat sink may be larger than shown here.*

The tag is bolted to the heat sink to keep them firmly in contact (Fig. 8.5) but the nut must not be tightened too much, for this may distort or even crack the transistor inside and lead to early failure. This effect can be minimised by using a bolt that is just able to pass through the holes. A washer helps by distributing the pressure more evenly.

With certain types of transistor or other device the tag is in electrical contact with one of the terminals of the device. Sometimes the metal case is the only connection to the terminal. An example is the 2N3055 transistor in its metal TO66 case. It has two wire terminals connecting to the base and emitter. The connection to the collector is the case itself. With devices such as these it is usually necessary to place an insulating washer between the device and the heat sink. This places another thermally resistive layer between the tag and the heat sink. Washers made of mica or of plastic have thermal resistances of about 2–3 °C/W, and about half of this if they are greased on both sides. There are also silicone rubber washers which have even lower thermal resistance and are used without greasing. Washers may be needed when two or more transistors are mounted on the same heat sink.

As well as the insulating washer, it is usual to employ a plastic bush around the bolt to prevent it from making an electrical connection between the tag and the heat sink (Fig. 8.5).

Example

A transistor running at a maximum power dissipation of 40 W, has a junction-to-tag thermal resistance of 0.8 °C/W. It is mounted on a large heat sink with resistance 2.4 °C/W, and is insulated from it by a greased mica washer with resistance 0.1 °C/W. The heat sink is mounted outside the enclosure of the amplifier in an ambient temperature of 27°C. What is the junction temperature and is it safe?

The three thermal resistances are in series so the solution is:

Junction temp. = power (sum of resistances) + ambient temp.
= 40 (0.8 + 0.1 + 2.4) + 27
= 159°C

This is higher than the maximum safe operating temperature of 150°C. Use a heat sink with lower thermal resistance, such as 1.2 °C/W. This reduces the junction temperature to 40 × 2.1 + 27 = 111 °C, allowing ambient temperature to be higher with safety.

Integrated circuits

Power amplifiers have so many applications, especially in the audio and TV field, that it is economic to produce them in large numbers as integrated circuits. Only a few external components, mainly capacitors, are required. Using an integrated circuit amplifier simplifies wiring, saves circuit-board space, and is generally cheaper. Having all the active components on the same chip, so that they are all at the same temperature, leads to greater stability and reliability.

Of the many power amplifiers available, each with its own distinctive features, we have chosen one as representative of this class of integrated circuit. This is the SGS Thompson TDA2040V. It is packaged as a 5-pin device with a metal tag for bolting to the heat sink. Fig. 8.6 shows how the TDA2040V is connected to build a Class AB audio amplifier capable of driving a 4Ω speaker at 22 W, when operating on a ±16 V supply.

On looking at Fig. 8.6 it is obvious that the circuit has many of the features of an inverting op amp circuit. There are two inputs, (+) and (–), with feedback from the output to the (–) input. The input stage of this amplifier is a differential amplifier. The external resistors and capacitors are there to act as filters. Others are there to hold the supply voltage steady against feedback through the speaker to the 0 V supply line.

The description on the data sheet makes it clear that this is not just a simple op amp. Like most integrated circuit audio amplifiers, it has several useful features, including:

- Short circuit protection. If the output terminal is short-circuited, the output current is automatically limited to prevent the transistors from being destroyed by over-heating.
- Thermal shutdown. If the device becomes too hot for any reason, it is automatically shut down to limit current and further heating.

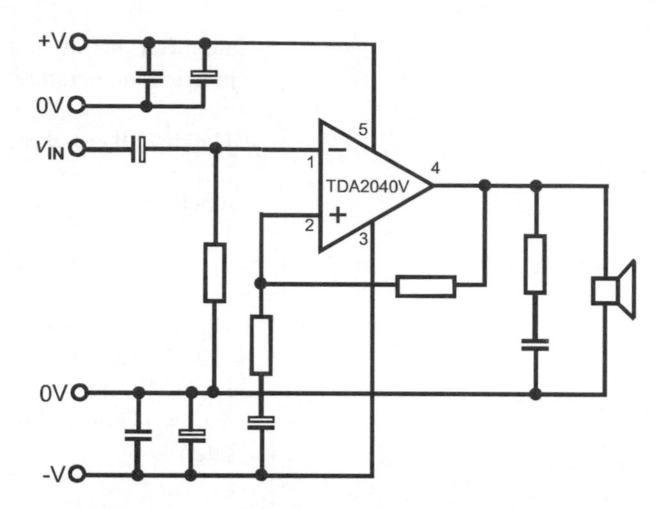

Figure 8.6 *A power amplifier in integrated circuit form may need only a few external resistors and capacitors to complete it.*

- The tag is connected internally to the 0V rail, so no insulating washer is needed on the heat sink.

Heat sinks are sometimes made of copper. This metal has higher thermal conductivity than aluminium, so copper heat sinks are used if there is a large amount of heat to be conducted away. Their disadvantage is that copper heat sinks are heavier than aluminium ones of the same size, and copper is more expensive. If there is space on the circuit board, a heat sink is made by leaving an area of copper unetched. Transistors are bent over on their leads and bolted to this.

Activity 8.1

Build and test a Class B power amplifier based on a complementary pair of BJTs, using the basic circuit shown in Fig. 8.1. Then modify the circuit as in Fig. 8.3. In each case, use an oscilloscope to observe the waveform of the output, when the input signal is a 500 Hz sinusoid, amplitude 1 V.

Problems on power amplifiers

1 Describe an example of a Class A amplifier and explain why its action is inefficient.

2 Describe a complementary push-pull power amplifier and how it works. In what way is it more efficient than a Class A amplifier?

3 What is meant by *crossover distortion* in Class B amplifiers and what can be done to avoid it?

4 Describe the precautions you would take to prevent the transistors of a power amplifier from overheating.

5 Explain why it may be necessary to place an insulating washer between the tag of the transistor and the heat sink. In what way may the thermal resistance between the tag and the heat sink be made as small as possible?

6 A transistor in a power amplifier has 12 V across it and passes 4.4 A. It is greased and mounted on a heat sink with a thermal resistance 1.2 °C/W. The junction-to-tag thermal resistance is 1.1°C/W and the tag-to-heat sink thermal resistance of the grease is 0.1 °C/W. What is the maximum ambient temperature under which the amplifier can safely operate?

7 With the amplifier described in question 6, explain giving reasons what measures you could take to make the amplifier suitable for operating at higher ambient temperatures?

8 Using information from the manufacturer's data sheet, design a power amplifier based on an integrated circuit.

Multiple choice questions

1 A basic type of amplifier used in power amplifiers is the:

A common-base amplifier.
B common-emitter amplifier.
C common-collector amplifier.
D common-drain amplifier.

2 The voltage gain of a typical Class B power amplifier is:

A 100.
B less than 1.
C more than 1.
D 20.

3 If the tag of a transistor is at 60°C, t_A is 25 °C, and the transistor is running at 3.5 W, the thermal resistance of the heat sink is:

A 10 °C/W.
B 35 °C.
C 10 W/°C.
D 0.1 °C/W.

4 The maximum allowable temperature of a transistor junction is:

A the ambient temperature.
B 150 °C above ambient temperature.
C 100 °C.
D 150°C.

9 Thyristor and triac circuits

Summary

A thyristor is a four-layer device acting like a diode but with a triggerable gate. Once triggered it continues to conduct until the current passing through it falls below a certain level. It can be used for switching direct current or alternating current. When switching AC it acts as a controlled rectifier. A triac is a similar device but conducts in both directions. There are several circuits for generating the firing pulses to obtain phase control. These include using an opto isolator to pass the firing signal from the low-power control circuit to the high-power thyristor or triac circuit. False triggering may occur due to rapid increase of voltage across the thyristor or triac, but this in minimised by using a snubber. Thyristors and triacs generate radio frequency interference but this is minimised either by wiring a radio frequency lowpass filter between the device and the supply mains, or by using a zero-crossing switch. Zero-crossing switches are used with integral cycle control.

Thyristor structure

A thyristor is also known as *silicon controlled switch* (SCS) and as *a silicon controlled rectifier* (SCR). The word *switch* tells us that it can be turned on and off. The word *rectifier* indicates that current flows through it in only one direction.

A thyristor consists of four semiconductor layers (Fig. 9.1) with connections to three of them. Its structure is easier to understand if we think of the middle layers being split into two, though still electrically connected. Then we have two three-layer devices, which are recognisable as a pnp transistor and an npn transistor. The base of the pnp transistor (n-type) is connected to the collector of the npn transistor (also n-type). The collector of the pnp transistor (p-type) is connected to the base of the npn transistor (also p-type). These connections show more clearly in Fig. 9.2, where the two transistors are represented by

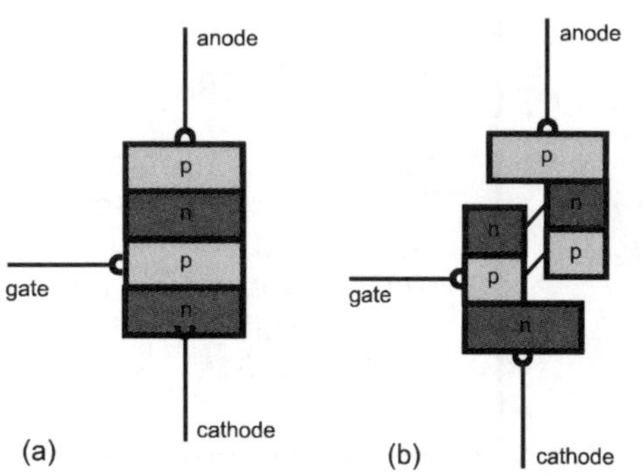

Figure 9.1 *A thyristor is a four-layered device, but can be thought of as a pnp transistor joined to an npn transistor.*

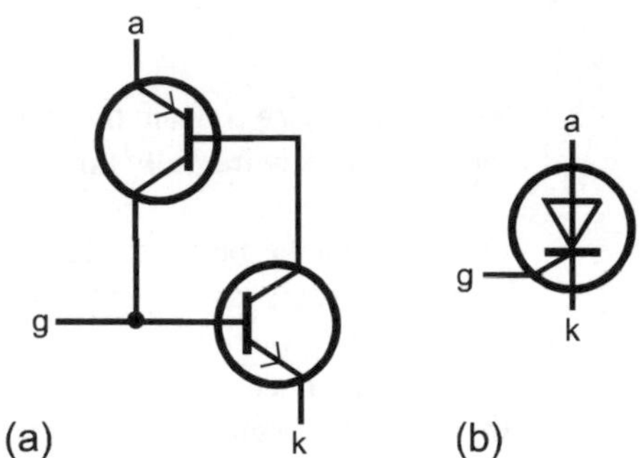

Figure 9.2 *It is easier to understand thyristor action if we think of it as a pnp transistor and an npn transistor combined. Its symbol shows that it has the one-way conduction property of a diode.*

their usual symbols. In this figure the terminals have been given the names that apply to a thyristor:

- anode: the emitter of the pnp transistor; current enters the thyristor here.
- cathode: the emitter of the npn transistor; current leaves the thyristor here.
- gate: the base of the npn transistor.

Test your knowledge 9.1

In what direction does current flow through a diode?

The names *anode* and *cathode* are the same as those given to the terminals of a diode. This shows that the thyristor has the same kind of one-way action as a diode and, like a diode, may be used as a rectifier. We will look more closely at this action to see how the voltage present at the gate terminal triggers it.

Thyristor action

Base-emitter voltage, v_{BE}: An npn transistor begins to turn on when this exceeds 0.6V.

This kind of action is described as *regenerative.*

When the gate is at the same voltage as the cathode, the npn transistor is off. No current can flow through the thyristor. If the gate voltage is made 0.6 V higher than that of the cathode, the npn transistor begins to turn on. Current begins to flow through the npn transistor. This current is drawn from the base of the pnp transistor. Drawing current from its base begins to turn it on. Current begins to flow through the pnp transistor. This current goes to the base of the npn transistor, adding to the current already flowing there and turning the npn transistor more strongly on. Its collector current increases, drawing more from the base of the pnp transistor and turning it more strongly on. This is positive feedback and results in both transistors being turned on very quickly.

Mains voltage

In the United Kingdom the mains voltage is quoted as 230 V. This is its *root mean square* value. Its amplitude is therefore $230 \times \sqrt{2} = 325$ V. Devices that operate in circuits that are directly powered from the mains need to be able to withstand a reverse voltage of at least 325 V. Usually they are rated to withstand 400 V.

The whole process, from the initial small gate current (base current) to the transistors being fully on, takes only about 1 microsecond.

Once current has begun to flow, it continues to flow even if the original positive voltage at the gate is removed. The current from the pnp transistor is enough to keep the npn transistor switched on. This means that the gate need receive only a very short positive pulse to switch the thyristor on. Typically, a gate voltage of about 1 V is enough to start turning the npn transistor on, and the current required may be only a fraction of a milliampere.

Once the thyristor has been triggered, it continues to pass current for as long as the supply voltage is across it. If the supply voltage is removed, current ceases to flow and it requires another pulse on the gate to start it again. There is a minimum current, known as the *holding current,* which must be maintained to keep the thyristor in the conducting state. If the supply voltage is reduced so that the current through it becomes less than the holding current, conduction ceases and a new gate pulse is needed to re-trigger it. Holding current is typically 5 mA but may be up to 40 mA or more with certain types of thyristor.

The amount of heat depends on how many watts of power are dissipated in the thyristor. This depends on the voltage between anode and cathode and the current through the thyristor: $P = IV$.

The voltage drop across a conducting thyristor is only about 1 V, though may be up to 2 V, depending on the type. Low voltage results in relatively low power, even if current is high. This means that the thyristor may pass a very high current without dissipating much energy and becoming excessively hot. Most thyristors can pass currents measured in amperes or tens of amperes. Yet a relatively minute gate pulse triggers such currents. It is the high gain of thyristors, combined with their extremely rapid switch-on which makes them popular for power control circuits. A simple npn or pnp power transistor is much less satisfactory for this purpose. If it is to have sufficient gain to switch a large current, its base layer needs to be very thin. This makes it unable to withstand high reverse voltages. By comparison, thyristors can withstand a reverse voltage of several hundred volts without breaking down. Many thyristors have a reverse *breakover voltage* of 400 V, and some can withstand a reverse voltage of more than 1000 V. For this reason thyristors are widely used for switching mains or higher voltages, as described later in this chapter.

Activity 9.1 Demonstrating thyristor action

Assemble the circuit of Fig. 9.3. Apply the 12 V DC supply from a bench PSU. Record and explain what happens when:

- power is first applied.
- S1 is pressed and held.
- S1 is released.
- S2 is pressed and held.
- S2 is released.

Measure the current flowing to the gate when S1 is pressed and held. Measure the current flowing through the lamp when it is lit.

Add a 22 kΩ variable resistor VR1 to the circuit where indicated. Begin with the variable resistor set to its minimum resistance so that maximum current flows. The lamp lights when S1 is pressed. Measure the current flowing through it. Repeat with VR1 set to various higher resistances and find the minimum holding current of the thyristor. The lamp may not light when current is small but the meter shows if a current is flowing.

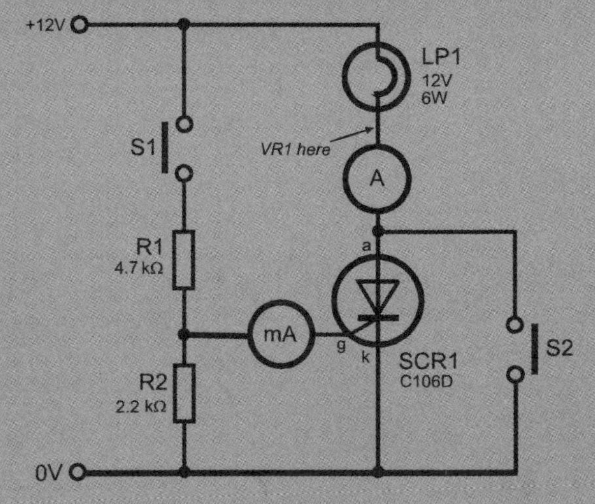

Figure 9.3 *This circuit is used for investigating the action of a thyristor. Note that the supply is 12 V DC.*

Operating voltage

Most of the thyristor and triac circuits in this chapter are shown supplied from the mains. This is because the control of mains-powered devices is the most important application of these devices. To investigate the action of these circuits practically but safely, **use a supply of lower voltage**, such as **12 V** AC or **20 V** AC taken from a bench power unit. Or run the circuits on a simulator, and set the simulated supply voltage to 230 V.

Switching DC

The switching action demonstrated in Activity 9.1 is used to turn on a high-power device when only a small triggering current is available. For example, the light-dependent resistor (LDR) in Fig. 9.4 triggers a thyristor to switch an alarm. The resistance of an LDR decreases as the amount of light falling on it decreases. When the light level falls below the level set by adjusting VR1, the LDR resistance increases, voltage at the gate rises and triggers the thyristor. Once triggered, the thyristor conducts for an unlimited period, even if light increases again. The circuit is sensitive to very short interruptions in the light, such as might be caused by the shadow of a passing intruder falling on the LDR.

This kind of switching action is mainly limited to alarm-type circuits with a once-for-all action. The alarm sounds until the reset button S1 is pressed or the power is switched off. A lamp can replace the alarm so that the thyristor turns on the lamp when light level falls. With some circuits, the light coming on could cause negative feedback and turn the lamp off again. The lamp would flicker on and off. There is no such problem in this circuit because of the once-for-all action.

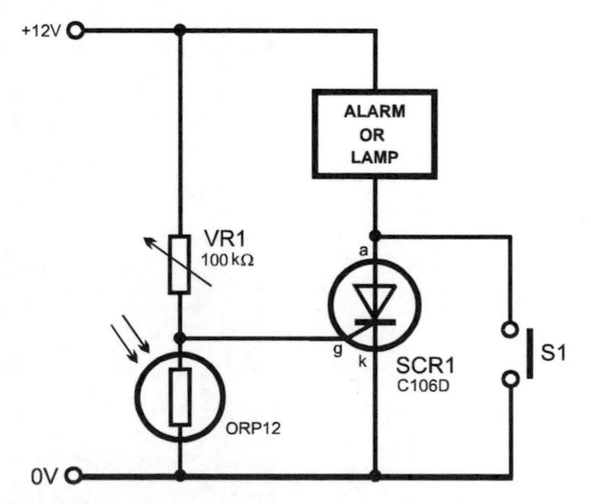

Figure 9.4 *Making use of the thyristor's high gain and rapid action in an alarm circuit.*

Switching AC

Most thyristor applications are concerned with AC circuits. Fig. 9.5 supplies power to a device represented by a 100 Ω resistance. The supply comes from the mains at 230 V AC and 50 Hz. On the positive half-cycle there is a positive voltage across *R*2, which acts to turn on the thyristor. It conducts for as long as the voltage across it is in the positive direction, from anode to cathode. The thyristor is turned off at the end of the half-cycle. On the negative half-cycle the voltage across *R*2 is negative, so the thyristor is not turned on. The result is that the thyristor conducts only during the positive half-cycle. It behaves just like a diode, and gives half-wave rectified current, as shown in Fig 9.6. In the figure we see the curve for current (black) following the input voltage curve (grey) during the positive half-cycle only.

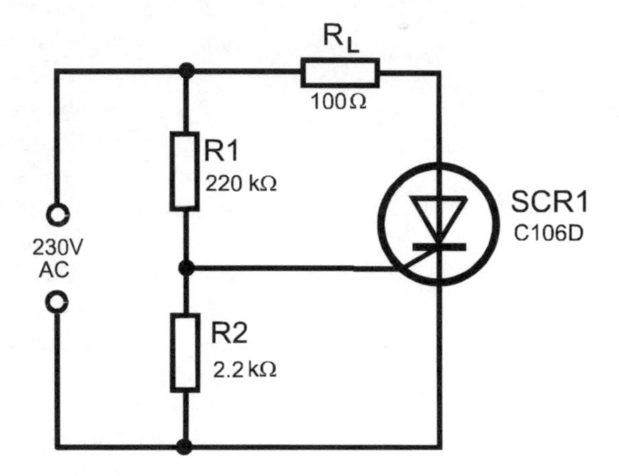

Figure 9.5 *A basic half-wave controlled rectifier, using a thyristor.*

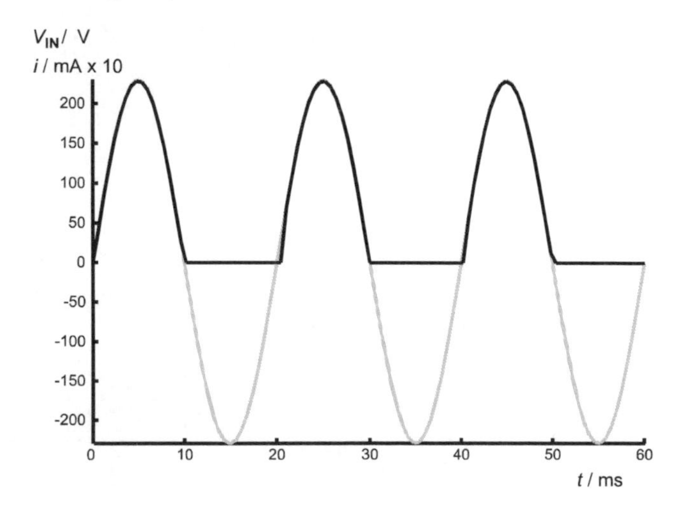

Figure 9.6 *The output (black) of the half-wave rectifier of Fig. 9.6 follows the input (grey) only during the positive half-cycles. There is a very slight delay at the beginning of every cycle.*

In this circuit, the thyristor is turned on at the beginning of every half-cycle. The resistor network provides the necessary pulses, which are synchronised with the alternating supply. The pulse occurs at the same point or phase early in every cycle. Because the timing of the pulses is related to the phase of the cycle, we call it *phase control.*

There can be no pulse at the very beginning of the cycle because the supply voltage is zero at that time. But very soon the input voltage has reached a level which produces a voltage across R2 that is sufficient to trigger the thyristor. The phase at which the pulse is generated is the *firing angle*. In Fig. 9.6 the pulse begins about 0.3 ms after the cycle begins, and the length of the cycle is 20 ms, so the firing angle is $360 \times 0.3/20 = 5.4°$. Current flows for the remainder of the half-cycle so the *conduction angle* is $180 - 5.4 = 174.6°$.

If R2 is a variable resistor it is possible to reduce its value and so reduce the amplitude of the triggering pulses. It then takes longer for the pulse to reach a value that will trigger the thyristor. Firing is delayed until later in the cycle. Fig. 9.7 shows what happens if R2 is reduced to 560Ω. The delay at the start of the cycle is now 2.2 ms. This gives a firing angle of $360 \times 2.2/20 = 40°$. The conduction angle is $180 - 40 = 140°$. The value of R2 controls the delay at the beginning of each cycle, which controls the firing angle and conduction angle. The conduction angle determines how much power is delivered to the load during each cycle. The circuit is a *half-wave controlled rectifier* which can be used for controlling the brightness of mains-powered lamps and the speed of mains-powered motors. There is also the fact that the thyristor rectifies the alternating current to produce pulsed direct current, which can be used for driving DC motors. But this is a

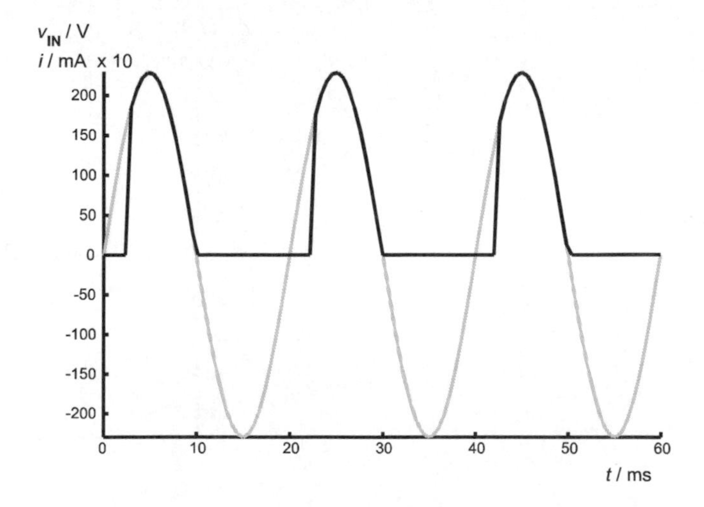

Figure 9.7 *Reducing the value of R2 increases the delay at the beginning of each cycle, so that the conduction angle is smaller, and the amount of power supplied to the load is reduced.*

half-wave circuit so, at the most, the lamp or motor is driven at only half power. This is a basic circuit and several improvements are possible, as will be explained later.

Full-wave controlled rectifier

There are various ways of using thyristors in full-wave controlled rectifiers. One of these has two thyristors in parallel but with opposite polarity so that one conducts on the positive half cycle and the other on the negative half cycle. They each have a pulse generator synchronised to switch on the thyristor during the appropriate half-cycle.

A different circuit, which requires only one thyristor, is illustrated in Fig. 9.8. The current is rectified by a conventional diode bridge, and the thyristor acts to turn the current on and off during both positive and negative half-cycles. The thyristor is able to act during both half-cycles because it is located in the rectified part of the circuit so that both half-cycles are positive. This circuit is suitable for loads that operate only on DC.

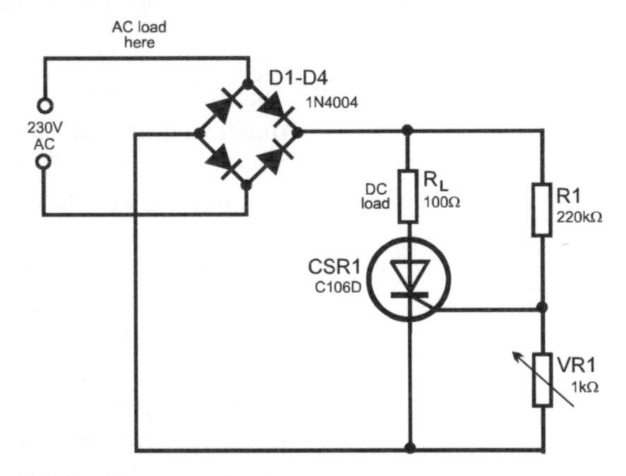

Figure 9.8 *A full-wave controlled rectifier has the same circuit as Fig. 9.5, but the power is first rectified by a full-wave diode bridge.*

As indicated in Fig. 9.8, the load R_L can be placed in the mains side of the diode bridge instead. The current still has to flow through the thyristor to complete its path through the circuit. The thyristor turns on the current on and off during both positive and negative half-cycles (Fig. 9.9).

Activity 9.2 Thyristor controlled rectifiers

Set up the half-wave and full-wave circuits illustrated in Figs. 9.5 and 9.8. Note the remarks in the box headed 'Operating voltage' and use a low-voltage AC supply, NOT the mains. Use an oscilloscope to monitor the input and output waveforms.

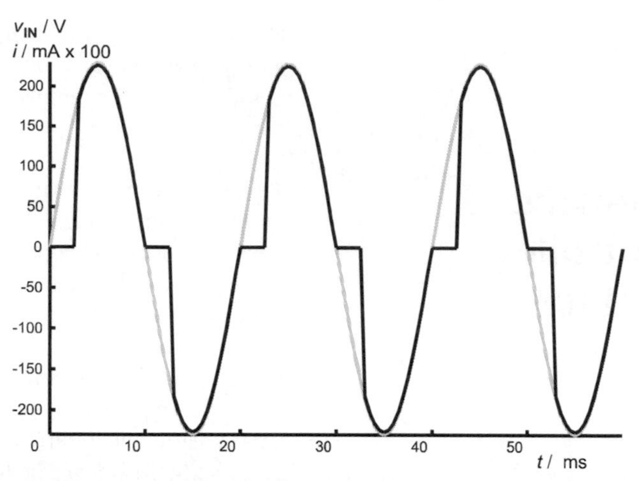

Figure 9.9 *If the load on a full-wave controlled rectifier is placed on the mains side of the diode bridge, the current through the load (black) is alternating and is controlled during both positive and negative half-cycles.*

Triacs

One of the chief limitations of thyristors is that they pass current in only one direction. At the most, they can operate only at half-power. If we require power during both half-cycles we need to wire two thyristors in parallel with opposite polarity, or rectify the supply, as in Fig. 9.8. The triac is a device in which two thyristors are combined in parallel but with opposite polarity in a single block of silicon. Gating is provided for one of the thyristors, but when the thyristors are combined in this way, the gate switches either one of them, whichever one is capable of conducting at that instant. Another useful feature is that either positive or negative pulses trigger the device. Like a thyristor, the device switches off whenever the supply voltage falls to zero, and then required re-triggering. It also switches off if the current through it falls below the minimum holding current.

Although triacs have advantages over thyristors, they have a disadvantage too. Since they conduct on both cycles they have little time to recover as the current through them falls to zero and then increases in the opposite direction. If the load is inductive, the voltage across the triac may be out of phase with the current through the triac. There may

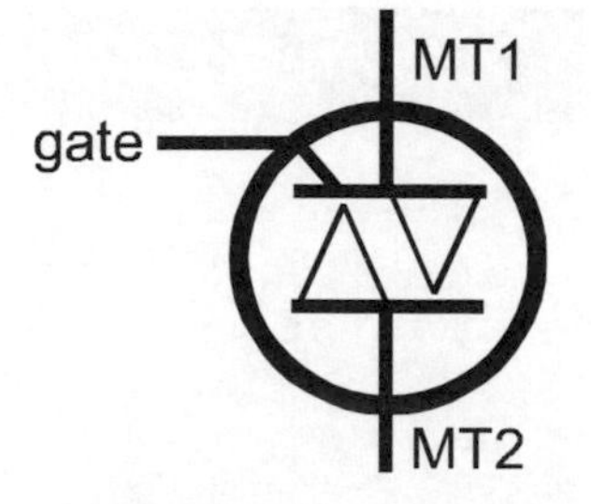

Figure 9.10 *Because the triac conducts in both directions, its main terminals are simply named MT1, MT2. The third terminal is the gate.*

still be a voltage across the triac when the current is zero. This may cause the triac to continue conducting and not switch off.

Firing pulses

In the thyristor and triac circuits described so far, we have used a pair of resistors as a potential divider to derive the firing pulse from the mains voltage. This simple system ensures that the pulse occurs at a fixed firing angle. The pulse can be made to occur at any time during the first 90° of the half-cycle (as voltage increases) but the conduction angle must always be between 90° and 180°. If we want a conduction angle between 0° and 90°, we use a capacitor to delay the pulse for more than 90°. The lamp-dimming circuit in Fig. 9.11 illustrates this technique. The triggering network consists of a resistor and capacitor. If VR1 is increased in value, the charge across the capacitor rises to its maximum later in the cycle. Fig. 9.12 shows firing at 126° and 306°. The triac is triggered by a positive pulse during the positive half-cycle and by a negative pulse during the negative half-cycle.

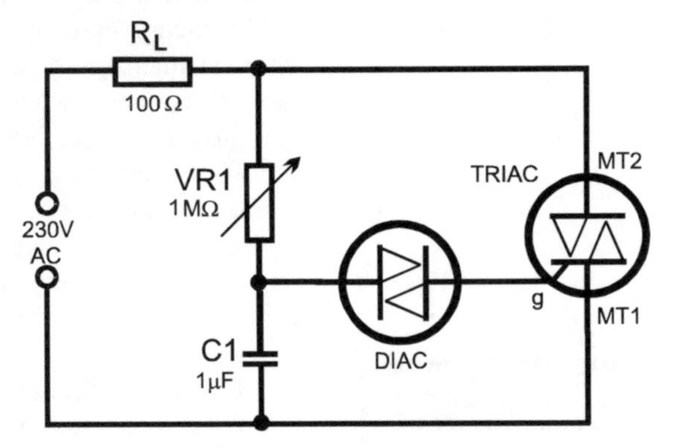

Figure 9.11 *A lamp dimming circuit using a triac, with a resistor-capacitor network to trigger the triac at any stage during the cycle.*

Diacs

The circuit of Fig. 9.11 has a component with a symbol like that of a triac, but without the gate terminal. This device is known as a *diac*. It is similar to a triac in structure but has no gate. A diac does not conduct in either direction until the voltage across it exceeds a given value (often 32 V). Then avalanche breakdown occurs and a burst of current passes through it. The effect of the diac is to produce a clear-cut pulse of current to trigger the triac.

Some triacs are made with the diac already built in, and are known as *quadracs*.

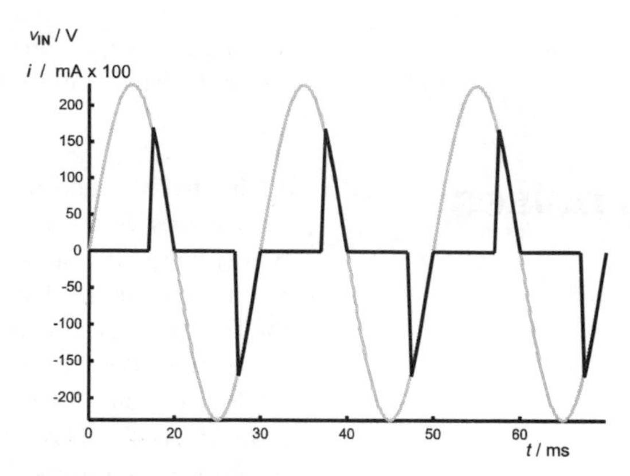

Figure 9.12 *The circuit of Fig. 9.11 dims the lamp completely if the triac is triggered late in the cycle.*

Although the resistors and capacitors in triggering networks pass very small currents, and so are not wasteful of power, there is always the risk of their eventually breaking down. They are directly connected to wires at mains voltage and there is danger in this. In commercially produced, manually operated circuits the control knobs are sufficiently well insulated to prevent the operator from receiving an electric shock, but when a circuit is automatically operated, perhaps by a computer, we need to take precautions to prevent high voltages reaching circuit where they can do expensive damage. Automatic thyristor and triac circuits often employ one of the following devices to isolate the control circuit from the thyristor or triac:

Test your knowledge 9.2

If a 4V pulse is passed into the primary coil of a 1:1 transformer, what will be the amplitude of the output pulse from the secondary coil?

- **Pulse transformer:** This is usually a 1:1 transformer so that the pulse generated in the control circuit is transferred virtually unchanged to the gate of the thyristor or triac. The insulation between the primary and secondary windings withstands voltages of 2000 V or more.
- **Opto isolator:** Opto isolators are often used when it is necessary to isolate two circuits, one at low voltage and one at high voltage, yet to pass a signal from one to the other. A typical opto-isolator consists of an integrated circuit package containing an infrared LED and a phototransistor (Fig. 9.13). A pulse is passed through the LED causing it to emit light. The phototransistor receives the light energy and is switched on. Current flows through the transistor. Fig. 9.14 shows how this current is used to generate a pulse at the gate of a triac. The opto-isolator is constructed so that its insulation can withstand voltages of 1000 V or more between the diode and the transistor.
- **Opto triac isolator:** This comprises an LED and a triac with a photosensitive gate (Fig. 9.15). A pulse of current applied to the LED causes the triac to be triggered into conduction.

Infrared LED: a light emitting diode that emits infrared radiation (invisible to the eye) instead of visible red, yellow, orange, green or blue light.

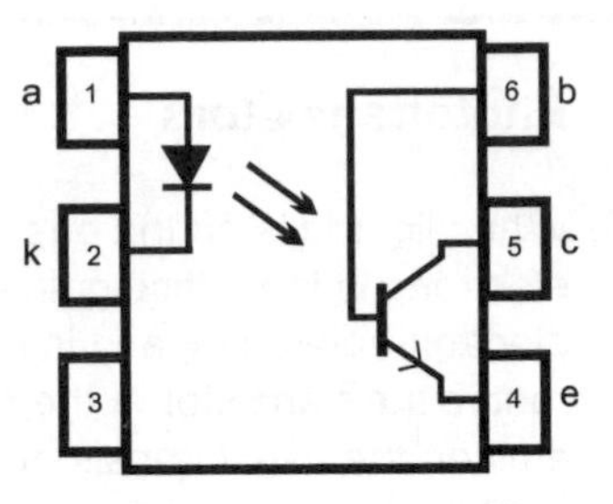

Figure 9.13 *The phototransistor in this opto-isolator is turned on by infrared light from the LED. It is not necessary to use the base connection, though it can be used to bias the base so that the transistor is ready to be turned on more quickly.*

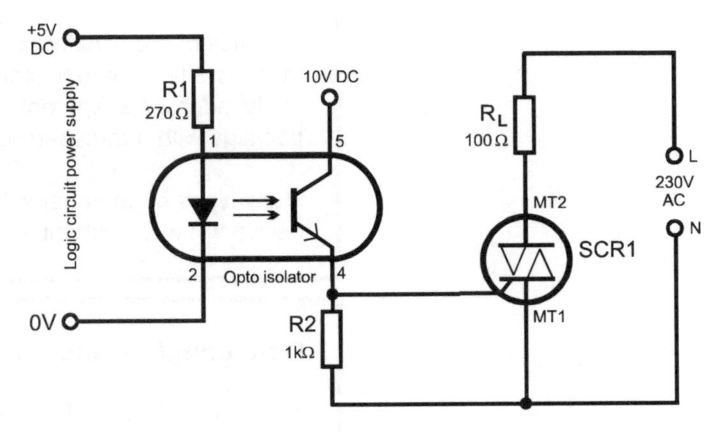

Figure 9.14 *The light from the LED provides sufficient energy to turn on the phototransistor. The current through R2 then generates a positive pulse to turn on the triac.*

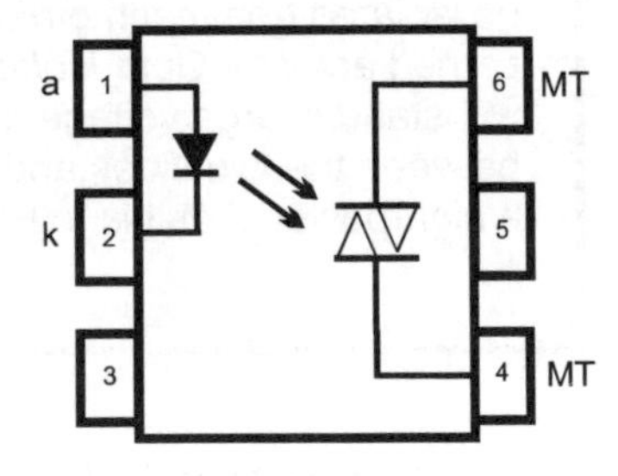

Figure 9.15 *In an opto triac isolator, the light from the LED triggers the triac directly.*

UJT pulse generator

A unijunction transistor consists of a bar of n-type semiconductor with a p-type region diffused into it about halfway along. The ends of the n-type are called base-1 and base-2, and the connection to the p-type region is known as the emitter. The *base-to-base resistance* (or *inter-base resistance*) is typically 500 Ω. The UJT is connected with b2 positive of b1, as in Fig. 9.16. There is a voltage gradient along the n-type bar ranging from 0 V at b1 to the voltage applied at b2. Current flows through R1 and gradually charges C1. Voltage across C1 rises until it reaches the *peak point* voltage of the circuit. This is the point at which the voltage of the emitter equals the voltage in the region of the bar in which the emitter is located. As the emitter voltage begins to

Phototransistors

When light falls on the base of a transistor, some of the electrons in the lattice gain energy and escape. This provides electron-hole pairs, and is equivalent to supplying a base current to the transistor. If the collector is positive of the emitter, the usual transistor action then occurs and a relatively large collector current flows through the transistor.

Transistors are usually enclosed in light-proof packages so that this effect can not occur, but a phototransistor is enclosed in a transparent plastic package or in a metal package with a transparent window.

Many types of phototransistor have no base terminal, as the device will work without it.

Opto couplers and opto isolators

Both kinds of device have an LED and an amplifying semiconductor device enclosed in a light-proof package. Their function is to pass a signal from one circuit to another. Opto *couplers* are used when, for *electronic* reasons, it is not suitable to connect the circuits directly (perhaps voltage levels are not compatible, or we want the two 'ground' rails to be isolated from each other to prevent interference being carried across). Opto *isolators* are specially designed to withstand a large voltage difference (usually 1000 V or more) between the two sides and are used for *safety* reasons, when a high-power circuit is to be controlled by a low-power circuit.

Test your knowledge 9.3

What is the main type of charge carrier in p-type semiconductor?

exceed the voltage in the bar, holes are injected into the bar from the emitter.This decreases the resistance of the bar between the emitter and b1. In turn, this reduces the voltage in the region of the bar next to the emitter Even more holes are injected and resistance is further decreased. The result is a high current flowing from emitter to b1, discharging C1 rapidly. This current may reach several amperes, generating a pulse of high voltage as it passes through R3. The current flows until it falls below the *valley point* current, which is only a few milliamperes. Once current has stopped flowing, the bar between the emitter and b1 regains its former high resistance, and the capacitor gradually charges again.

As a result of this action a series of pulses is generated across R3.

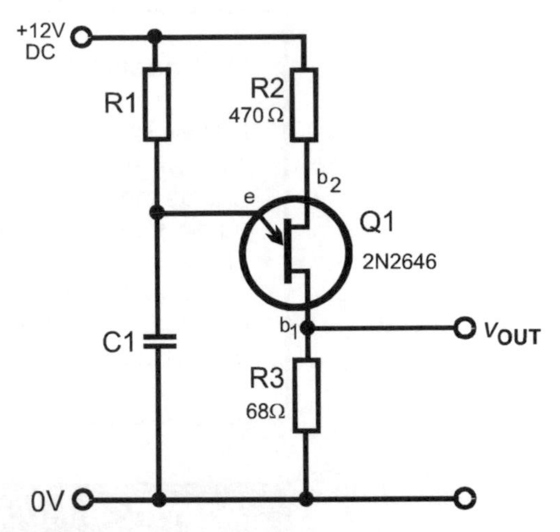

Figure 9.16 *A UJT provides the basis for building an oscillator with the minimum of components.*

These may be used to trigger a thyristor or triac, often after being passed through an isolating pulse transformer. In most UJT triggering circuits the UJT oscillator runs at several hundred kilohertz. There are fifty or more pulses during each half-cycle of the mains, which results in the thyristor or triac being turned on early during the half-cycle. Once triggered on, it stays on for the remainder of the half-cycle. In this way the UJT oscillator is used as a low-power pulse source for switching a high-power circuit.

Peak point and valley point voltages

Peak point voltage (V_P) depends on the voltage across the UJT (V_S) and its *intrinsic stand-off ratio* (η):

$$V_P = \eta V_S$$

Both voltages are referenced to b1, which is taken to be at 0 V. In most UJTs, η ranges from 0.56 to 0.75, with a typical value of 0.6.

Example

If $V_S = 9$ V, and $\eta = 0.6$, then $V_P = 9 \times 0.6 = 5.4$ V.

In the oscillator of Fig. 9.16, C_1 charges until the emitter is 5.4 V positive of b1.

It then discharges until the emitter voltage reaches the *valley point*, V_V, which is typically 2 V.

Test your knowledge 9.4

A UJT operating with 7V across it has a standoff ratio of 0.65. At what emitter voltage does it trigger?

UJT oscillators

UJT oscillators like those in Fig. 9.16 have other applications such as audio-frequency buzzers in lamp-flashing circuits. As well as the pulsed output from R3, we can take the sawtooth output from the connection between C1 and the emitter. There is a steadily rising voltage (not truly linear, but exponential) followed by a rapid fall as the peak point is reached. If the oscillator has low frequency, this output can be used as a positive ramp generator.

Activity 4.3 UJT oscillator

Build a UJT oscillator as in Fig. 9.16. To begin with, make R1 = 10 kHz and C1 = 1 μF. Use an oscilloscope to monitor the outputs from base-1 and from the emitter. Find the value of the stand-off ratio, η. Try the effect of changing the values of R1 and C1.

False triggering

If the voltage across a thyristor or triac increases too rapidly, this may lead to false triggering and loss of control. This may occur when:

- power is first applied to the circuit.
- there is an inductive load, such as a motor, causing voltage and current to be out of phase.
- there are spikes on the supply lines.

The cure for this is to wire a resistor-capacitor network across the device, to prevent the voltage across it from rising too rapidly. This is known as a *snubber* network and is illustrated by R1 and C2 in Fig. 9.17. This network could be added to any of the circuits described in this chapter. A relatively new development is the *snubberless triac*. This can be subjected to voltage change rate as high as 750 V/ms without causing false triggering. A snubber network is not required with this device.

RFI

When a thyristor or triac switches on or off during a half-cycle, there is a rapid change in current. Rapid changes in current are equivalent to a high-frequency signal. The result is the generation of *radio frequency interference*. You can demonstrate this by holding a portable AM radio receiver close to a domestic lamp dimmer switch. RFI is radiated from the circuit and also appears on the mains supply lines which are providing power to it. It is conducted along the mains lines to other

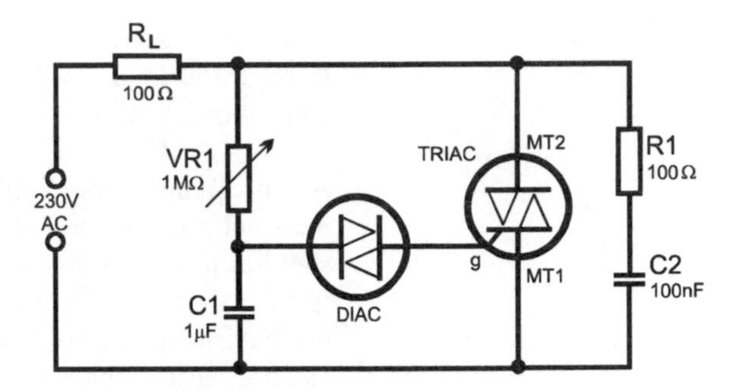

Figure 9.17 *A snubber network (R1 and C2), when added to the circuit of Fig. 9.11, prevents the triac from being triggered by rapid rise in the voltage across it.*

equipment, where it may interfere with their operation. The larger the currents being switched by the thyristor or triac, the worse the interference.

To reduce RFI it is usual to place an RF filter on the supply lines, as shown in Fig 9.18. This is wired as close as possible to the thyristor or triac so as to reduce the amount of RFI being radiated from the lines joining the thyristor or triac to the load. In Fig. 9.18, L1 and C3 form a lowpass filter with the cutoff frequency in the RF band.

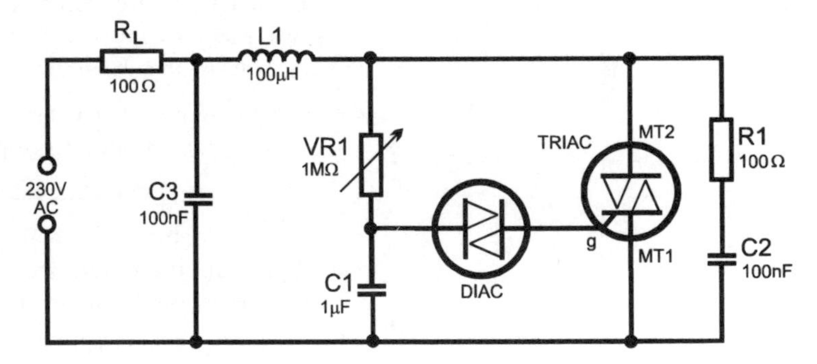

Figure 9.18 *A lowpass RF filter (L1 and C3), when added to the circuit of Fig. 9.17, prevents radiofrequency interference produced by the triac from spreading to the supply mains.*

Integral cycle control

All the circuits described so far in this chapter are examples of *phase control.* The thyristor of triac is switched on at a particular phase of the half-cycle and remains on until the half-cycle ends. The amount of power supplied to the load depends on for how large a fraction of the half-cycle the thyristor or triac is switched on. In integral cycle control the device is switched on for complete half-cycles. The amount of power supplied to the load depends on for how many half-cycles it is switched on and for how many cycles it is switched off. The advantage of this technique is that switching occurs when the supply voltage is at or close to zero. There is no large increase or decrease of current through the load, so RFI is reduced to the minimum. Integral cycle control is sometimes referred to as *burst fire* control

As an example, we could supply power to a motor for 5 half-cycles, then cut off the power for the same number of half-cycles. This supplies power for half the time. Power is interrupted for 5 half-cycles (for 0.1 s on a 50 Hz mains supply) but the inertia of the motor and the mechanism it is driving keeps it rotating during the off periods. An electric heater may also be controlled by this technique because interruptions of a fraction of a second or even several seconds makes little difference to the temperature of the heating element. On the other hand, this form of control is not suitable for controlling the brightness of lamps because it makes them flicker noticeably.

Integral cycle control relies on being able to detect the instant at which the alternating supply has zero value and is changing direction. This is the zero-crossing point at the beginning of every half-cycle. A zero-crossing detecting circuit can be built from discrete components but special integrated circuits are more often used. Some triac opto-couplers (Fig. 9.15) have a built-in zero-crossing detector that allows the triac to be switched on only at the beginning of the half-cycle.

A typical example of a zero-crossing switch IC is the CA3059, which can be used to control either thyristors or triacs. Fig. 9.19 shows it being used to provide on/off control of a triac. The integrated circuit contains four main sub-circuits:

- A regulated 6.5 V DC power supply for the internal circuits. This has an outlet at pin 2 for supplying external control circuits.
- Zero-crossing detector, taking its input from the mains supply at pin 5.
- High-gain differential amplifier, with its inputs at pins 9 and 13. Depending on the application, this can accept inputs from:
    - a switch (as in Fig. 9.19), for simple on/off control
    - a sensing circuit based on a device such as a thermistor or light-dependent resistor, for automatic control of a heater or lamp.
    - a transistor or opto-isolator, for control by automatic circuits, including computers.
    - an astable for burst-fire control.
- Triac gate drive with output at pin 4. It produces a pulse when the mains voltage crosses zero volts. Pulses are inhibited if the pin 9 input of the differential amplifier is positive of the other input at pin 13. Pulses may also be inhibited by a signal applied to the 'inhibit' input at pin 1 and the 'fail-safe' input at pin 14. The drive can be triggered by an external input at pin 6.

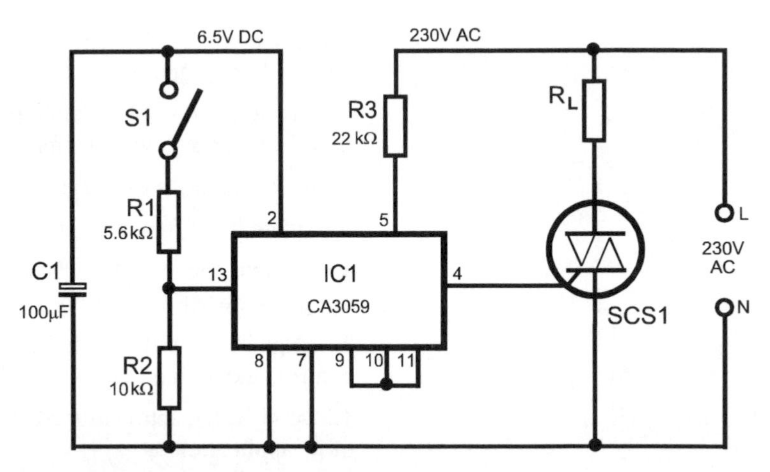

Figure 9.19 *The integrated circuit is a zero-crossing switch, used to minimise RFI and to interface the triac to the control circuitry, represented here by a simple on/off switch.*

As can be seen from Fig. 9.19, the CA3059 needs very few external components. These comprise:

- The triac, with its load R_L.
- A snubber network connected across the terminals of the triac (not shown in the figure).
- A dropper resistor R_3, which should be a high-wattage type (about 5 W).
- A smoothing capacitor C_1 for the 6.5 V supply.
- A control circuit.

In the figure, the control circuit consists of S1 and a pair of resistors. Pin 9 is connected to pins 10 and 11 and this puts it at mid-rail voltage (approx 3 V). When S1 is open, R2 pulls pin 13 down toward 0 V. The gate drive circuit is inhibited because pin 13 is negative of pin 9 and the triac does not fire. When S1 is closed, pin 13 is raised to just over 4 V, making it more positive than pin 9 and enabling the gate drive to produce pulses. The triac is fired at the beginning of every half-cycle and full power is applied to the load.

Problems on thyristor and triac circuits

1 A thyristor has its anode several volts positive of its cathode. A positive pulse is applied to its gate. Describe what happens next.

2 Once a thyristor is conducting, what can we do to stop it?

3 Describe how a thermistor and a thyristor may be used to control an electric heating element powered by the mains.

4 Explain why it is preferable to use a thyristor rather than a transistor for controlling large currents.

5 Explain what is meant by *firing angle* and *conduction angle*.

6 Draw a circuit diagram of a thyristor half-wave controlled rectifier and explain its action.

7 Draw a diagram of a simple phase-control triac circuit that could be used in a mains-powered lamp-dimming circuit.

8 List the methods used to isolate a triac circuit from its control circuit.

9 Explain the difference between an opto isolator and an opto coupler.

10 Explain the possible causes of false triggering in a triac power circuit and what can be done to prevent it.

11 Explain the differences between phase control and integral cycle control, and their advantages and disadvantages.

12 Describe the action of the CA3059 zero crossing switch ic and some of its applications.

Multiple choice questions

1 A thyristor is triggered by:

A a negative pulse applied to the gate.
B a negative pulse applied to the anode.
C a positive pulse applied to the cathode.
D a positive pulse applied to the gate

2 Starting from the anode, the semiconductor layers in a thyristor are:

A pnpn.
B npn.
C pnp.
D npnp.

3 Triacs are triggered by:

A positive or negative pulses.
B positive pulses only.
C negative pulses only.
D pulses produced as the mains voltage crosses zero.

4 A snubber network is used to reduce:

A overheating of the triac.
B RFI.
C the length of the trigger pulse.
D false triggering.

5 The voltage at which the capacitor is discharged in a UJT oscillator is:

A the peak point.
B the trigger point
C the valley point.
D the time constant.

6 As well as producing a pulsed output, a UJT oscillator can also produce a:

A sine wave.
B sawtooth wave.
C triangular wave.
D square wave.

10 Logic families

Summary

The three main logic families are transistor-transistor logic, complementary MOS, and emitter coupled logic. Each has its own features that suit it for particular applications.

Gates

All the elementary logical operations (NAND, NOT and the like) can be performed electronically. A circuit that performs these operations is known as a *gate,* which is not the same thing as the gate of a thyristor or an FET. A gate may be constructed from individual diodes, transistors and resistors, but usually even the simplest of logic circuits needs so many gates that it is quicker, cheaper and more reliable to use the ready-built gates available as logic integrated circuits.

There are three main logic families:

- Transistor-transistor logic (TTL).
- Complementary MOSFET logic (CMOS).
- Emitter coupled logic (ECL)

Each of these families has its own versions of the standard logic gates such as NAND and NOR. Each family includes a range of ics with more complex functions such as flip-flops, adders, counters, and display drivers. The families differ in the ways the basic gates are built, and this gives rise to family differences in operating conditions and performance. We will look at each family in turn.

Transistor-transistor logic

This was the first family to become widely used and set the standards for logic circuits for many years. The family is based on the TTL NAND gate. Fig. 10.1 shows the circuit of a NAND gate with two inputs. It comprises four npn transistors, four resistors and two diodes. One of the transistors is unusual in having two emitters, but this can be

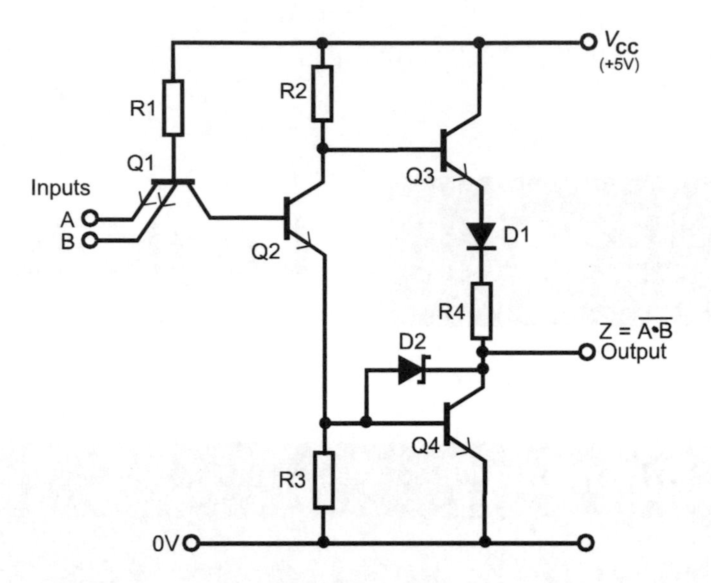

Figure 10.1 *A TTL NAND gate is built on the chip from transistors, resistors and diodes. The Schottky diode (D2) is characteristic of the low-power Schottky series (74LSXX).*

NAND operation: the output is low only when all inputs are high.

AB or A•B is the symbolic way of writing A AND B.

A bar over a symbol or group of symbols indicates negation, the inverse quantity.

thought of as two ordinary npn transistors wired in parallel.

If inputs A and B are both at logic high (+5 V) or are unconnected, the whole of Q1 is high and Q2 is switched fully on. This puts Q3 off and Q4 on, resulting in a low output level Z. But if either one or both of A and B are connected to 0V, current flows through Q1 and the base of Q2 is pulled low, turning it off. Then Q3 is on and Q4 is off, and Z is high. Writing these levels in a truth table we obtain:

Inputs		Output
A	B	Z
0	0	1
0	1	1
1	0	1
1	1	0

This is recognisable as the truth table for NAND, in which 0 = low or false and 1 = high or true. Adding more emitters to Q1 can provide more inputs. These operate in the same way as the two already described, so producing NAND gates with 3 or more inputs.

The output stage of this gate has two transistors operating in opposition. The arrangement is called a *totem pole* output, likening it to the

7400 and others

The devices in the original TTL family were all given type numbers beginning with '74'. This was followed by two or three digits to distinguish the types. For example 7400 is a quadruple 2-input NAND gate, and 7493 is a 4-bit counter/divider. To refer to the family itself we substitute 'XX' for the last two or three digits.

The original 74XX series is now virtually obsolete, being used mostly for repairing old equipment, and is becoming much more expensive than it used to be. There are over 20 more recent versions of the series differing in various features, such as power, speed, immunity from noise, EMI reduction, and supply voltage. Some of the newer versions operate with low power on a 3.3 V supply, which makes them suitable for battery-powered portable equipment, as well as being more economical of power generally.

One of the most popular series is the low power Schottky series, with device numbers 74LSXX. The 74HCXX and 74HCTXX series comprise CMOS versions of the 74XX devices.

stacked icons on the traditional totem pole of North American Indians. In operation there is a changeover stage during which both of the transistors are switched on. This causes a large current to flow as the gate changes state. This may overload the supply line, causing a fall in voltage which may interfere with the action of nearby ics. For this reason the supply to TTL ics must be decoupled by wiring capacitors between the supply lines. Decoupling usually requires a 100 nF disc ceramic capacitor for every 5 ics.

Certain TTL logic devices are made with an *open collector* output. This has a transistor with its base and emitter connected to the gate circuit but its collector left unconnected (Fig. 10.2). The output must have a pull-up resistor connected to it (1 kΩ is often suitable). The resistor is connected to a voltage higher than the standard 5 V operating voltage. Depending on the type of gate, the higher voltage can be as much as 15 V or 30 V. Provided that the gate can sink sufficient current, this allows high-voltage low-current devices to be switched under logical control.

Fig. 10.1 shows a Schottky diode connecting the base of the output transistor Q4 to its collector. The purpose of this is to prevent the transistor from becoming saturated when switched on. A saturated transistor takes longer to switch off than one that is only just switched on, so saturation makes the gate operate more slowly.

Schottky diode: has a forward voltage drop of only 0.25 V.

Test your knowledge 10.1

A 74LS05 inverter gate has its open collector output pulled up to 15 V. When it has a low output, it sinks 8 mA. What value of resistor should be used to give a low output of 1V?

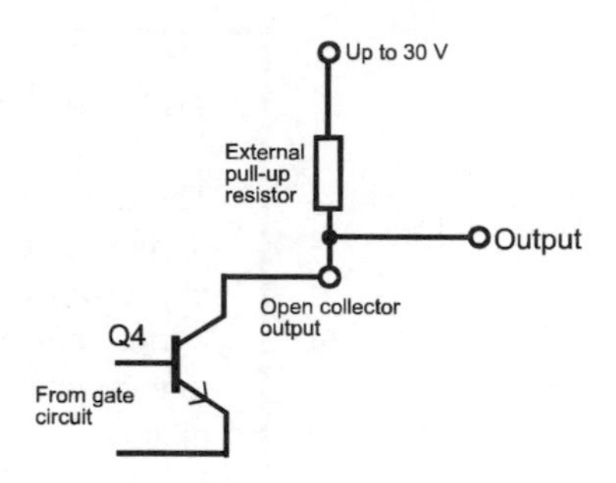

Figure 10.2 *The output of TTL gates with an* open collector *output can be pulled up to voltages higher than the standard 5 V operating voltage.*

Standard TTL does not have the Schottky diode. It is present in the low power Schottky version of TTL (74LSXX). The diode gives it faster operating times than standard TTL. The 74LSXX series is widely used, partly because of its speed but also because of its low power requirements.

CMOS logic

This is based on complementary MOSFET transistors. By complementary, we mean that the transistors operate in pairs, one of the pair being n-channel and the other p-channel. Both are enhancement MOSFETs. Fig. 10.3 shows the structure of a CMOS two-input NOR gate. It consists of two p-channel gates (Q1 and Q2) and two n-channel gates (Q3 and Q4). If input A is low Q1 is on and Q3 is off. Similarly if B is low, Q2 is on and Q4 is off. If both A and B are high, the output terminal is connected through Q1 and Q2 to the positive V_{DD} line and the output is high.

NOR operation: the output is high only when all inputs are low.

If one or both of A and B are low, one or both of Q1 and Q2 are off. There is no connection to the V_{DD} line. But one or both of Q3 and Q4

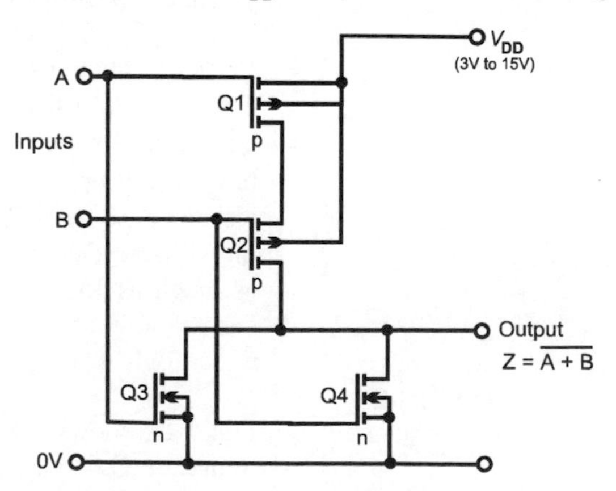

Figure 10.3 *A CMOS gates consists of complementary pairs of MOSFETs. This is a two-input NOR gate.*

are on. There is a connection to the 0V line and the output is low. Putting these results into a table:

Inputs		Output
A	B	Z
0	0	1
0	1	0
1	0	0
1	1	0

This is the truth table for the NOR operation. Other logical operations can be performed in a similar way and more complex devices can be built up from these basic logic gates. One of the advantages of CMOS technology is that it is that the gates are physically very small. It is easy to fabricate hundreds or even thousands of them on a single chip. Highly complex logical devices are readily made in CMOS, which is one of the reasons for its popularity.

CMOS is available in two main series, in which the type numbers range from 4000 and 4500 upwards. The most commonly used series, the 4000B series, has buffered outputs. The 4000UB series has unbuffered outputs. The gate shown in Fig. 10.3 is an unbuffered gate. In the 4000B series the buffer consists of *two* inverter stages (Fig. 10.4) so the truth table is the same. Although the inclusion of buffers adds to the time that the gate takes to operate, the buffers make the output more symmetrical, with equal and faster rise and fall times. This is important when CMOS gates are used in pulse circuits and timing circuits.

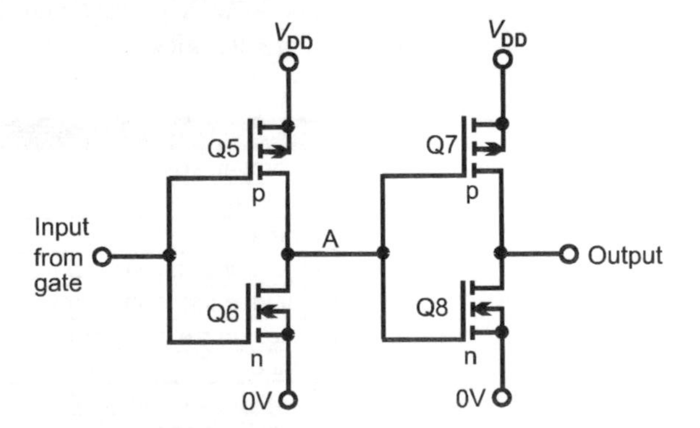

Figure 10.4 *The output buffer of the 4000B series consists of two complementary pairs acting as inverters. If the gate output is high, Q5 is off and Q6 is on, making point A low. This puts Q7 on and Q8 off, making the output high. Similarly, if the gate output is low the output of the last buffer is low. The output is always the same as the gate output.*

TTL and CMOS compared

For all except the fastest logic, the choice of family usually lies between TTL and CMOS. Here we compare the characteristics of the most popular TTL series, the low power Schottky series, and the CMOS 4000B series. Data sheets list many characteristics but the ones of most interest to the designer are the following:

- Operating voltage.
- Power consumed, usually specified as the power per gate.
- Input voltage levels, the level above which the gate recognises the input as logic high, and the level below which it is taken as logic low.
- Output voltage levels, the minimum high level, the typical high level, the maximum low level and the typical low level produced by the gate.
- Input currents, the maximum when a gate is receiving a high input and the maximum when it is receiving a low input.
- Output currents, the minimum when a gate has a high output and the minimum when the gate has a low output.
- Fanout, the number of inputs that can be driven from a single output (can be calculated from input and output currents).
- Propagation delay, the time, usually in nanoseconds, between the change of input voltages and the corresponding change in output voltage.
- Speed, as indicated by the maximum clock rate it is possible to generate.

The 74LSXX series and the CMOS 4000B series have the following characteristics:

Characteristic	74LSXX	CMOS
Operating voltage	5 V ± 0.25 V	3 V to 15 V (max 18 V)
Power per gate	2 mW	0.6 μW
Input current, high (max)	20 μA	0.3 μA
Input current, low (max)	–0.4 mA	–0.3 μA
Output current, high (min)	–0.4 mA	–0.16 mA (–1.2 mA)
Output current, low (min)	8 mA	0.44 mA (3 mA)
Fanout	20	50 (nominal)
Propagation delay (typical)	9 ns	125 ns (40 ns)
Fastest clock rate	40 MHz	5 MHz

Their input and output voltages are plotted in Fig. 10.5.

The 74LS series, like all TTL, operates on 5 V DC which must be held to within 5%. This is the voltage commonly used in computers and

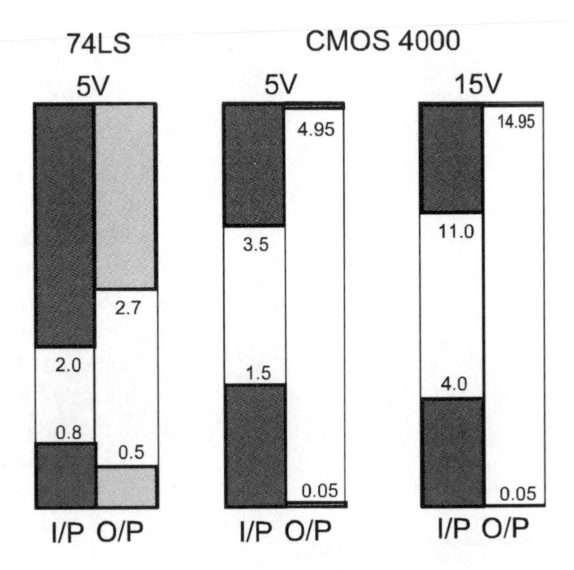

Figure 10.5 *The shaded areas indicate the voltage ranges that are defined as logic low and logic high for TTL and for CMOS with a 5 V or a 15 V supply voltage. The output voltage levels of gates are set to be well within the limits accepted at the inputs.*

similar equipment. By contrast, CMOS is able to work on a wide voltage range, which makes it much more suitable for battery-powered equipment. It is also easier to interface with a range of different circuits and devices. Even through the 74LS is a low-power series compared with standard TTL (which uses 10 mW per gate), 74LS series needs much more power than CMOS. Again, this gives CMOS the advantage for battery-powering and for portable equipment in general. Although the *average* power requirement of CMOS is low, the amount of power used depends on the amount of activity in the circuit. If the circuit is quiescent, with no logical changes occurring, it uses almost no power. But when it is changing state at high frequency the amount of power used may rise almost to the level used by TTL.

CMOS has insulated input gates (using 'gate' in the other sense) like any other FET circuit, so it is not surprising to find that input currents are very low in CMOS. The output currents are lower than those of TTL, the figures being quoted for a 5 V supply and (in brackets) for a 15 V supply.

CMOS versions of TTL devices are available in the 74HCXX and 74HCTXX series. They have input and output features similar to CMOS, but the operating speeds of TTL. The supply voltage of 74HC is 2 V to 6 V, but 74HCT is limited to 5 V ± 0.5 V.

When calculating fanout we have to consider what happens under both high and low logic conditions.

Example

Consider a 74LS gate, gate A which is feeding its output to an input of a second 74LS gate, gate B. If A is high, then, according to the table on p. 146, it can supply at least 0.4 mA, or 400 μA. Only 20 μA are needed to supply the input of B, so A can supply 400/20 = 20 gates. Now suppose that the output of A is low. It can draw at least 8 mA from any outputs to which it is connected. But gate B needs only 0.4 mA drawn from it to qualify as a low input. So the number of gates that A can bring down to a low level is 8/0.4 = 20. At both high and low levels A can drive the inputs of 20 gates. Its fanout is 20. This is far more inputs than a designer would normally need to connect to a single output so, for practical purposes, the fanout is unlimited.

Test your knowledge 10.2

What is the fanout when:
(a) a 74LS gate drives a 4000B series gate?
(b) a 4000B gate drives a 74LS gate?

Similar calculations can be done for CMOS but here we have the extra complication that output currents depend on operating voltage. The high input resistance of CMOS gates means that the fanout is large. A fanout of 50 is taken as a rule-of-thumb number, and is one that will very rarely be needed.

Looking at Fig. 10.5 we note a major difference between 74LS and CMOS. On the left we see that a 74LS gate takes any input voltage (I/P) greater than 2.0 V to be logic high. Any input lower than 0.8 V is low. Inputs between 0.8 V and 2.0 V are not acceptable and give unpredictable results. These inputs are provided by 74LS outputs as shown in the O/P column. A 74LS gate is guaranteed to produce an output of 2.7 V or more when it goes high. Thus any high output will register as a high input, with a margin of 0.7 V to spare. This allows for spikes and other noise in the system and gives a certain degree of *noise immunity*. Provided that a noise spike takes a high output level down by no more than 0.7 V, it will still be taken as a high level. In this case the noise immunity is 0.7 V. At low level the maximum output voltage is 0.5 V, though typically only 0.25 V. This is 0.3 V (typically 0.55 V) below the input level that is registered as logic low, so again there is a useful amount of noise immunity.

In the case of CMOS, the high and low input levels vary with operating voltage. But the output levels are always within 0.05 V of the positive or 0 V supply lines. This is because the family is built from MOSFETs. When turned on, they act as a very low resistance between the positive supply line and the output. When turned off, they act as a very low resistance between the negative line and the output. This gives CMOS very high noise immunity.

The table shows that 74LS is much faster than CMOS and this allows clocks and other time-dependent circuits to run at much higher rates with 74LS. Propagation delay is reduced in CMOS if the supply voltage is made higher. The values quoted are for the more frequently used 4000B series. Part of the delay is due to the output buffers. If speed is important, the 4000UB series has a propagation delay of only 90 ns when used at 5 V.

Activity 10.1 Logic levels

Investigate the input and output logic levels of CMOS and 74LSXX gates using, say, CMOS 4011 and 74LS00 dual-input NAND gates. You need:

- A breadboard.
- A supply of short lengths of single-core wire with insulation stripped from both ends.
- A power supply, 5 V DC for 74LS series ics, or any DC voltage between 3 V and 15 V for CMOS.
- Integrated circuits to provide the gates.
- Data sheets to show the pin connections of the ics.
- Output indicator, either a testmeter or a logic probe. You can also use an LED in series with a resistor to limit the current to a few milliamperes.

Connect the two inputs of a gate together and supply them with input from the wiper of a variable resistor connected across the supply lines as a potential divider. Measure the input voltage with a testmeter and use an LED wired in series with a 270 Ω resistor as an output indicator. If NAND or NOR gates are used with inputs wired together they act as INVERT gates. Increase the voltage slowly from 0 V up to the supply voltage. Record the level at which the output becomes firmly low. Decrease the voltage slowly from supply voltage to 0 V. Record the level at which the output becomes firmly high. Use a testmeter to measure high and low output voltages.

The CMOS ic can be operated on any supply voltage in the range +3 V to +18 V. Repeat this investigation at two or three different supply voltages in this range. The LSTTL ic must be operated at +5 V. Unused CMOS inputs must be connected to the supply or 0 V before testing. Unused TTL inputs should be connected to the supply through a 1 kΩ resistor; several inputs can share a single resistor.

Record your results in a table to compare the two families.

Emitter coupled logic

The prime advantage of this family is its high speed, which is due to the fact that the transistors are never driven into saturation. This makes it the fastest of all logic families with a propagation delay of 1 ns per gate and a maximum clock rate of 500 MHz.

As well as avoiding transistor saturation, the circuits have low impedances to avoid the speed-reducing effects of capacitance. Because of these measures, power consumption is high, as much as 30 mW per gate. Also, the logic output levels are close together at –0.8 V and –1.6 V. This leads to low noise immunity. Consequently, an ECL circuit needs an ample power supply carefully protected against voltage spikes. Another factor which makes ECL difficult to work with is that the fast edges on the signals mean that connecting wires must be treated as transmission lines (see Chapter 16), with precise routing on the circuit board.

Because of the difficulties associated with ECL, its use is restricted to large computer systems in which its high speed is an indispensable benefit.

Activity 10.2 Truth tables

Review the notes at the beginning of Activity 10.1. Investigate the behaviour of the basic kinds of logic gate, using CMOS and 74LS gates. For each gate, draw up a blank truth table and mark in all possible combinations of inputs, using 1 and 0 or H and L to indicate the two possible logic states. For each combination of inputs, connect the input terminals to the supply line or the 0 V line. Use an LED or logic probe to find the state of the output and record this in the table. Use the two-input versions of these gates (except for NOT):

NOT, AND, NAND, OR, NOR, exclusive-OR, and exclusive NOR.

Connect two or more logic gates as in the circuits of Fig. 10.6. Test the combination as described above and identify the function that the combination performs.

Problems on logic families

1 Describe the structure and action of a TTL 2-input NAND gate. In what way is this modified as a Schottky NAND gate, and what effect does this have?

2 Write out the 2-input truth tables for AND, NAND, OR and NOR.

3 Explain why a totem pole output may cause large voltage changes on the supply lines, and what can be done to minimise these.

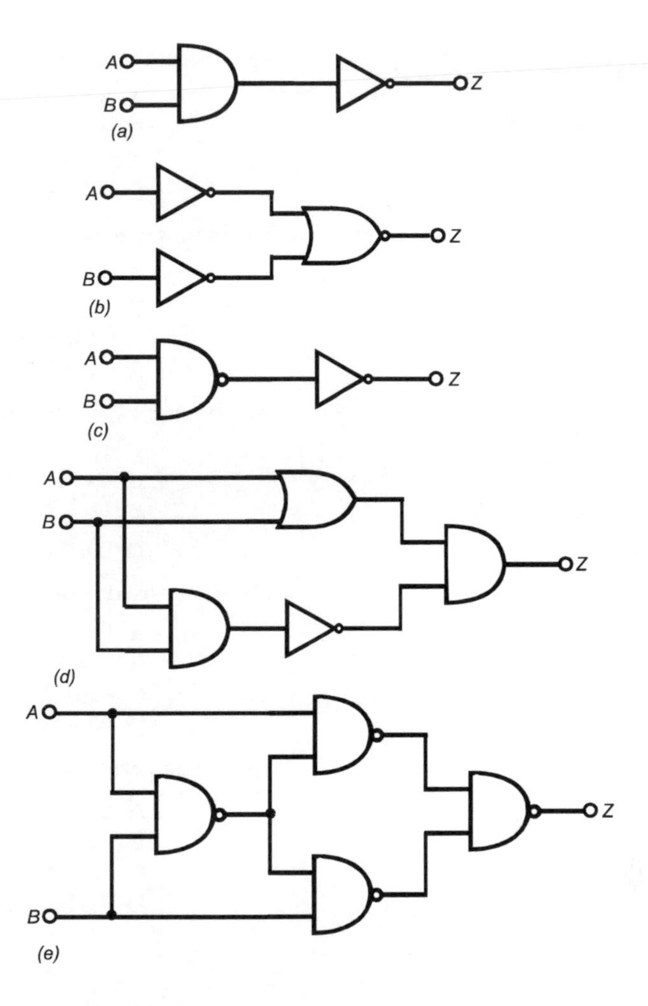

Figure 10.6 *Logic circuits for Activity 10.2.*

4 Describe the structure and action of a CMOS 2-input NOR gate.

5 What is the difference between the CMOS 4000B series and the 4000UB series?

6 Design a CMOS gate for the NAND operation.

7 Compare the 74LS series with the 4000B series from the viewpoint of power supplies.

8 Compare the 74LS series with the 4000B series from the viewpoint of operating speed and noise immunity.

9 List 10 applications of logic ics, and name those which would be most suitable for TTL logic and those most suitable for CMOS logic.

10 Explain how ECL achieves its very high speed and what are the problems that this raises.

Multiple choice questions

1 The forward voltage drop of a Schottky diode is:

A 0.6 V.
B 0.7 V.
C 0.3 V.
D 0.25 V.

2 The logic series with the fastest operating time is:

A 74XX.
B ECL.
C CMOS 4000B.
D 74LSXX.

3 The fanout of 74LS devices is:

A 10.
B 50.
C 16.
D 20.

4 The power used by a CMOS gate is:

A less than 1 μW.
B 2 mW.
C more than 5 mW.
D 30 mW.

5 The fanout of a CMOS gate is very high because of:

A the high output current from the gate.
B the high noise immunity.
C the high gain of complementary MOSFETs.
D the high input resistance of MOSFETgates.

11 Logical combinations

Summary

The output of a combinational logic circuit depends only on its present inputs, and can be predicted by the use of truth tables. These are based on the four basic logical operators and operators derived from them. Tools and techniques for designing logic circuits, simplifying them, and predicting their output include truth tables, Karnaugh maps, breadboarded circuits and logic software that simulates logical devices. Most combinational circuits can be built solely from NAND and NOR gates. Basic gates can be combined to give more complicated functions, such as exclusive-OR and exclusive-NOR gates, majority logic gates, half adder and full adder. Combinational logic circuits are manufactured as integrated circuits for use in data processing. These include encoders, priority encoders, decoders, data selectors (multiplexers) and data distributors (demultiplexers). ROM chips and programmable array logic (PAL) are also used to perform complex logic functions.

In this chapter we look at the output from logical circuits when that output depends only on the *present* inputs to the circuit. Previous inputs have no effect at all. First we summarise the essentials of the logic that we are representing by our electronic circuits.

Logical operators

There are four basic logical operators, TRUE, NOT (or INVERT), AND and OR. Their action is shown in *truth tables*, in which A and B are inputs and Z is the output. Logic 'true' is represented by 1, and 0 represents 'false'. In practical circuits logical operations are performed by gates. In *positive logic* circuits, we represent 'true' by a high voltage (fairly close to the positive supply voltage) and 'false' by a low voltage (close to zero volts). All the circuit descriptions in this chapter assume positive logic.

TRUE has only one input and the output is equal to it:

A	*Z*
0	0
1	1

There does not seem to be much use for a gate which performs this operation, but TRUE gates are usually buffer gates with a greater output *current* than ordinary gates. We use them for driving large numbers of other gates, for switching transistors, for flashing LEDs, and for similar tasks.

NOT has only one input and the output is its inverse:

A	*Z*
0	1
1	0

Some NOT gates have the same current output as other logic gates, but some can produce a greater output current and are known as *inverting buffers*.

The truth tables for the other two basic gates and some other commonly used gates are shown below side by side for comparison:

Inputs		Output *Z*					
B	*A*	$A \bullet B$	$A+B$	$A \oplus B$	$\overline{A \bullet B}$	$\overline{A+B}$	$\overline{A \oplus B}$
0	0	0	0	0	1	1	1
0	1	0	1	1	1	0	0
1	0	0	1	1	1	0	0
1	1	1	1	0	0	0	1

NOT-AND and NOT-OR gates are more often called NAND and NOR gates.

EX-OR is short for 'exclusive-OR'.

Reading from left to right, the outputs are *A* AND *B*, *A* OR *B*, *A* EX-OR *B*, *A* NOT-AND *B*, *A* NOT-OR *B* and *A* NOT-EX-OR *B*. The outputs for the last three gates are simply the inverse of the outputs of the first three. The truth tables are shown with two inputs, but gates can have more than two inputs. The truth tables for these can be worked out from the table above.

Test your knowledge 11.1

If A = 0 and B = 0, what are A+B and $\overline{A}$•B?

Test your knowledge 11.2

Write the truth table for a NOR gate with 4 inputs.

Example

In the AND column there is only one way to get a 1 output. This is by having *all inputs at 1*. This rule applies to any AND gate. For example an 8-input AND gate must have *all inputs at 1* to get a 1 output. Any other combination of inputs gives a 0 output.

Logical symbols

The symbol for AND is • , but it may be omitted. So A•B and AB both mean 'A AND B'.

The symbol for OR is +, so A+B means 'A OR B'.

The symbol for exclusive-OR is ⊕.

The symbol for NOT or INVERT is a bar over the variable, so $\overline{A}$ means NOT-A. It makes a difference whether two variables *share* a bar or have *separate* bars. For example: $\overline{A \bullet B}$ means '*A* NAND *B*' but $\overline{A} \bullet \overline{B}$ means 'NOT-*A* AND NOT-*B*'. Write out the two truth tables to see the difference.

Predicting output

Truth tables are an effective way of predicting the output from a circuit made up of gates. Usually we need to try all possible combinations of inputs and work out the output for each combination. If there are two inputs there are four combinations, as listed in the table above. If there are 3 inputs there are 8 combinations. If there are 4 inputs there are 16 combinations. With even more inputs the number of combinations becomes too large to handle easily and it is simpler to use logic simulator software on a computer. Sometimes there are combinations of inputs that we know in advance can not occur. These can be ignored when working out the outputs.

Example

Predict the outputs of the circuit shown in Fig. 11.1.

First work from left to right, entering the logical expressions at the output of each gate, as has already been done in Fig 11.1. The output of gate 1 (NOT) is the invert of *A*. The inputs of gate

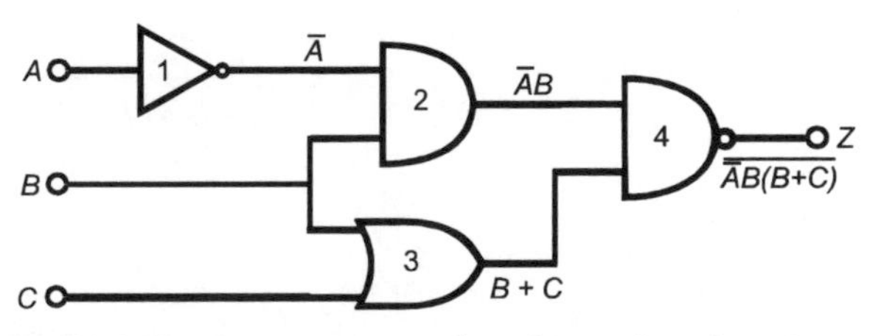

Figure 11.1 *This logic circuit is the subject of analysis in the first part of this chapter.*

2 (an AND gate) are $\overline{A}$ and *B*, so the output is $\overline{A} \bullet B$. The output of gate 3 (an OR gate) is $B+C$. The inputs to gate 4 are $\overline{A} \bullet B$ and $B+C$ so its output is their NAND.

Now follow this through in a truth table, for all 8 possible combinations of *A*, *B*, and *C*. You can get all the combinations simply by writing out the eight 3-digit binary numbers from 0 to 7.

Inputs			Outputs			
			(1)	(2)	(3)	(4)
C	*B*	*A*	$\overline{A}$	$\overline{A} \bullet B$	$B+C$	*Z*
0	0	0	1	0	0	1
0	0	1	0	0	0	1
0	1	0	1	1	1	0
0	1	1	0	0	1	1
1	0	0	1	0	1	1
1	0	1	0	0	1	1
1	1	0	1	1	1	0
1	1	1	0	0	1	1

The numbered columns show what each gate does for each combination of inputs. Column (1) is the inverse of the *A* column. Column (2) is obtained by ANDing column (1) with the *B* column. Column (3) is the OR of columns *B* and *C*. Finally column (4) is the NAND of columns (2) and (3).

Column (4) shows the output for all possible combinations of inputs.

Identities

A second way of solving this problem is to use the logical identities.

Example

Find the output of the circuit in Fig. 11.1 by labelling the output of each gate with the logical result of its action. This gives the logical expression for the output shown on the extreme right of the figure. Next simplify this, using De Morgan's Therorem and other logical identities:

$$\overline{\overline{A} \bullet B \bullet (B+C)} = \overline{\overline{A} \bullet B} + \overline{B+C} \quad \text{De Morgan}$$

$$= \overline{\overline{A}} + \overline{B} + \overline{B} \bullet \overline{C} \quad \text{De Morgan again}$$

$$= A + \overline{B} + \overline{B} \bullet \overline{C} \quad \text{NOT-NOT A = A}$$

$$= A + (\overline{B} + \overline{B} \bullet \overline{C}) \quad \text{Bracket } \overline{\text{B}} \text{ expressions}$$

$$= A + \overline{B} \quad \text{Eliminate } \overline{\text{C}}$$

De Morgan's Theorem

This comprises two rules that are probably the most useful of all the logical identities:

$$\overline{A \bullet B \bullet C \bullet D} = \overline{A} + \overline{B} + \overline{C} + \overline{D}$$

$$\overline{A + B + C + B} = \overline{A} \bullet \overline{B} \bullet \overline{C} \bullet \overline{D}$$

The first rule states that the NAND of two or more variables is equivalent to the OR of their inverts. The second rule states that the NOR of two or more variables is equivalent to the AND of their inverts. The rules are used for converting NAND expressions to OR, and NOR expressions to AND.

The last line shows that the result is independent of the value of *C*. The output has been simplified to *A* OR NOT-*B*.

Looking again at the truth table, we can see that the last four lines of the table ($C = 1$) repeat the first four lines ($C = 0$). The output is 1 whenever *A* is 1 or *B* is 0. This confirms our result.

Karnaugh maps

A third way of solving this problem is to use a Karnaugh map. This has the advantage that it tells us if it is possible to simplify the logic circuit and, if so, how to do it.

Example

Taking column (4) of the previous truth table, we enter these results into a 3-variable Karnaugh map (Fig. 11.2). It is possible to encircle two overlapping groups of four 1's. The groups are A and $\overline{B}$, so the circuit simplifies to A OR $\overline{B}$. Fig. 11.3 shows the circuit that has the same action as the circuit of Fig. 11.1. It does not have a *C* input, as the state of *C* makes no difference to the output.

Breadboards and simulators

A fourth way to solve the circuit is to connect gates together as in Fig. 11.1, run through all possible input states, as listed in the truth table, and see what output is produced at each stage. You can do this either by connecting integrated circuits on a breadboard (*see box*, Activity 11.1) or by setting up the circuit on a simulator.

Example

Fig. 11.4 is the result of a simulation of the circuit of Fig. 11.1

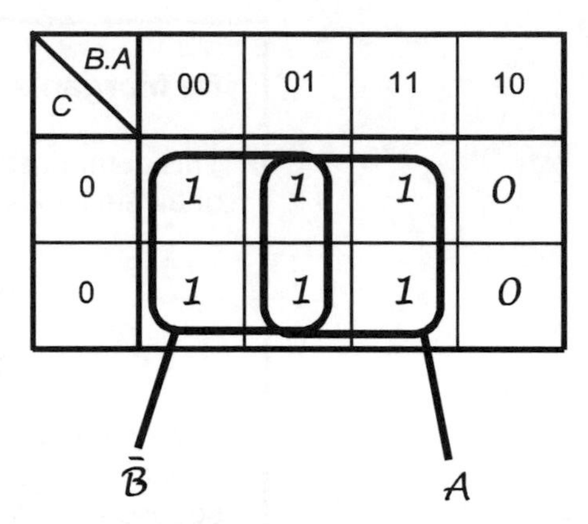

Figure 11.2 *The Karnaugh map of the logic of Fig. 1.1 shows us how to perform the same logic with fewer gates. There are two groups of four '1's, representing* A = *1 and* B = *0 (or* $\overline{B}$ = *1). The output is 1 when A = 1 OR when* $\overline{B}$ *= 1. In symbols, Z = A +* $\overline{B}$*.*

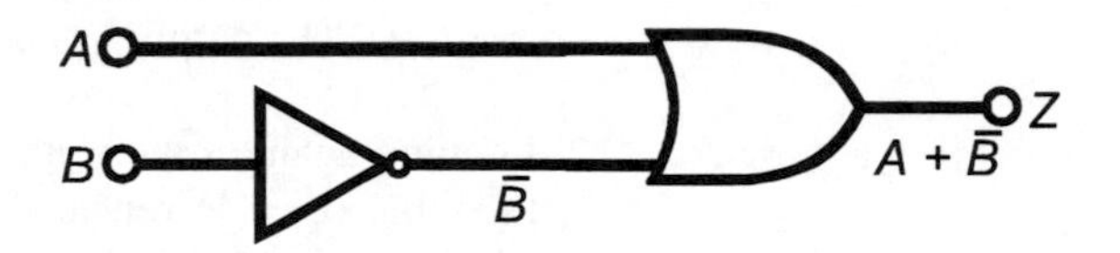

Figure 11.3 *The original circuit (Fig. 11.1) uses four gates. This minimised version uses only two gates, a NOT gate and an OR gate. An input is not required for variable* C *in the minimised version.*

using 74LS series devices. To provide the input we used a 5 MHz clock and fed its signal to a 3-stage binary counter. These are seen on the left of the screen in Fig 11.4. Times are in nanoseconds, the count incrementing every 200 ns. The vertical dashed lines on the output display represent 200 ns intervals. The inputs (plots A, B and C) begin at the count of 000 and the trace runs while they increase to 111 and return to 000 again. The count then runs through 000 to 100 at the extreme right side of the screen. The output from the NAND gate is plotted as 'Out' on the display. This is *Z* and equals 1 *except* when *A* is low and *B* is high (*A* = 0 and *B* = 1). The state of *C* has no effect on output. This result agrees with the truth table on p. 156.

Fig. 11.4 shows an unexpected effect. There is a *glitch* in the output, shown as a brief low pulse occurring as input changes from 011 to 100. The reason for this is discussed in the next chapter. It is not a fault in the logic of the circuit of Fig. 11.1.

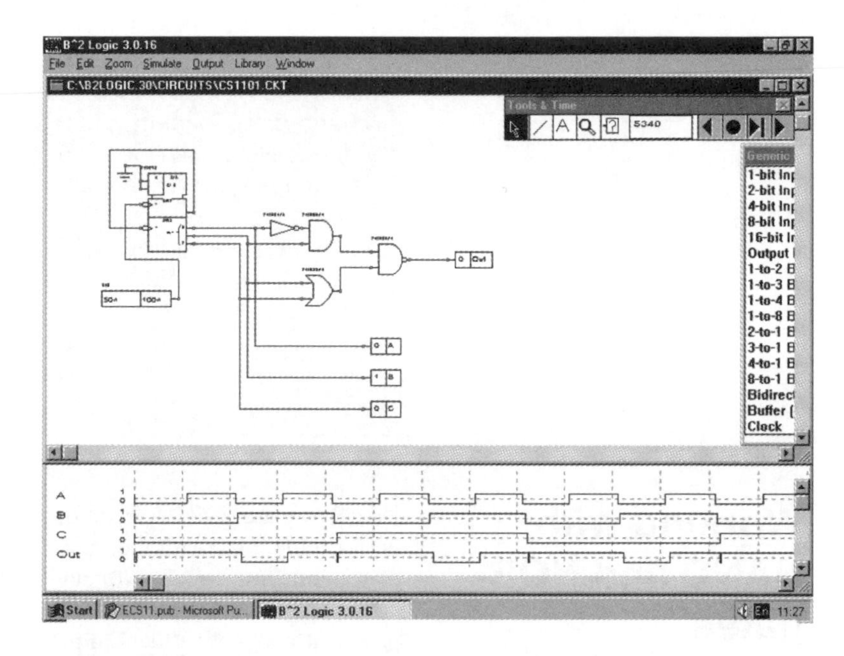

Figure 11.4 *A run on a simulator confirms that the truth table for Fig. 11.1 is correct.*

Activity 11.1 Logical circuits

Any of the logical circuits in this chapter may be set up on a breadboard for testing. You need:

- A breadboard.
- A supply of short lengths of single-core wire with insulation stripped from both ends.
- A power supply, 5 V DC for 74LS series ics, or any DC voltage between 3 V and 15 V for CMOS.
- Integrated circuits to provide the gates.
- Manual to show the pin connections of the ics.
- Output indicator, either a testmeter or a logic probe. You can also use an LED in series with a resistor to limit the current to a few milliamperes.

Complete the wiring and check it *before* applying power. With CMOS, all unused inputs *must* be connected to *something* before switching on, usually either to the positive or to the 0V supply line. Unused outputs are left unconnected. Take care to avoid pins becoming bent beneath the ic when inserting the ics into the sockets in the breadboard. The pin may not be making contact but you will not be able to see that there is anything wrong.

CMOS ics need careful handling to avoid damage by static charges. Keep them with their pins embedded in conductive plastic foam until you are ready to use them. It is good practice to remove charges from your body from time to time by touching a finger against an earthed point, such as a cold water tap. Follow any other static charge precautions adopted in your laboratory.

Apply 1 or 0 inputs by connecting the appropriate pins to the positive or 0V supply lines.

Working with NAND and NOR gates

Usually we are given a circuit or a truth table and we design a circuit to perform the logic with the minimum of gates. Fig. 11.3 is an example of a minimised circuit. Although it requires the minimum of components, takes up the least possible board space and uses the minimum of power, it may not be ideal in practice. When a circuit is built from ics there is the practical point that an ic usually contains several identical gates. A single ic may contain six NOT gates, or four 2-input gates performing the other basic logical operations. It is uneconomical of cost, board space and power to install a 6-gate or 4-gate ic on the circuit board if only one or two of its gates is going to be used. It is better to restrict the design to a few different types of gate.

This circuit requires only two gates, NOT and AND but unfortunately they are of different kinds. To set up this circuit on its own requires two ics containing 10 gates, of which we use only 2. If this circuit is part of a larger logic system, we may find that the system does not need any other AND gates. Or, if it does need a few AND gates, they are situated a long way from the NOT gates. This means having long tracks wandering all over the board. Tracks take up space and, if they are long, may delay the signal sufficiently to cause the logic to fail. The more long tracks there are, the more difficult it becomes to design the board, and the more we need to resort to wire links and vias to make the connections.

Via: on a printed circuit board a via is a connection (usually a double-headed pin soldered at both ends) that goes through the board, linking a track on one side of the board to another track on the other side.

Circuit board design can be simplified by keeping to two or three types of gate. Better still, use only one type. All logic functions can be obtained with NOT, AND and OR, but these are not the most useful gates in practice. NAND and NOR gates too can be used to perform all the basic logic functions and are also useful in building flip-flops, latches and other more complicated logical circuits. For this reason we prefer to work only with NAND and NOR gates. There are some circuits, such as decoders, that need a lot of AND gates, which makes it worth while to use ics of this type. Some kinds of circuit need a lot of NOT gates. But, on the whole, NAND and NOR gates are preferred.

Example

The first step in converting the circuit of Fig. 11.1 to NAND and NOR is to note that the second half of the truth table repeats the first half, so it is unnecessary to take *C* into account. This leaves two inputs, *A* and *B*, so there are four possible combinations of input state. Now work backwards from the output. The outputs for the four combinations comprise one 0 and three 1's. The truth tables (p. 154) show that this mix of outputs is a feature of NAND gates. Therefore the output gate of the converted circuit must be NAND. Compare the normal output of NAND with the output of Fig. 11.1:

Inputs		NAND	Fig. 11.1
B	*A*		
0	0	1	1
0	1	1	1
1	0	1	0
1	1	0	1

Inverting *A* in the bottom two lines makes the NAND output agree with the required output. Inverting *A* in the top two lines makes no difference as both lines have output 1. So the solution is to invert *A* before feeding it to the NAND gate. We can use a NOT gate if there is one to spare. Otherwise, build the circuit from two NAND gates, as in Fig. 11.5.

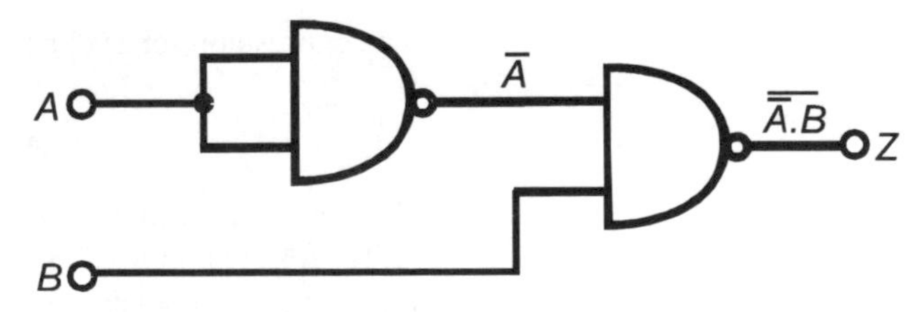

Figure 11.5 *It is sometimes useful to be able to work only with NAND gates. This is the NAND gate equivalent of Fig. 11.1.*

We could also use De Morgan's Theorem to help us design the circuit in NAND or NOR gates. Fig 11.6 illustrates the hardware equivalents of the logical equations in the box on p. 157. Here we work from right to left to arrive at NAND or NOR expressions or gates. The output of the minimal circuit in Fig. 11.3 is given as an OR expression, *A* OR $\overline{B}$. Using De Morgan's NAND/OR rule (Fig 11.6a) and going from right to left, we see that the equivalent of the OR expression is a NAND expression with the variables inverted. We replace A OR $\overline{B}$ with $\overline{A}$ NAND *B*).

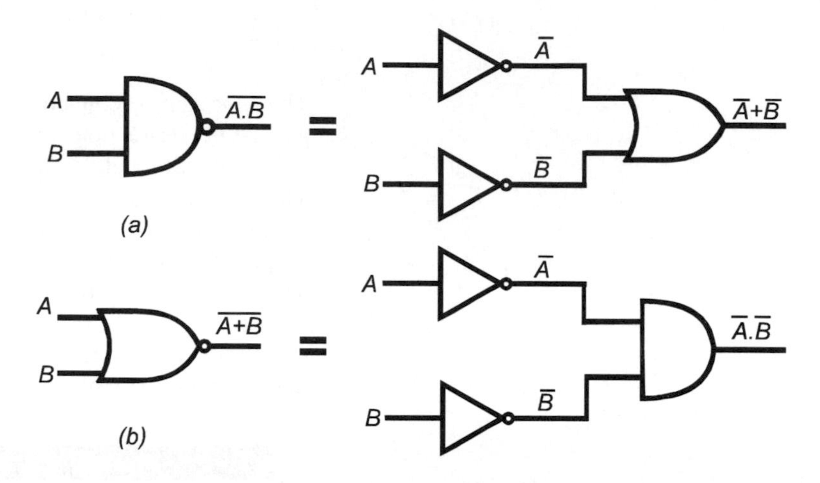

Figure 11.6 *De Morgan's Theorem. In (a) to get from left to right, remove the NAND, invert the inputs, then OR them. In (b) remove the NOR, invert the inputs, then AND them.*

Exclusive-OR

Exclusive-OR and exclusive-NOR ics are available, usually four gates to an integrated circuit, and sometimes a circuit needs several ics (see Fig. 11.13). But often we need such a gate only once or twice in a circuit and it is better then to build it from other basic gates. The Karnaugh map technique is no help as there are no entries from the truth tables that can be grouped together.

The expression for exclusive-OR from the truth table itself is:

$$A \oplus B = A\overline{B} + \overline{A}B$$

This can be translated directly into gates, using three of the basic types, NOT, AND and OR (Fig. 11.7). In practice we might not find all three types already available as spare gates, and they might be too far apart on the circuit board to make it convenient to connect them. Instead we will try to produce the same function using only NAND gates.

Starting with the expression already given above, the aim is to eliminate all the ORs and to replace them with NANDs. Obviously the NAND/OR rule of de Morgan will be applied in reverse. Before we do that we introduce two extra terms into the equation but since these both have the value 0 (false), they make no difference to the logic.

The result has only NAND operations in it. It looks formidable but can easily be put into words: NAND *A* with *B* (twice) then NAND one of these with *A* and the other with *B*. Finally NAND the two expressions so obtained. Fig. 11.8 shows this as a circuit diagram.

TRUE/INVERT gate

An exclusive-OR gate is sometimes known as a true/invert gate. If one input receives the data and the other is connected to a control line (C_i in Fig. 11.12), the output of the gate follows the input (TRUE) when the control line is low. The output is the inverse of the input when the control line is high.

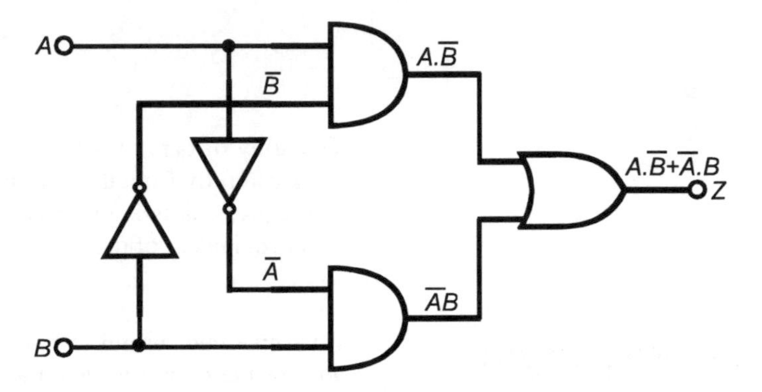

Figure 11.7 *An exclusive-OR gate is built from NOT, OR and AND gates, requiring three different ics.*

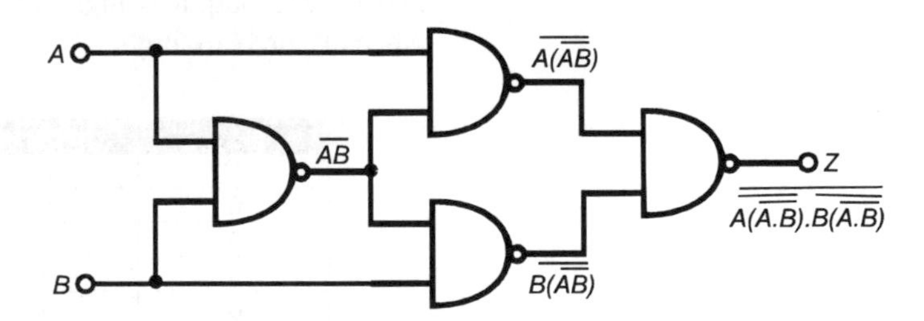

Figure 11.8 *This version of the exclusive-OR gate uses only NAND gates to build it. It is assembled from a single ic containing four NAND gates.*

When trying to solve any particular logical problem, it is useful to adopt more than one of the four possible approaches. The results from one approach act as a check on those produced in another way. So, to confirm the all-NAND version of the exclusive-OR gate we will try to arrive at the same result by using logical equations. As usual, De Morgan's equations (p. 157 and Fig 11.6) are important in the proof.

We begin with the equation quoted on p. 162:

$$A \oplus B = A\overline{B} + \overline{A}B$$

The proof continues by including two expressions that are certain to be false, for they state that *A* AND NOT-*A* is true, and also that *B* AND NOT-*B* is true. Their value is zero so they do not affect the equality. Next, common factors are taken to the outside of brackets. The final two lines rely on De Morgan:

Test your knowledge 11.3

Write the truth table for the circuit of Fig. 11.8 if the NAND gates are replaced with NOR gates.

$$= A\overline{B} + \overline{A}B + A\overline{A} + B\overline{B}$$

$$= A(\overline{A} + \overline{B}) + B(\overline{A} + \overline{B})$$

$$= A(\overline{AB}) + B(\overline{AB})$$

$$= \overline{\overline{A(\overline{AB})}.\overline{B(\overline{AB})}}$$

The array of bars over the final expression is complicated but, if you work carefully from the bottom layer of bars upward, you will see that this expression is is made up entirely of NAND operations and is the exact equivalent of the circuit in Fig 11. 8.

Majority logic

It is sometimes useful to have a circuit with an odd number of inputs and for the output to take the state of the majority of the inputs. For example, with a 5-input majority circuit, the output is high if any 3 or 4 or all 5 inputs are high.

The truth table of a 3-input majority circuit requires three inputs, *A*, *B*, and *C*. The output is high when any two of *A*, *B*, and *C* are high or when all three are high:

Inputs			*Z*
C	*B*	*A*	
0	0	0	0
0	0	1	0
0	1	0	0
0	1	1	1
1	0	0	0
1	0	1	1
1	1	0	1
1	1	1	1

Stating this as a logical equation we obtain:

$$Z = \overline{A}BC + A\overline{B}C + AB\overline{C} + ABC$$

To implement this as written would take three NOT gates, four 3-input

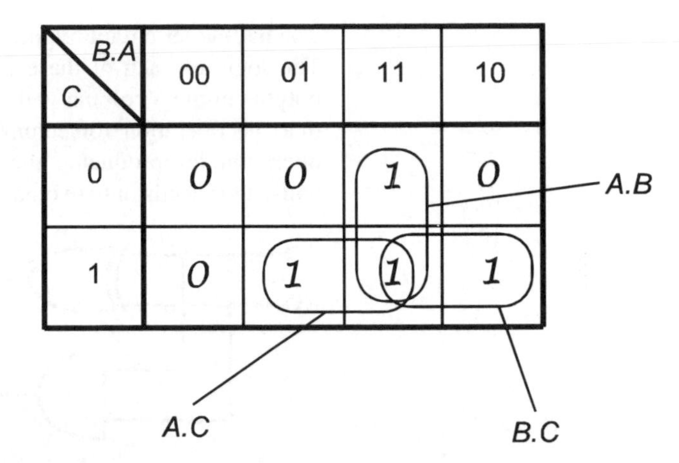

Figure 11.9 *A Karnaugh map indicates that best-of-three majority logic can be built from three AND gates and an OR gate.*

AND gates and a four-input OR gate, a total of 8 gates. Fig. 11.9 is the Karnaugh map of the circuit. There are three groups of 1's, giving:

$$Z = A \bullet B + A \bullet C + B \bullet C$$

Three two-input AND gates and one 3-input OR gate, a total of 4 gates can obtain the same logic. This is only half the number of gates used by the original logic and these have fewer inputs.

The all-NAND version is easy to find because the equation above is converted to the required form by a single step, using De Morgan's theorem:

$$Z = \overline{\overline{A \bullet B} \bullet \overline{A \bullet C} \bullet \overline{B \bullet C}}$$

NAND the three pairs of inputs, then NAND the three NANDS so obtained. Again we need just 4 gates.

Half adder

Arithmetical operations are often an essential function of logic circuits. This truth table describes the simplest possible operation, adding two 1-bit binary numbers *A* and *B*:

Inputs		Outputs	
A	B	Sum (S)	Carry (Co)
0	0	0	0
0	1	1	0
1	0	1	0
1	1	0	1

The half-adder circuit requires two outputs, sum (S) and carry-out (C_o). The logic of each of these is worked out separately. Comparing the outputs in the S column with those in the truth tables on p. 156 shows that the operation for summing is equivalent to exclusive-OR. The operation for producing the carry-out digit is AND. The complete half-adder needs only two gates, as in Fig. 11.10.

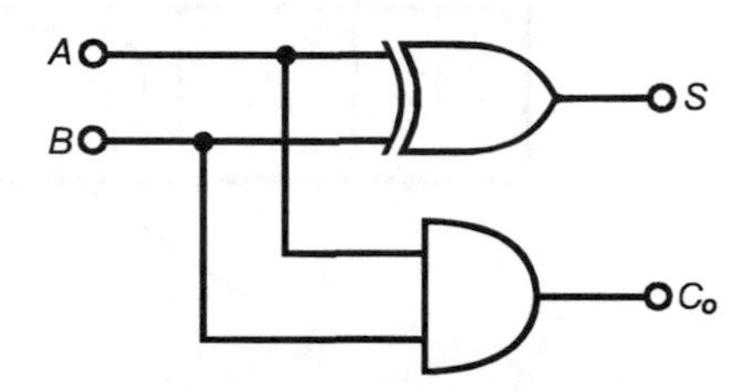

Figure 11.10 *A half-adder needs only two gates to find the sum of two 1-bit numbers and the carry out digit.*

If required, the half-adder can be built entirely from NAND gates by using the exclusive-OR circuit of Fig. 11.8 to produce S. This circuit already produces the NAND of A and B. It is necessary to use only one more NAND gate as inverter to obtain the AND function. This provides C_o, as in Fig. 11.11.

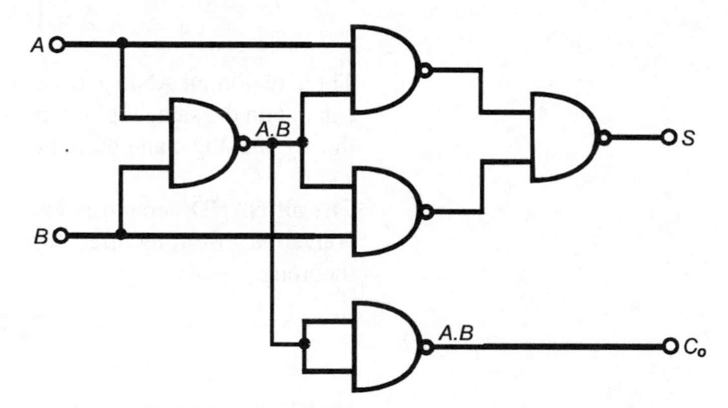

Figure 11.11 *The half-adder can also be built entirely from NAND gates.*

Full adder

The half adder is so-called because it is incomplete. It is not able to accept a carry in C_i from a previous stage of addition. The truth table of a full adder is shown on the opposite page. The first four lines show the addition of A and B when the carry-in is zero. The entries for A, B and C_o are identical with those in the half-adder table on p. 165. The last four lines show the addition when carry in is 1.

The first thing to notice is that, for S, the first four lines have an exclusive-OR output, as before. The last four lines have an exclusive-NOR output. In Fig. 11.12 we use a second exclusive-OR gate to perform the true/invert function on the output from the exclusive-OR

C_i	A	B	S	C_o
0	0	0	0	0
0	0	1	1	0
0	1	0	1	0
0	1	1	0	1
1	0	0	1	0
1	0	1	0	1
1	1	0	0	1
1	1	1	1	1

gate of the half-adder, Fig. 11.10. When C_i is low the second exclusive-OR gate does not invert the output of the first one, so the output is exclusive-OR, and gives S for the first 4 lines of the table. When C_i is high, the second gate inverts the output of the first one, giving exclusive-NOR, which gives S for the last four lines.

SSI: small scale integration; logic gates and devices of equal complexity integrated on a chip.

MSI: medium scale integration; flip-flops, counters, registers and similar devices integrated on a chip.

Looking at the column for C_o, this is a majority logic function. C_o is high whenever two or more inputs are high. We use majority logic but base it on 4 NAND gates. If the two exclusive-OR gates are built from NAND gates, as in Fig. 11.8, the whole circuit is then in NAND gates. A gate is saved because the first exclusive-OR gate already has a NAND gate to provide the NAND of A and B so the output of this is used in producing the carry-out. Altogether, a total of 11 NAND gates is required. Although it is interesting to build a complex circuit from the basic SSI units, this is the point where we begin to use MSI devices such as the CMOS 4008 and the 74LS283 binary 4-bit full adders.

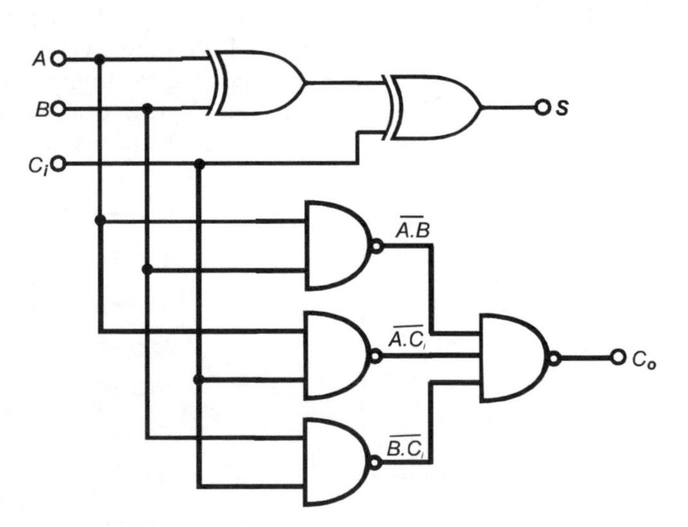

Figure 11.12 *A full adder allows for a carry-in digit from a previous stage of addition.*

Parity tree

Bit: a binary digit, 0 or 1.

Byte: a group of bits, usually 8.

Parity trees are used for checking if a byte of data is correct. In an 8-bit system, 7 bits are reserved for the data and the eighth bit is the *parity bit*. When the data is prepared, the numbers of 1's in each group of 7 bits is counted. The parity bit is then made 0 or 1, so as to make the total number of 1's odd. This is called *odd parity*. Later, perhaps after having been transmitted and received elsewhere, the data can be checked by feeding it to a parity tree (Fig. 11.13). If the data is correct it still has odd parity and the output of the tree goes high. If one of the bits has changed, the number of 1's is even and the output of the tree goes low, warning that the data is corrupted.

Test your knowledge 11.4

Write truth tables for Fig. 11.13 to confirm that the parity tree has a high output only when its input contains an odd number of 1's.

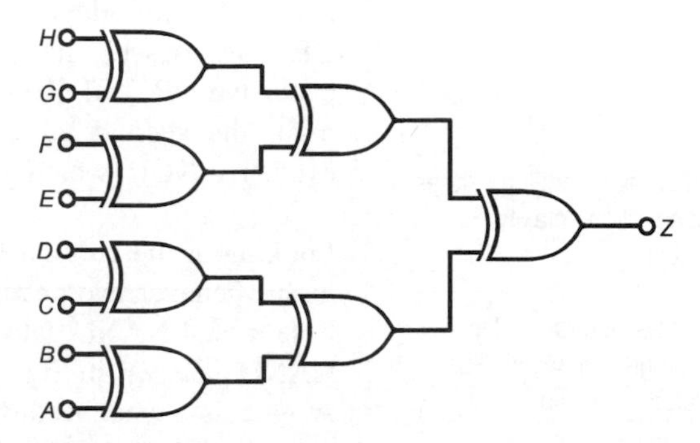

Figure 11.13 *The output of this parity tree goes high if the 8-bit input contains an odd number of 1's.*

Magnitude comparator

This circuit (Fig. 11.14) compares two 4-bit words, and gives a high output if they are identical. The two inputs of each gate receive the corresponding bit from each word. MSI versions of this circuit have additional logic to detect which of the two words is larger (supposing them to be 4-bit binary numbers) if they are unequal.

Test your knowledge 11.5

Write truth tables for Fig. 11.14 (or investigate it, on a breadboard or simulator, to find the output when:
(a) both words are 1011, and
(b) word A is 1100 and word B is 0101.

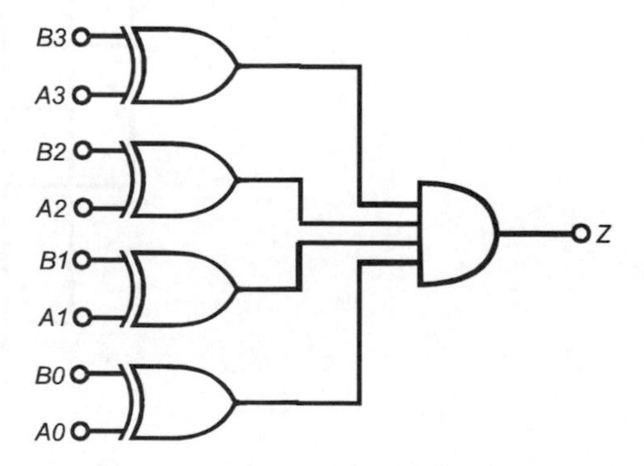

Figure 11.14 *The output of a magnitude comparator goes high if the bits of word A are all equal to the corresponding bits of word B.*

Decimal to binary converter

Assuming that only one input is high at any one time, this circuit (Fig. 11.15) produces the 2-bit binary equivalent of the decimal numbers 1 to 3. An extended converter is built with three 3-input OR gates to convert numbers from 1 to 7. A circuit that converts decimal to binary is also called an *encoder*, which is sometimes shortened to *coder*.

The circuit does not allow for a zero (0) input. This is because, in any given application, a zero input may or may not have a meaning different from 'no input'. In either case the output should be 00. If the range of possible inputs starts at 1, a zero input can not occur and there is no confusion. '00' means 'no input'. If a zero input is possible, the circuit needs an extra line, known as an 'input present' line. It has an output that is driven to 1 whenever there is an input to the circuit. If the output is 00 and the 'input present' output is 1, this is interpreted as an input of 0. If the output is 00 but the 'input present' output is 0, it means there is no input.

Test your knowledge 11.6

Write the truth table for Fig. 11.15 (note the table has *two* output columns).

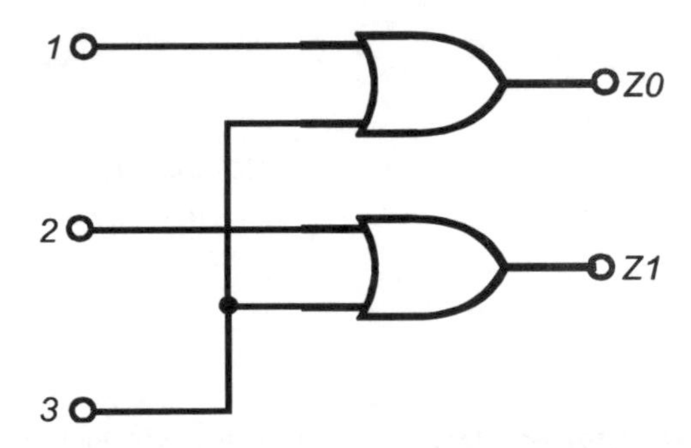

Figure 11.15 *When one of the inputs is made high, the outputs of this decimal-to-binary converter (or encoder) show the binary equivalent.*

Binary to decimal converter

This performs the reverse function to that of Fig. 11.15. Given a binary input, the *one* corresponding decimal output goes high. A circuit such as this, which converts binary to decimal, is also known as a *decoder*.

The truth table is:

Input		Output			
B	A	3	2	1	0
0	0	0	0	0	1
0	1	0	0	1	0
1	0	0	1	0	0
1	1	1	0	0	0

Looking at the output columns, each is 1 for only one set of inputs. This suggests that NOR gates might be suitable as the output gates. Fig. 11.16 shows how each gate is fed with either the TRUE or the NOT form of the two input digits.

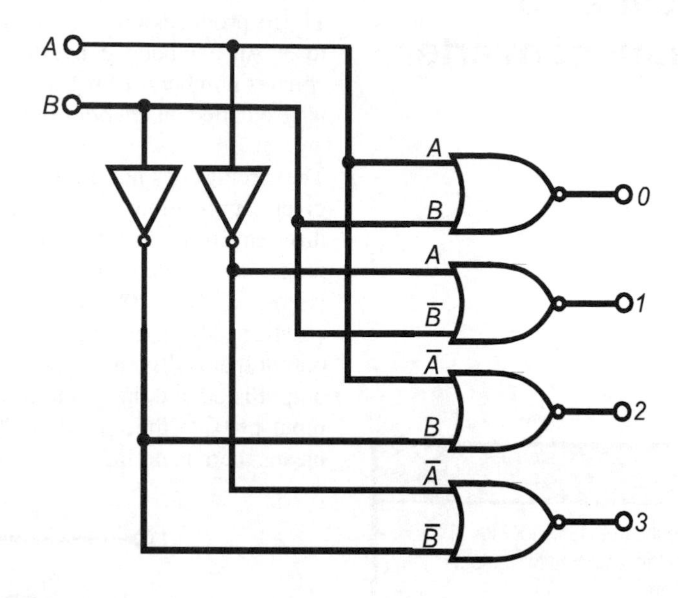

Figure 11.16 *A binary number is fed to this binary-to-decimal converter (or decoder), making the corrresponding one of its outputs go high.*

Priority encoder

A priority encoder is similar to the converter of Fig. 11.15 but does not assume that only one input is high at any time. If two or more inputs are high, the outputs show the binary equivalent of the larger-numbered (or priority) input. The truth table is:

Inputs			Outputs		Priority
3	*2*	*1*	*Z1*	*Z2*	
0	0	0	0	0	0
0	0	1	0	1	1
0	1	0	1	0	2
0	1	1	1	0	2
1	0	0	1	1	3
1	0	1	1	1	3
1	1	0	1	1	3
1	1	1	1	1	3

To follow the operation we start with the highest-numbered input. If input *3* is high, both OR gates receive at least one 1, so both *Z0* and *Z1* are high, coding binary 11 and indicating priority *3*. Whatever else is on the other lines makes no difference.

If input *2* is high (but *3* is low), *Z1* is high. The output of the NOT gate is low, so the AND gate output is low. Inputs to the *Z0* gate are both low, so *Z0* is low. Outputs are binary 10, indicating priority *2*. The state of input *1* makes no difference.

If only input *1* is high, the NOT gate has a high output. Both inputs to the *Z0* gate are high. *Z0* is high. Outputs are 01, indicating priority *1*.

With all three inputs low, both outputs are low. Whether this is to be taken as a zero input or no input depends on the application. As before, we can add extra logic if we need to distinguish between these two conditions.

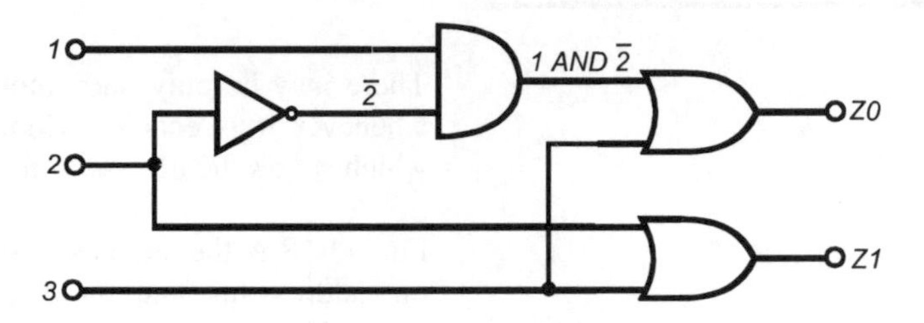

Figure 11.17 *When two or more of the inputs of this priority encoder are made high the outputs show the number of the higher ranking input.*

Data selector

A data selector (or *multiplexer*) has several inputs but only one or two outputs. The data selector circuit in Fig 11.18 has two inputs, *A* and *B*, and one output, *Z*. The action of the selector depends on the logic level at the select input, *S*. If *S* is low, as in the figure, output *Z* follows the state of input *A*. Conversely, *Z* follows *B* if *S* is high. So by setting *S* to 0 or 1 we can select data *A* or data *B*. The select input is the equivalent of an address, the two possible addresses in this circuit being 0 or 1.

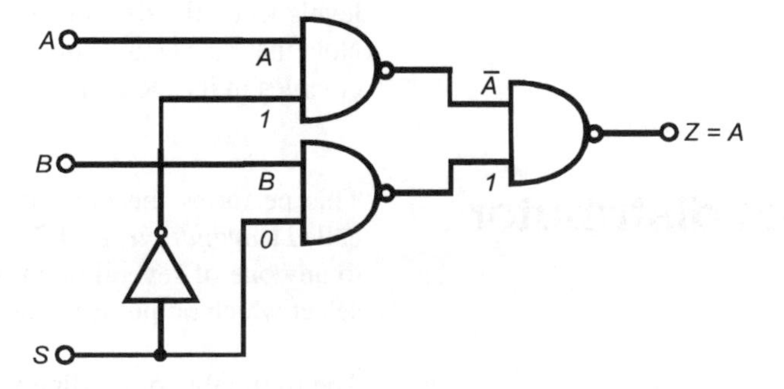

Figure 11.18 *The logic levels in this data selector (or multiplexer) are shown for a low select (S) input. The current value of A appears at the output. The value of B appears if S is made high.*

Many data selectors are more complicated than Fig 11.18, but they all have the same three kinds of input:

- Data inputs, of which there are usually two, four, eight or sixteen.
- Address inputs, used to select one of the data inputs to appear at the output. Depending on the number of data inputs, there are one, two, three, or four address inputs, which are the equivalent of 1-bit 2-bit, 3-bit and four-bit addresses .
- An enable input, to switch the output on or off. This is also known as the *strobe* input. This is optional and Fig 11.18 does not have one.

Test your knowledge 11.7

How many address inputs are needed to select between 8 data inputs?

There may be only one output, which shows the selected data input whenever it is enabled. Some data selectors have a second output which shows the inverse of the data.

Fig. 11.18 is the simplest possible data selector, with two data lines, one address line and one output. It does not have an enable input. Its truth table is:

Inputs			Output
B	A	Select	Z
X	0	0	0
X	1	0	1
0	X	1	0
1	X	1	1

The 'X' means 'Don't care'. In other words, the state of that input has no effect on the output. When select is low, the output shows the state of input *A*. When it is high, it shows the state of input *B*. This action is obtained by using a pair of NAND gates. The figure shows the logic levels when the data on A is selected by making the *Select* input low. Note that the signal is inverted twice as it passes through the circuit and emerges in its TRUE form.

Data distributor

This performs the opposite function to a data selector, and is often called a *demultiplexer*. It has one data input and the data may be routed to any one of several outputs (Fig. 11.19). There are address inputs to select which output is to show the data.

The truth table of a 2-line to 4-line data distributor is shown opposite. The first line states that if the data is 0 all outputs are 0, whatever the state of the select inputs. As before, 'X' means 'Don't care'. If the data is 1, it appears as a 1 output at the *one* selected output. The unselected outputs remain at 0.

Digits in a binary word or byte are numbered from 0 for the least significant digit (on the right) to the most significant digit (on the left). For example, if A = 1011 then A0 = 1, A1 = 1, A2 = 0 and A3 = 1.

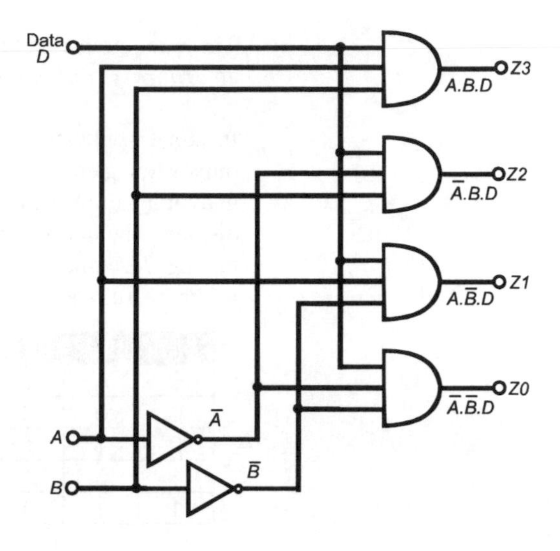

Figure 11.19 *A data distributor (or demultiplexer) routes the data to one of its outputs. The output is selected by supplying its binary address to inputs A and B.*

Inputs			Outputs			
Data	Select		Z3	Z2	Z1	Z0
	B	A				
0	X	X	0	0	0	0
1	0	0	0	0	0	1
1	0	1	0	0	1	0
1	1	0	0	1	0	0
1	1	1	1	0	0	0

Complicated truth tables

Very often a logic circuit has to perform a complicated function that is not simply related to one or more of the standard logical or mathematical functions.

Example

Consider a logic circuit controlling the hot water valve in a domestic washing machine. Whether it is to be open or not depends on many factors: the water temperature required, the level of the water in the wash-tub, the washing cycle selected, whether or not a 'half-load' version of the cycle has been selected, the stage in the washing cycle, and whether or not an operating fault has been detected requiring the water supply to be cut off.

These requirements produce a very complicated truth table, known as an *arbitrary truth table.*

In some applications, a Karnaugh map may solve the problem, and only a few gates are needed. In other cases a large number of gates is unavoidable. One instance of this is a circuit for driving a 7-segment display. The display (Fig. 13.1) represents the decimal digits 0 to 9 by having 7 segments which are turned on or off. This action has an arbitrary truth table, of which the first 4 lines are:

Inputs		Outputs						
B	A	a	b	c	d	e	f	g
0	0	1	1	1	1	1	1	0
0	1	0	1	1	0	0	0	0
1	0	1	1	0	1	1	0	0
1	1	1	1	1	1	0	0	1

We will follow the steps for designing a circuit to decode the first 4 digits, 0 to 3. Column *a* shows when segment *a* is to be on. This is the output column of the truth table for:

$$a = \overline{A \bullet \overline{B}}$$

The reasoning is the same as for the example on p. 161, except that *A* and *B* are exchanged. The next segment, segment *b,* is on for all combinations of inputs (though not if we continue the table to include digits 4 to 9). Its equation for the table above is $b = 1$. The equations for the remaining segments are:

Test your knowledge 11.8

Confirm that the equations for digits *c* to *g* are correct.

$$c = \overline{\overline{A} \bullet B}$$
$$d = a$$
$$e = \overline{A}$$
$$f = \overline{A + B}$$
$$g = A \bullet B$$

The 0-3 decoder has the circuit of Fig. 11.20. This is fairly complicated but the decoder 0-9 is even more so, because there are two more inputs (*C, D*) to operate with. We have to add these to the 0-3 decoding because we must make sure that digits 0 to 3 appear only when $C = 0$ and $D = 0$. Extra logic is needed for driving segment *b* because it is not lit for digits 5 and 6. Segment *d* needs its own logic because it is not equal to *a* for digit 7. In fact, all of the segments need more complicated logic to cover the additional digits 4 to 9. To make things even more difficult we could decode for hexadecimal digits A to F when the input is 1010 to 1111.

Hexadecimal: a counting system based on 16 digits, 0 to 9 and A to F, the letters representing decimal values 10 to 15. This covers all the 16 binary values 0000 to 1111.

Designing a binary-to-seven-segment decoder is an interesting project

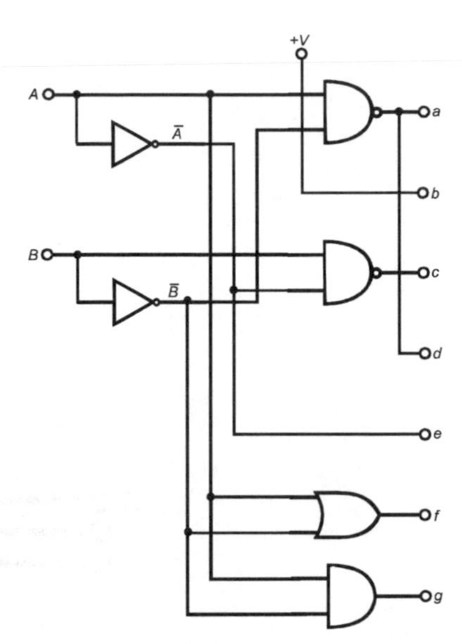

Figure 11.20 *This circuit decodes 2-bit binary inputs 0 -3 and produces outputs for driving a 7-segment display. Segment b is switch on in all of digits 0-3, so it is powered directly from the positive supply rail.*

and helps you to learn about simple gates and how they work. But, in practice, we do not attempt to build decoders from individual gates. Instead, we use a ready-made decoder such as the 74LS47 or the CMOS 4511. These inexpensive ics contains all the logic for decoding a binary input, and also such features as *leading zero blanking*. In a display of several digits, this switches off all the segments on the left-most digits when the digit receives an all-zero input. A 4-digit display shows '63', for example, instead of '0063'. Another feature is a 'lamp test' input that switches all segments on to check that they are in working order.

Data selector = multiplexer

When truth tables are so long and complicated there are other ways in which we can obtain the same function, if it is not already available as one of the standard ics. One of these ways is to use data selectors. Fig. 11.21 shows how we might use a data selector in the control circuit of a car radio. The radio has 7 pre-tuned stations, some of which transmit in mono and the others in stereo. In the truth table, a 0 output means mono and a 1 output means stereo. Stations are numbered 1 to 7 in binary.

The truth table is on p. 176. In the first line we see that there is no button for selecting Station 0. We can ignore this possibility when working out the logic. In practice, since the '0' input of the ic must be connected somewhere and the adjacent pin is to be connected to the positive supply, the simplest course is to connect it to the positive supply. The ic has an enable ($\overline{EN}$) input that must be held low, as

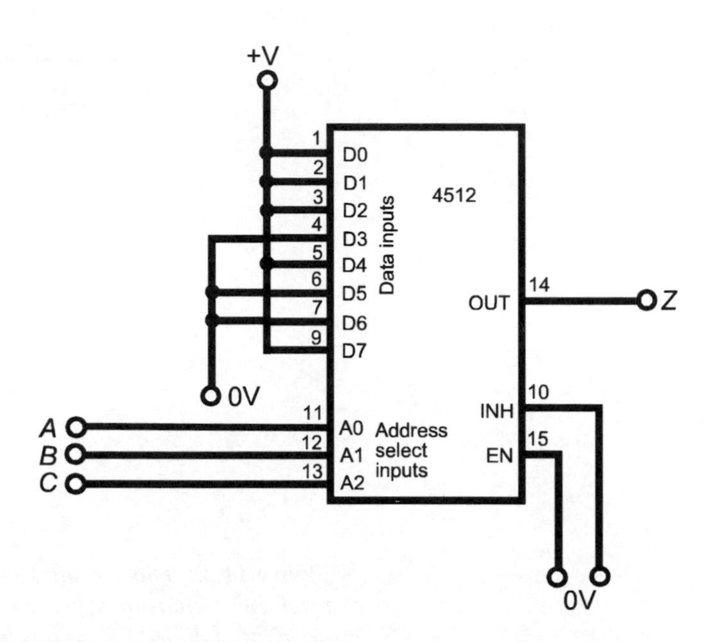

Figure 11.21 *A data selector can be used in place of a complicated network of gates to produce any logic function. Inputs are wired to the positive supply or to the 0 V line to give the required pattern of highs and lows in the output.*

Inputs			Output
C	B	A	Z
0	0	0	Can't happen
0	0	1	1
0	1	0	1
0	1	1	0
1	0	0	1
1	0	1	0
1	1	0	0
1	1	1	1

shown in the diagram. If it is made high the output goes low independently of the data inputs.

Another advantage of this technique is that the output for any particular combination of inputs can easily be altered, simple by connecting the corresponding input to the other supply rail. If we were using an array of gates, a change of logic is likely to require gates of a different type, and re-routing the connections.

Another useful feature of the 4512 data selector is that it has *tri-state outputs*. When the inhibit (INH) input is low the output pins behave in the normal way, outputting either high or low logical levels. When the inhibit input is made high, *all* the outputs go into the *high impedance* state. In this state a high impedance exists between the output gate and the external line to which it is normally connected. In effect, all outputs become disconnected from the system and no signals come from the device. This feature is of great use in computers and other data processing systems, where several ics may be wired to a *data bus*. This carries data to many other parts of the system. But other parts of the system may receive data from only one device at a time. All the devices that put data on to the bus have tri-state outputs, but only the one device that is supposed to be transmitting data has its inhibit input made low. All the others have their inhibit inputs high and are, in effect, disconnected from the bus at that time. We look more closely at this in Chapter 20.

Data bus: a set of conductors between any two or more parts of a data-processing system, often each carrying the individual bits of a data group.

Using ROM

ROM: a read-only memory device.

Another approach to complicated combinational logic is to use a ROM. As explained in Chapter 20, a memory chip has address inputs by which any item of data stored in the memory can be read. Chips vary in the number of data items stored and the number of bits in each item. An example of the smaller ROMs is the 82S123A, which stores 256 bits arranged in 32 bytes, each consisting of 8 bits. Any particular byte is read by inputting its address (00000 to 11111, equivalent to 0 to 31 in decimal) on five address input pins. When the 'chip enable' input is made low, the byte stored at this address appears on the 8 tri-state output pins. A device such as this can control up to eight different devices such as lamps, displays, solenoids, or motors. They can be switched on in 32 different combinations according to the address selected.

The 82S123A is programmable only once, but many other ROMs may be reprogrammed if the logic has to be changed. There is more about ROMs in Chapter 20.

Programmable array logic

If a logic design is highly complicated it may be best to create it in a logic array. An array has several hundreds or even several thousands of gates, all on one chip. They may be connected together by programming the device, using a personal computer. The chip is usually contained in a PLCC package, which is square in shape with a large number (as many as 84) pins arranged in a double row around the edge of the device. Depending on the type of PAL, some of the pins are designated as inputs, some as outputs, and some may be programmable to be normal inputs or outputs or to be tri-state outputs. Many PALs also contain an assortment of ready-made flip-flops and other useful sequential logic units.

Programming PALs is an intricate matter. It is done on a development board or programmer, which provides the programming signals. This is under the control of a personal computer, running special software to produce the required logical functions. Most types of PAL are erasable and reprogrammable.

Logic in software

Mention of software reminds us that there is an increasing tendency to use microprocessors and microcontrollers (such as the PIC and the BASIC Stamp) to replace complicated logic circuits. We use *software* logic instead of *hardware* logic. This allows even more complicated logical operations to be performed, yet requires no greater expense, board space or power. The microcontroller receives inputs from sensor circuits and control switches. The logic is performed by software running on the microcontroller. The truth table is built into the logic of the software. This determines what the outputs must be. Then the appropriate signals are sent to the microcontroller's output terminals. If the logic needs to be changed or updated, it is relatively simpler and quicker to re-write software than it is to redesign and reconstruct hardware.

Activity 11.2 MSI devices

Investigate some of the MSI devices that perform the combinational functions described in this chapter. Work on a breadboard or use a simulator.

Devices to investigate include:

Adder (4-bit, full): 4008, 74LS283.

Magnitude comparator: 4063 (4-bit), 74LS85 (4-bit), 4512 (8-bit).

Data selector/multiplexer: 4019 (2 line), 74LS157 (2-line), 74LS153 (4-line), 74LS151 (8-line), 4512 (8-line).

Data distributor/demultiplexer/decoder: 74LS138 (3-to-8 line), 74LS139 (2-to-4 line), 4555*(2-to-4 line).

Decoder: 4028 (1 of 10), 4511 (7-segment), 74LS47 (7-segment).

Priority encoder 4532 (8-to-3 line), 74LS148 (8-to-3 line), 74LS147 (10-to-4 line).

Verify that the circuit operates according to the truth table (function tables) in the manufacturer's data sheet.

Problems on logical combinations

1 Simplify this expression and write its truth table:

$$\overline{A \bullet B + \overline{A} \bullet \overline{B}}$$

To what logical operation is it equivalent?

2 Draw a circuit diagram to represent this expression:

$$Z = \overline{A} \bullet B \bullet C + A \bullet \overline{B} \bullet C + A \bullet B \bullet \overline{C} + A \bullet B \bullet C$$

Use a breadboard or simulator to investigate its logic.

3 Design a half-adder, using (a) only NAND gates and (b) only NOR gates.

4 Design a decimal to binary decoder for numbers 1 to 3, using only NAND gates.

5 Obtain the logical equations from the Karnaugh maps in Fig. 11.22. Draw the logic circuit for each equation.

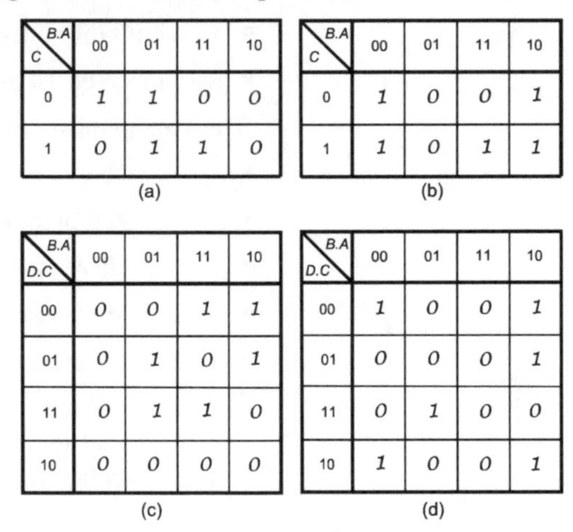

(a)

C \ B.A	00	01	11	10
0	1	1	0	0
1	0	1	1	0

(b)

C \ B.A	00	01	11	10
0	1	0	0	1
1	1	0	1	1

(c)

D.C \ B.A	00	01	11	10
00	0	0	1	1
01	0	1	0	1
11	0	1	1	0
10	0	0	0	0

(d)

D.C \ B.A	00	01	11	10
00	1	0	0	1
01	0	0	0	1
11	0	1	0	0
10	1	0	0	1

Fig 11.22 *Karnaugh maps for Question 5.*

6 Convert the circuits of Question 5 into an all-NAND or all-NOR form.

7 Design a 3-bit binary to decimal converter, using only NOR gates.

8 Explain the action of a 2-input data selector.

9 Design a 3-input data selector, with an enable input and TRUE and NOT outputs.

10 Design a 4-input data selector, using only NAND gates.

11 Design and verify (on a breadboard or simulator) a circuit for driving a 7-segment display for (a) the digits 0 to 7, and (b) the even digits 0 to 8.

12 Explain how a binary-to-decimal decoder can be used as a data distributor.

13 In a shop, a siren is to be sounded automatically if:

- the emergency button is pressed, or
- smoke is detected, or
- a person is detected inside the shop between dusk and dawn, or
- if someone enters the storeroom without first closing the hidden 'storeroom access' switch, or
- the shopkeeper leaves the premises without first locking the storeroom door.

Design the logic for this system. State what outputs each sensor and switch gives to indicate the 'alarm' condition.

14 In a greenhouse, the irrigation system is to be turned on automatically when:

- the temperature rises above 25°C, or
- humidity falls below 70%, or
- between 1430 hrs and 1530 hrs every day.

The system must never be turned on when:

- it is between dusk and dawn, or
- (b) temperature falls below 7°C,
- (c) there is someone in the greenhouse, or
- (d) the manual override switch has been closed.

Design the logic for this system. State what outputs each sensor and switch gives to indicate the 'irrigate enable' condition.

15 Outline the techniques available for building circuits for arbitrary truth tables when the logic is complicated.

Multiple choice questions

1 In this list, the operator that is not a basic logical operator is:

A AND.
B OR.
C NAND.
D NOT.

2 The operator represented by the ⊕ symbol is:

A NAND.
B NOT.
C NOR.
D Exclusive-OR.

3 The output of a 3-input NAND gate is low when:

A all inputs are low.
B one input is high.
C more than one input is low.
D all inputs are high.

4 The number of input combinations which can make the output of a 4-input OR gate go low is:

A 8. C 15.
B 16. D 1.

5 The output of a majority logic circuit goes high when:

A more than half the inputs are 1.
B all inputs are 0.
C some inputs are 1 and some are 0.
D all inputs are 1.

6 The inputs to a half adder are $A = 1$ and $B = 1$. Its outputs are:

A $S = 0$, $C_o = 1$.
B $S = 1$, $C_o = 1$.
C $S = 1$, $C_o = 0$.
D $S = 0$, $C_o = 0$.

7 The outputs of a full adder are $S = 0$, $C_o = 1$. Its inputs are:

A $A = 1, B = 0, C_i = 1$.
B $A = 0, B = 0, C_i = 0$.
C $A = 1, B = 1, C_i = 1$.
D $A = 0, B = 0, C_i = 1$.

8 An 8-input parity tree has odd parity. Its output goes high when the input is:

A 11111111.
B 10011000.
C 10000001.
D 00000000.

12 Logical sequences

Summary

Logical circuits with feedback or delay in them go through a sequence of states when triggered. Set-Reset *flip-flops* may be built from NAND or NOR gates. They are used for temporary storage of data. Arrays of S-R flip-flops are used as memories in large logical systems such as computers. Pulse generators and timers rely on the delay occurring while a capacitor is charged or discharged. There are special timing ics such as the 555 timer, which can be used as a monostable or an astable multivibrator. Astables can also be built from NOT gates, and their timing may be made more precise by including a quartz crystal in the circuit. MSI integrated circuits include clocked-logic devices such as D-type and J-K flip-flops, which operate on the master-slave principle. These can be used for making toggle flip-flops and connected to form counters, storage registers and shift registers. Synchronous counters avoid the output glitches of ripple counters. RAMs are used for storage of large amounts of data and are of two types, static RAM, in which data is stored in flip-flops, and dynamic RAM, in which data is stored as quantities of elecric charge.

Flip-flops

If a logic circuit has *feedback* in it, then its response to a given input may not be the same the second time the input is applied. The circuit is in one state to begin with and then goes into a different state. An example of this is the *set-reset flip-flop,* or SR flip-flop (Fig. 12.1). Its name suggests that it can exist in two states, *Set* and *Reset*, and that it can be made to flip into one state and then flop back into the original state. The essential point is that the circuit is *stable* in either state. It does not change state unless it receives the appropriate input. Because it has *two* stable states it is also known as a *bistable* circuit.

Fig. 12.2 shows how the flip-flop works:

NAND: Output is low when all inputs are high.

(a) The flip-flop is in its *Set* state, with Q=1 and $\overline{Q}$=0. The inputs must be held high by the circuit that provides input. In the Set state, both gates have inputs and outputs that conform to the NAND truth table, so the circuit is *stable*.

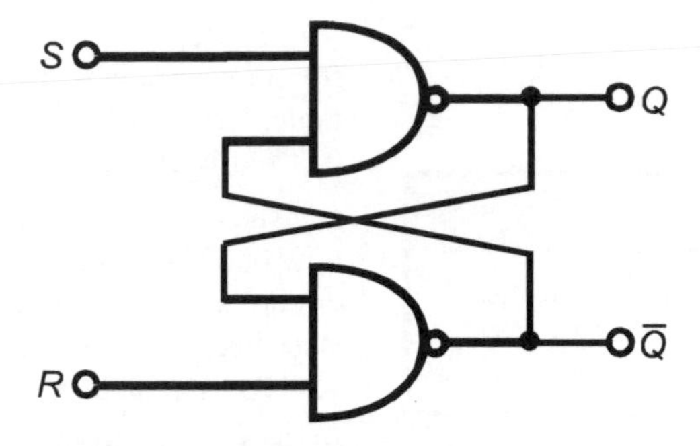

Figure 12.1 *A set-reset flip-flop has two stable states. It is made to change from one state to the other by applying a low pulse to its normally high inputs.*

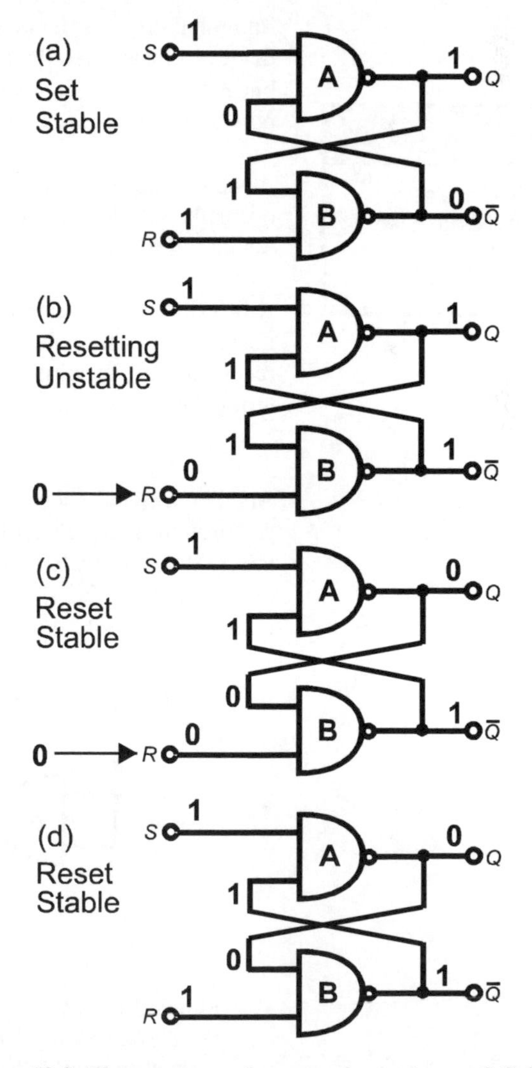

Figure 12.2 *The sequence of stages of resetting an S-R flip-flop.*

(b) The Reset input has been made low. The input to Gate B is now 01. Its output changes from 0 to 1, in accordance with the NAND truth table. Gate A then has a 11 input, and its output is about to change to 0. The flip-flop can not stay in the state of Fig. 12.2b; it is unstable.

(c) The output of Gate A has gone low, so that the input to Gate B is now 00. This makes no difference to its output, which stays high. The circuit is now in the *Reset* state, with Q=0 and $\overline{Q}$=1. It is *stable* in this state because both gates have inputs and outputs that conform to the NAND truth table.

(d) There is no further change of output when the Reset input goes high again. This is because the output of B remains high when it has 01 input. The flip-flop remains stable.

Test your knowledge 12.1

Draw diagrams to illustrate what happens when we Set a flip-flop which is in the Reset state.

Summing up, $\overline{Q}$ is the inverse of Q in the stable states. The flip-flop is made to change state by applying a low pulse to its Set or Reset input, but it does not change state when an input receives two or more low pulses in succession.

Test your knowledge 12.2

Draw diagrams to illustrate the action of a flip-flop based on two NOR gates. Label the Set and Reset inputs.

A similar flip-flop is built from a pair of NOR gates. Its inputs are normally held low and it is made to change state by a high pulse to its Set or Reset inputs.

A flip-flop stays in the same state until it receives an input that changes its state. It can 'remember' which kind of input (Set or Reset) it received most recently. Flip-flops are the basic circuit used for registering data, particularly in computer memory circuits.

Delays

Another way in which the behaviour of a logic circuit can be made to go through a sequence of changes is by introducing a *delay*. Usually this is done by including a capacitor, which takes time to charge or discharge. Fig. 12.3 shows a simple example of this. The way it works is shown in Figs 12.4 and 12.5.

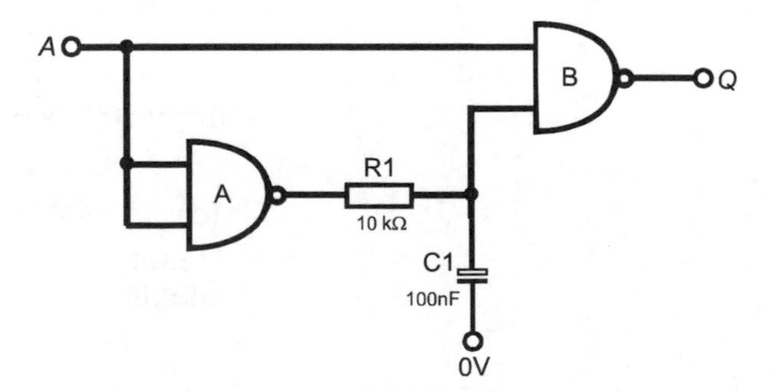

Figure 12.3 *This edge-triggered circuit produces a low output pulse when the input rises from low to high. Its output remains high when input falls from high to low.*

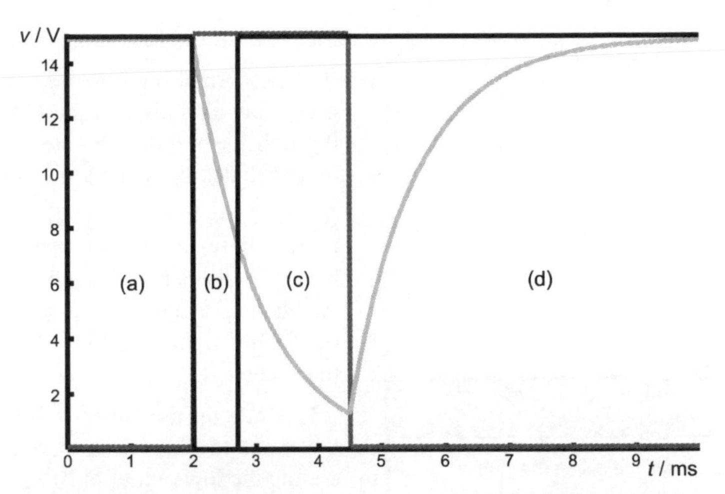

Figure 12.4 Changes of *voltage in the circuit of Fig. 12.3 show a rise of input at 2 ms (light grey), which immediately produces a low output pulse lasting 0.7 ms (black). Output remains high when input goes low at 4.5 ms. Changes of voltage across the capacitor are plotted in dark grey.*

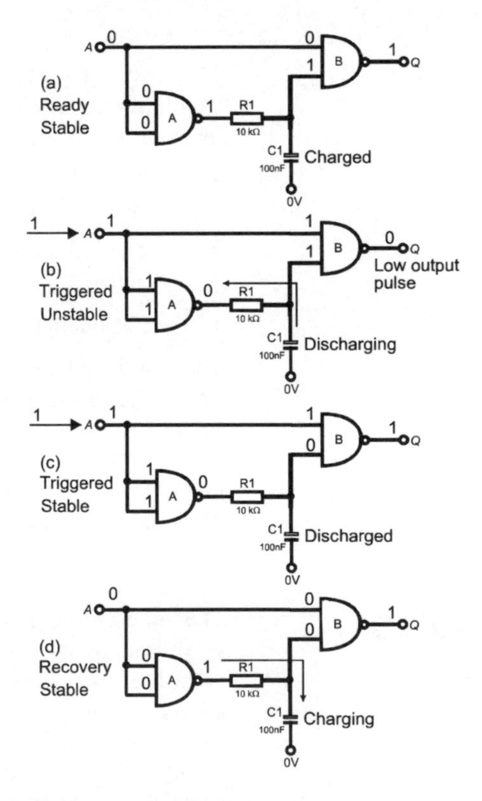

Figure 12.5 *The sequence of stages in the generation of a pulse by the circuit of Fig. 12.3. The letters (a) – (d) indicate the stages labelled in Fig. 12.4.*

Fig. 12.5 shows four stages in the operation of the delay:

(a) The normal input to the circuit (dark grey line) is low (0 V). This gives Gate A a high output (15 V). The capacitor has charged to 15 V through the resistor. So the input to Gate B is 01, giving it a high output (black line). This is a stable state.

(b) At 2 ms the input to the circuit (dark grey line) goes high. This immediately changes the output of Gate A from 1 to 0. The capacitor begins to discharge through R1, current flowing *into* the output of Gate A. The exponential fall in voltage across the capacitor can be seen in the figure (light grey line). Nothing happens to Gate B (black line) at this stage.

Test your knowledge 12.3

Describe the action of a pulse generator in which the gates of Fig.12.3 are replaced by NOR gates.

(c) At 0.27 ms the falling voltage across the capacitor reaches 7.5 V, so the input to Gate B is now equivalent to 01. Its output goes high, ending the low output pulse.

(d) When the input pulse ends (here at 4.5 ms but it could be at *any* time after the beginning of stage (c)) the state of Gate A is reversed and the capacitor begins to charge again. It makes no difference to the output of Gate B when its input changes from 00 to 01. So there is no output pulse when the input pulse ends.

This circuit is often used for generating pulses in logic circuits. The length of the pulse depends on the values of the resistor and capacitor. Its chief disadvantages are:

- It must be given time to completely recharge the capacitor before being triggered again.
- The input must go high for longer than the length of the output pulse.
- Pulse length depends on supply voltage.

The pulse generator is stable in its quiescent or 'ready' state (Fig. 12.5a). Once triggered it produces a pulse and then returns to the same stable state. It is stable in only one state and is therefore known as a *monostable*. More correctly, but rarely, it is called a *monostable multivibrator*. Monostables with higher precision are available in the 74LS and CMOS series.

Monostables

The 74LS221 (Fig. 12.6) is a typical monostable pulse generator. It is very similar to the older 74121. The pulse length is determined by the values of the resistor and capacitor and is approximately 0.7RC seconds. This is the time taken to charge the capacitor. Using a capacitor of maximum value 1000 μF and a resistor of maximum value 70 kΩ, the greatest pulse length obtainable is 70 s. Output Q is normally low and $\overline{Q}$ is normally high. The monostable is triggered by a falling edge on input A or a rising edge on input B, as indicated by the truth table on the opposite page.

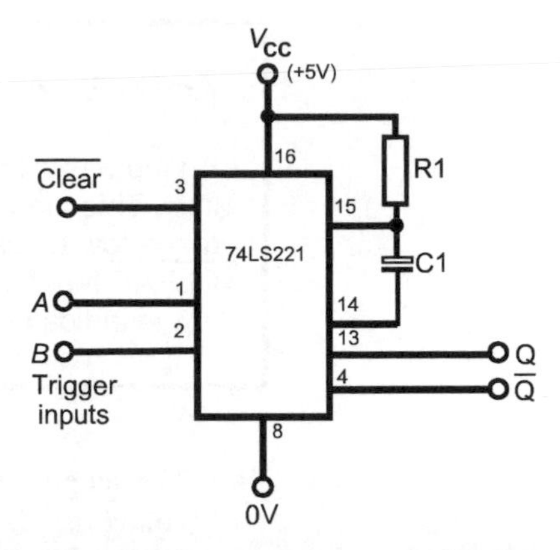

Figure 12.6 *The 74LS221 comprises two identical monostables. The pin numbers in this figure refer to only one of these.*

Inputs			Outputs	
$\overline{CLR}$	*A*	*B*	*Q*	$\overline{Q}$
0	X	X	0	1
X	1	X	0	1
X	X	0	0	1
1	0	↑	_⎍_	‾⊔‾
1	↓	1	_⎍_	‾⊔‾
↑	0	1	_⎍_	‾⊔‾

The $\overline{\text{CLEAR}}$ input is *active low* (see box). When it is made low, the values of A and B have no effect; output Q is low and output $\overline{\text{Q}}$ is high. The monostable can also be cleared by making A high or B low, whatever the state of the $\overline{\text{CLEAR}}$ input.

To generate a pulse the $\overline{\text{CLEAR}}$ input must first be made high. Then a falling edge on A (a downward arrow in the table) or a rising edge on B (a rising arrow) triggers the pulse to begin. Once the monostable is triggered, the length of the pulse depends only on the values of R and C. In other words, the monostable is *not retriggerable*. However, a low level on $\overline{\text{CLEAR}}$ can end the pulse at any time. The last line of the table shows that a pulse can also be triggered by the $\overline{\text{CLEAR}}$ input. This only occurs if certain inputs have already been made to A and B and is not part of the normal action of the circuit.

The 74LS221 has features that make it superior to the circuit of Fig. 12.3:

- It requires very little time to recover at the end of the output pulse. Under the right conditions it can have a duty cycle up to 90%. That is, it recovers in about one-ninth of the length of its output pulse.
- B is a Schmitt trigger input. This means that the monostable is reliably triggered on a *slowly changing* input voltage level.

Active Low

An input to a logic circuit may often be described as 'active low'. This means that the input is normally held high, but is made low to make a certain action occur. An example is the $\overline{\text{CLEAR}}$ input of the 74LS221. The fact that an input is active low is indicated by printing a bar over its name.

- The length of the timing period does not depend on the supply voltage.
- The length of the timing period does not depend on temperature.

The CMOS 4528 is a monostable with similar properties, except that it is re-triggerable, and its recovery time is usually several milliseconds.

Bipolar devices: conduction is by electrons and holes, for example, in npn and pnp transistors.

Sink or source current: to take in or to provide current.

One of the most popular monostable ics is the 555. It belongs neither to the TTL family nor the CMOS family but its inputs and outputs are compatible with both. It operates on a wide range of supply voltages, from 4.5V to 16V. The original 555 is based on bipolar circuitry, and there are several newer devices, notably the 7555 based on CMOS. The advantages of the CMOS versions is that they use far less current, produce longer timing periods, cause less disturbance on the power lines when their output changes state, and operate on a wider range of supply voltages (3 V to 18 V). The 7555 output is able to sink or source only 100 mA, in comparison with the 200 mA of the 555, but this is rarely a drawback. The 7555 has 2% precision compared with 1% for the 555.

Fig. 12.7 shows the 555 connected as a monostable. The length of the output pulse depends on the values of R1 and C1 according to the equation:

$$t = 1.1R_1 C_1$$

Test your knowledge 12.4

A 555 monostable circuit has R_1= 2.2 kΩ and C_1=1.5 μF. What is the length of the output pulse?

The control pin is sometimes used to vary the length of the period but usually it is either connected to the 0 V line through a capacitor or (in the 7555) is left unconnected. The 555 needs the decoupling capacitor C3 connected across the supply terminals, but this is not necessary with the 7555.

The trigger (set) and reset pins are normally held high. A low-going pulse (which need only fall to two-thirds of the supply voltage) triggers the monostable. The device is not retriggerable, but the output pulse can be ended at any time by a low-going pulse on the reset pin. In the figure, the reset pin is permanently wired to the positive supply (high) so that resetting is not possible.

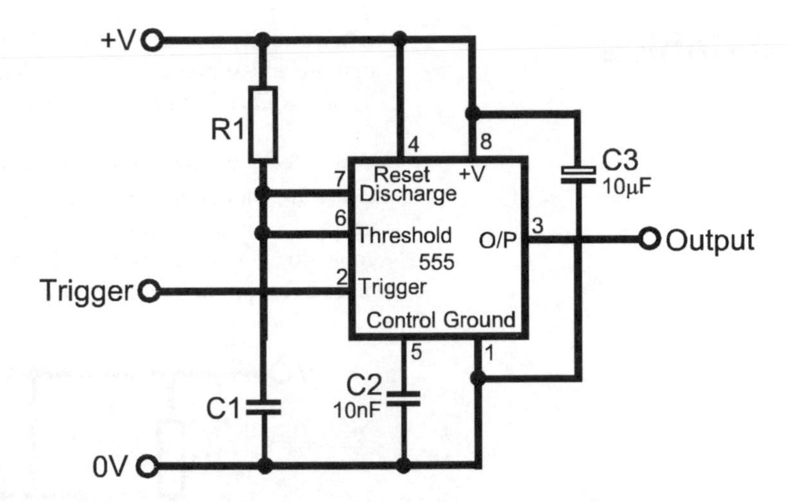

Figure 12.7 *The 555 timer ic provides an accurate output pulse length determined by the values of the timing capacitor and resistor, C1 and R1.*

Test your knowledge 12.5

Design a 555 monostable circuit using a 220 nF capacitor to give an output pulse of 25 ms.

The action of the monostable is as follows:

(a) In the quiescent state the timing capacitor (C1) is charged to one-third of the supply voltage. The output is low (0 V). Current flowing through R1 is passed to ground by way of the discharge pin. The threshold pin monitors the voltage across C_1.

(b) When the circuit is triggered, its output goes high and the discharge pin no longer sinks current. Current flows to the capacitor, which begins to charge.

(c) When the charge across the capacitor reaches two-thirds of the supply voltage, as monitored by the threshold pin, the output goes low. The charge on the capacitor is rapidly conducted away through the discharge pin until the voltage across the capacitor has fallen again to one-third of the supply voltage.

The 555 timer has the advantage that the voltage levels are precisely monitored by the internal circuitry, so that timing has high precision. Most often it is limited by the precision of the timing resistor and capacitor. This is particularly the case when an electrolytic capacitor is used to give a long period. Because the starting and finishing voltages are defined as one-third and two-thirds of the supply, the voltage across the capacitor has to rise by one-third *of the supply.* At every stage of charging, the *rate of rise* of voltage is proportional to the supply voltage. The *total increase* of voltage required is also proportional to the supply. So the time taken to charge is independent of the supply.

Astables

After circuits that are stable in one or two states, we look at circuits that are not stable in *any* state. These unstable circuits are known as *astable multivibrators*, or *astables* for short.

The 555 can be used as an astable simply by connecting the trigger pin to the threshold pin and adding another resistor. The circuit is triggered again as soon as the capacitor is discharged at the end of its cycle. The capacitor is charged and discharged indefinitely and the output alternates between high and low. Fig. 12.8 shows an astable 555 circuit.

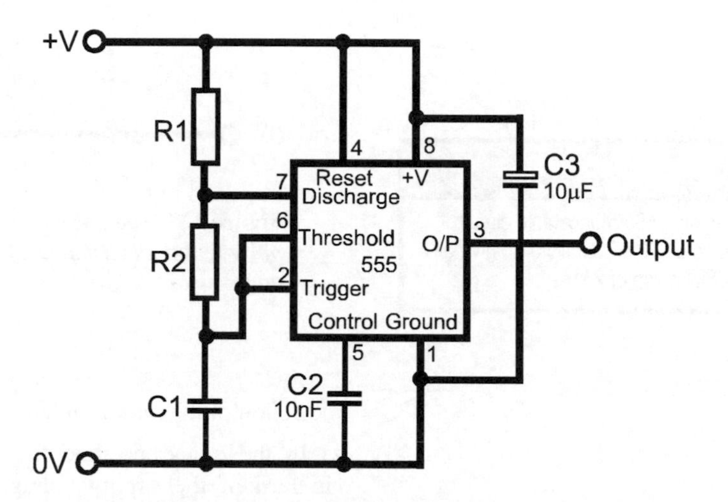

Figure 12.8 *The 555 timer ic can also be used as an astable.*

The action of the 555 astable is divided into two phases:

(a) Charging: output is high while the capacitor is charged from one-third to two-thirds the supply voltage, through R1 and R2 in series. The time taken is:

$$t_H = 0.69(R_1 + R_2)C_1$$

(b) Discharging: the output is low while the capacitor is discharged through r2, current flowing into the discharge pin. The time taken is:

$$t_L = 0.69R_2C_1$$

The total time for the two phases is:

$$t = t_H + t_L = 0.69C_1(R_1 + 2R_2)$$

The frequency of the signal is:

$$f = \frac{1}{t} = \frac{1}{0.69\,C_1(R_1 + 2R_2)} = \frac{1.44}{C_1(R_1 + 2R_2)}$$

Test your knowledge 12.6

A 555 astable has R_1=10 kΩ, R_2=6.8 kΩ and C_1=150 nF. What is the frequency and the mark-space ratio of the output signal?

The *mark-space ratio* is the ratio between the period when output is high and the period when it is low. Using the equations for t_H and t_L:

$$\text{Mark-space ratio} = \frac{t_H}{t_L} = \frac{R_1 + R_2}{R_2}$$

It follows from this result that the mark-space ratio is greater than 1 in all circuits connected as in Fig. 12.8. It is possible to obtain ratios equal to 1 or less than 1 by using diodes in the timing network to route the charging and discharging currents through different resistors.

CMOS astable

Astables can be built from TTL and CMOS gates. Fig. 12.9 shows one built from a pair of CMOS NOT gates. The third gate is not part of the astable but acts as a buffer between the astable and an external circuit such as an LED. R2 is not an essential part of the astable, but serves to give the output signal a squarer shape. It has a value about 10 times that of R1.

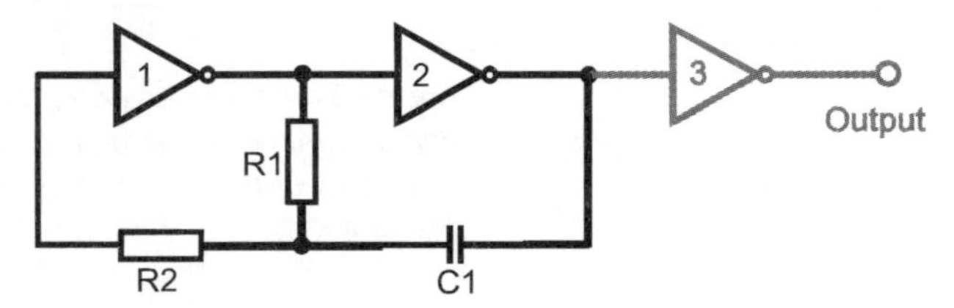

Figure 12.9 *A pair of NOT gates (or their equivalent made from NAND or NOR gates) make an astable multivibrator. The third gate (grey) acts as a buffer.*

The action of the astable is as follows (Fig. 12.10). To avoid having to repeat the words 'the supply voltage' several times, assume that it is 15 V:

(a) The input of Gate 1 is 0 V, its output is 15 V, and therefore the output of Gate 2 is 0 V. The capacitor is at a low voltage (actually negative, but see later). Current flows from Gate 1 through R1 and gradually charges the capacitor. The rate of charging depends on the values of R1 and C1.

(b) When the voltage across the capacitor reaches about 7.5 V, the input to Gate 1 (which is at almost the same voltage) is taken to be at logic high. Instantly the levels change to those shown in Fig. 12.10b. The sudden rise in the output of Gate 2 pulls *both* plates of the capacitor up by 15 V. This means that the plate connected to R1 rises to 22.5 V. Now current flows in the reverse direction through R1 and *into* the output of Gate 1. The voltage at this plate falls until it has reached 7.5 V again. The input of gate 1 is taken to be logic low and all the logic levels change back to state (a) again. The sudden fall in the output of Gate 2 pulls both plates of the capacitor down by 15 V. The plate connected to R1 falls to –7.5 V. The astable has now returned to its original state, stage (a).

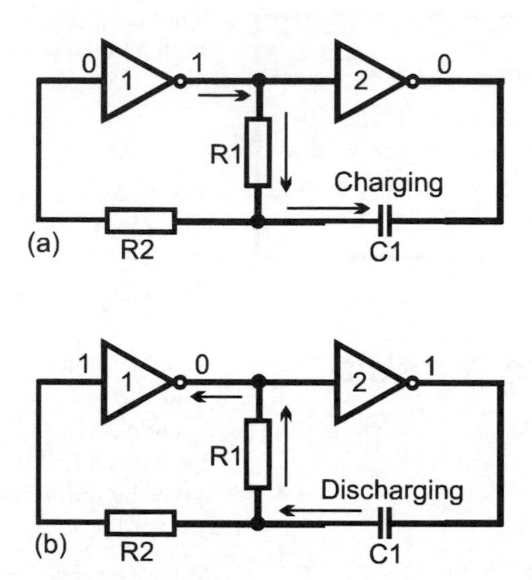

Figure 12.10 *The inverting action of the NOT gates is used to charge and discharge the capacitor. The time taken to do this, and therefore the frequency of the astable depends on the values of the capacitor and resistor R1.*

Crystal astable

If accurate frequency is important, a CMOS astable can be linked with a network containing a crystal as in Fig. 12.11. The circuit oscillates because of the negative feedback through R1. When output goes high, the input is pulled high too, making the output go low. The input is then pulled low, making the output go high. The gate alternates between two states but is not stable in either of them.

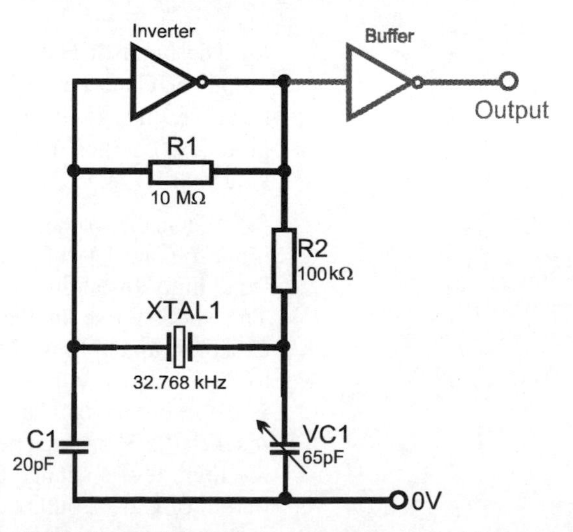

Figure 12.11 *An astable based on a quartz crystal has a precisely determined frequency.*

The circuit oscillates even if it consists only of the gate and R1. Adding the other components forces it to oscillate at the frequency of the crystal. R2, VC1, C1 and the crystal form a network that links the output of the gate to its input. The values of the resistor and capacitors are chosen so that, at the frequency of the crystal, the phase of the signal from the gate output is delayed by 180°. In other words, when the output is low the signal reaching the input is high and the other way round. As the crystal oscillates (see the description of the action in Fig. 5.5) the reversals of voltage across it are in step with and reinforce the negative feedback through R1. The gate alternates between its two states and the frequency of the crystal.

Phase: a stage reached in an oscillation, expressed as an angle. A complete cycle is equivalent to 360°.

The second gate is simply a buffer used to drive other circuits without drawing sizeable current from the astable.

Bistable latch

A latch is a circuit is which the output is normally equal to the data input but which can be held (latched) at any time. In Fig. 12.12 the gates on the right form an S-R flip-flop exactly as in Fig. 12.1. In that circuit the inputs are normally held high and a low pulse to the Set input sets the flip-flop. In this circuit the Set input of the flip-flop receives input $\overline{D \cdot S}$ while the Reset input receives $\overline{\overline{D} \cdot S}$. D is the Data input which at any instant is 0 or 1. S is the Store input. The truth table for D and S is:

Inputs		Outputs		Effect on Q
S	D	$\overline{D \cdot S}$	$\overline{\overline{D} \cdot S}$	
0	0	1	1	Latched
0	1	1	1	Latched
1	0	1	0	Equals D
1	1	0	1	Equals D

When $S = 0$ the outputs of the NAND gates on the left are 1. They hold the flip-flop in the latched state, since neither input can go low. When $S = 1$, and D goes low, the Reset input goes low, resetting the flip-flop so that Q goes low. When $S = 1$ and D goes high, the Set input goes

Test your knowledge 12.7

Design a data latch using NOR gates instead of NAND.

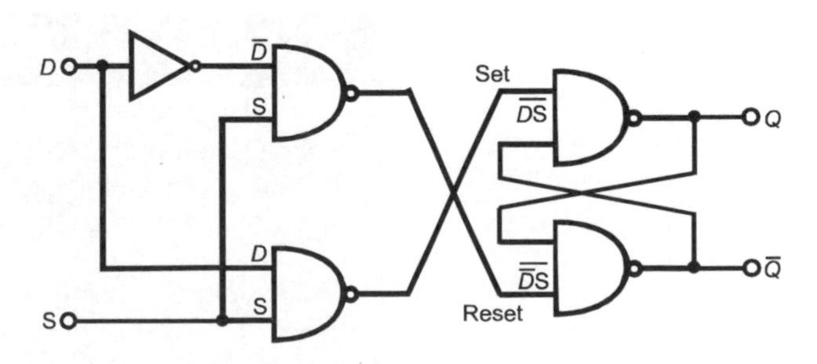

Figure 12.12 *With a data latch, the Q output follows the data input (D) when the STORE input (S) is high, or is latched when S is low.*

low, setting the flip-flop so that Q goes high. Whatever the value of D, output Q is equal to D. The other output, $\overline{Q}$, always is the inverse of Q and therefore of D.

There are ics that contain several data latches equivalent to the above. They have individual data inputs, Q outputs and sometimes $\overline{Q}$ outputs as well. Examples are the CMOS 4042 and the 74LS75, which latch 4 bits of data simultaneously.

Clocked logic

In the bistable circuits that we have looked at so far in this chapter, the circuit changes state immediately there is a suitable change of input. This can lead to difficulties in large logic circuits. Logic gates always take time to operate and different amounts of delay may arise in different parts of a circuit owing to the different numbers of gates through which the signals have to pass. It becomes very difficult to make sure that all the bistables change state in the right order and at the right times to respond to the data. The solution to this is having a *system clock.* This controls the exact instant at which the bistables change state. It is like the conductor of an orchestra, keeping all the players in time with each other by beating 'time'. We say that the players (or the bistables) are *synchronised*.

MSI: medium scale integration.

Clocked logic devices usually operate on the *master-slave* principle. They are available as MSI integrated circuits. Each flip-flop actually consists of two flip-flops, the *master* and the *slave*. The master flip-flop receives data and is set or reset accordingly. The state of the master flip-flop can not be read on the output terminals of the ic at this stage. Then, on the next positive-going edge of the clock input, the data on the master is transferred to the slave and appears at the output terminals of the ic. After this, new data may change the state of the master but this has no effect on the slave or on the data appearing at the terminals. The present output state is held unchanged until the next positive-going clock edge. Then the data currently on the master is transferred to the slave.

Activities 12.1 Flip-flops, and related circuits

Investigate each of the circuits listed below, either on a breadboard or by using a simulator. Use the suggested ics and consult data sheets for connection details.

Build a flip-flop using NAND gates as in Fig. 12.1(CMOS 4011 or 74LS00).

Design and build a flip-flop using NOR gates (CMOS 4001 or 74LS02).

Build a pulse generator using NAND gates as in Fig. 12.3.

Design and build a pulse generator built as in Fig. 12.3 but with NOR gates.

Design and build a pulse generator which produces a low pulse on a *falling* edge, using two NAND gates.

Investigate a pulse generator built from the 74LS221, with pulse length 1.5 ms. Check the pulse length with an oscilloscope.

Build a 555 or 7555 monostable (as in Fig. 12.7) with pulse length 10 s. Check the pulse length with a stopwatch.

Build a 555 or 7555 astable (as in Fig. 12.8) with a frequency of 1 Hz and a mark-space ratio of 2.

Design and build a CMOS astable as in Fig. 12.9, using either NAND (4011), NOR (4001) or NOT (4069) gates. Select components to give a frequency of 100 Hz and check the frequency by using an oscilloscope.

Build a CMOS astable using exclusive-OR gates (4070).

Construct the bistable latch shown in Fig. 12.12 and study its action by placing probes at various points in the circuit.

Investigate an MSI data latch (CMOS 4042, 74LS75).

Data-type flip-flop

The symbol for the data-type (or D-type) flip-flop is shown in Fig. 12.13. and its action in Fig. 12.14. For a *clocked mode* operation such as this the Set input (*S*) and Reset input (*R*) are held low. Some kinds of D-type flip-flop do not have the R input.

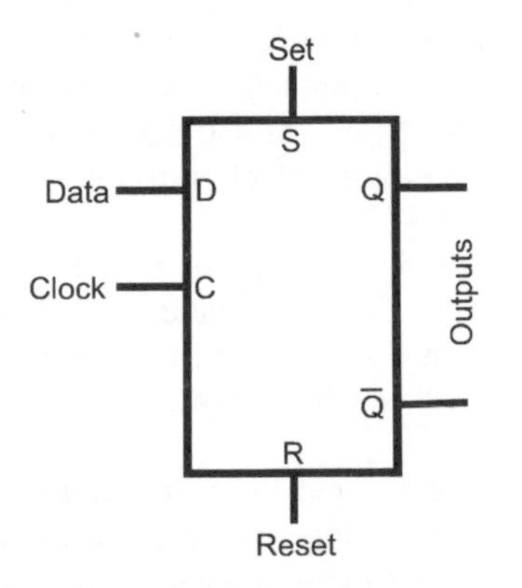

Figure 12.13 *A D-type flip-flop has a data input (D), a clock input (C), and a Q output. It may also have set and reset inputs and an inverted Q output. Data is transferred to the Q output on the rising edge of the clock.*

Latches and D-Type flip-flops

The *Q* output of a *data latch* follows the input until the $\overline{STORE}$ input goes low. Then it remains latched at that level until $\overline{STORE}$ goes high again.

The *Q* output of a *data-type flip-flop* changes only when the CLOCK input goes high. Then it changes to the same value as the data input at that instant.

The graphs of Fig. 12.14 are obtained from a simulator, modelling a 74LS74 D-type flip-flop. The clock is running at 10 MHz and the time scale is marked in nanoseconds. The lowest graph shows output *Q*. The simulator is stepped manually so that the data input can be made high or low at any desired time.

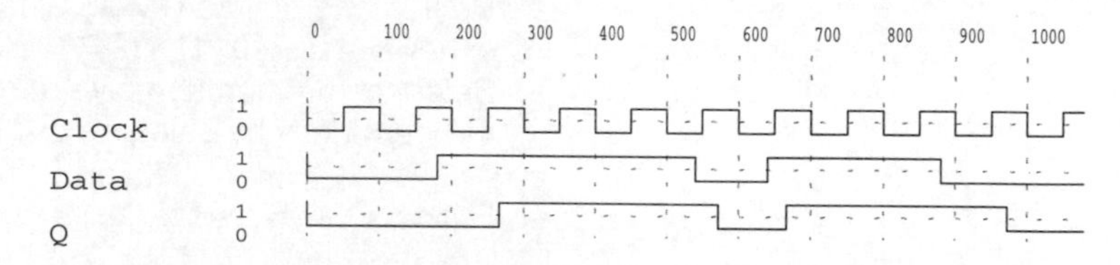

Figure 12.14 *In this plot of the action of a D-type flip-flop the data is changed at* any *time but the Q output changes only on the rising edge of the clock. The change is slightly delayed because of the propagation delay of the flip-flop (22 ns in this case).*

The run begins with both *D* and *Q* low. *D* is made high at 170 ms. At this time the clock is high. Nothing happens to the output when the clock goes low. But *Q* goes high at the next *rising edge* of the clock. This occurs at 250 ns. Looking more closely at the timing, we note that the rise in *Q* is slightly later than the rise in the clock. This is due to the propagation delay of the flip-flop, here set to a typical value of 22 ns. So *Q* eventually goes high at 272 ns. Subsequent changes in *D*, at 540 ns, 640 ns and 880 ns, always appear at output *Q* at the *next* rising edge of the clock (plus the propagation delay).

D-type flip-flops are used for sampling data at regular intervals of time. They then hold the data unchanged while it is processed. An example is the use of these flip-flops at the input of digital-to-analogue converters (see Chapter 14).

The Set and Reset inputs are used in the *direct mode* of operation of the flip-flop. They are also known as the Preset and Clear inputs. When the set input is made high, *Q* goes high instantly, without waiting for the rising edge of the clock. Similarly, Q goes low immediately the Reset input is made high.

Toggle flip-flop

The output of a toggle (or T-type) flip-flop changes every time the clock goes high. Toggle flip-flops are not manufactured as such. It is easy to make one from a D-type flip-flop, by connecting the $\overline{Q}$ output to the D input (Fig. 12.15). Assuming the D input at the most recent clocking was 0, then Q is 0 and $\overline{Q}$ is 1. So the D input is now 1. On the next clocking Q follows D and becomes high, so $\overline{Q}$ goes low. Now the D input is 0 and Q goes to 0 at the next clocking. The state of D and therefore of Q changes every time the flip-flop is clocked. Fig. 12.16 is a graph of the toggle action.

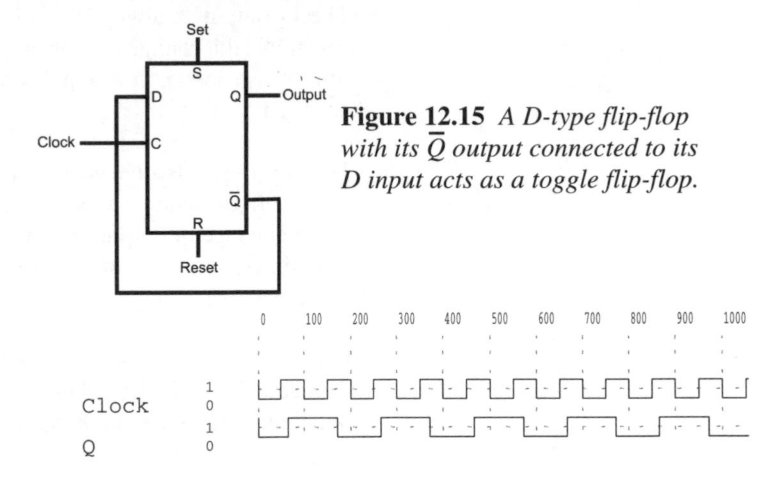

Figure 12.15 *A D-type flip-flop with its $\overline{Q}$ output connected to its D input acts as a toggle flip-flop.*

Figure 12.16 *In the toggle flip-flop of Fig. 12.15, Q changes state slightly after every rising edge of the clock.*

The toggle flip-flop demonstrates one of the advantages of clocked logic. After the flip-flop has been clocked, there is plenty of time before the next clocking for the circuit to *settle* into its stable state, and for its output to be *fed back* to the data input. This makes the operation of the circuit completely reliable. Unless the clock is very fast, there is no risk of its operation being upset by propagation time and other delays.

J-K flip-flop

The J-K flip-flop (Fig. 12.17) has its next output state decided by the state of its J and K inputs:

- If $J = K = 0$, the outputs do not change.
- If $J = 1$ and $K = 0$, the Q output becomes 1.
- If $J = 0$ and $K = 1$, the Q output becomes 0.
- If $J = K = 1$, the outputs change to the opposite state.

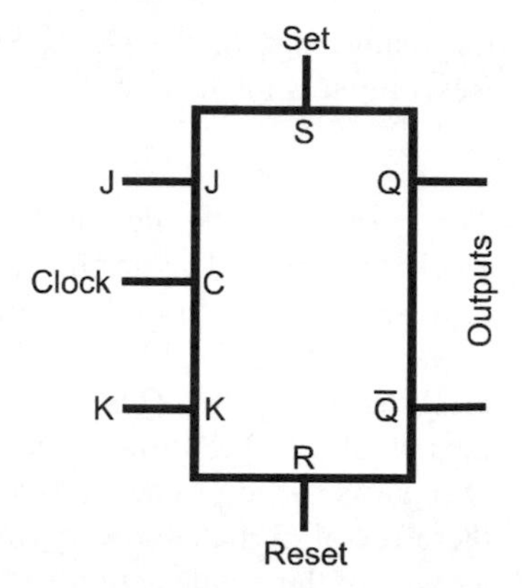

Figure 12.17 *The output of a J-K flip-flop changes state on the rising edge of the clock (on the falling edge in a few types). Whether it changes or not depends on the inputs to the J and K terminals.*

The $\overline{Q}$ output is always the inverse of the Q output. With most J-K flip-flops, the changes occur on the next rising edge of the clock. In the fourth case above, in which $J = K = 1$, we say that the outputs *toggle* on each rising edge.

J-K flip-flops also have Set inputs and (usually) Reset inputs. These operate in the same way as those of the D-type flip-flop, producing an instant change of output, without waiting for the clock. Several types of J-K flip-flops are clocked by a negative-going edge instead of the more usual positive-going edge.

Fig. 12.18 illustrates an important application of a negative-edge triggered J-K flip-flop. Three flip-flops are *cascaded,* the Q output of one being fed to the clock input of the next.

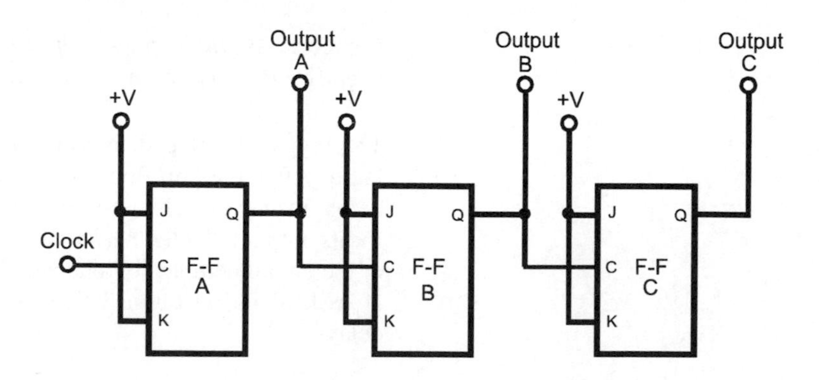

Figure 12.18 *Three J-K flip-flops wired as toggle flip-flops act as a 3-bit counter/divider.*

The J and K inputs are held high, so that the outputs toggle at each falling clock-edge. As each Q output falls from 1 to 0, this toggles the next flip-flop in the chain. The result of this can be seen in Fig. 12.19 and in the table opposite.

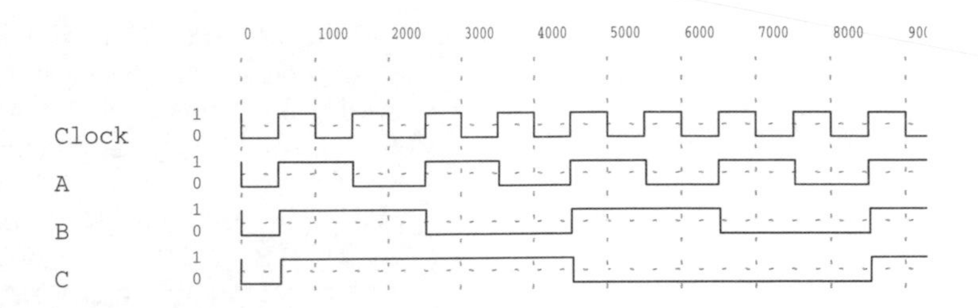

Figure 12.19 *When the clock is running at 1 MHz the output of the 3-bit counter of Fig. 12.18 apparently runs correctly through all stages from 000 to 111.*

Time (ns)	*C*	*B*	*A*	Count (decimal)
0000	0	0	0	0
1000	0	0	1	1
2000	0	1	0	2
3000	0	1	1	3
4000	1	0	0	4
5000	1	0	1	5
6000	1	1	0	6
7000	1	1	1	7
8000	0	0	0	0

The three outputs show binary values, equivalent to decimal values 0 to 7, and repeating. The chain of flip-flops is acting as a *counter.*

Another way of looking at the output of Fig. 12.18 is to consider the frequency of each output. The output of A is toggling every time the clock rises. It toggles back again the next time that the clock rises. This means that the frequency of output A is exactly *half* that of the clock. Similarly the frequency of output B is half that of A (1/4 the clock frequency), and C is half that of B (1/8 the clock frequency). The circuit acts as a *frequency divider.*

Activities 12.2 Flip-flops

Investigate each of the circuits listed below, either on a breadboard or by using a simulator, using the suggested ics. Consult data sheets for connection details. Most of the ics specified below contain two or more identical flip-flops, of which you will use only one. With CMOS it is essential for *all* unused inputs to be connected to the positive or 0 V rail.

Check out a D-type flip-flop (CMOS 4013, 74HC74, 74LS174, or 74LS175) as in Fig. 12.13. Use a slow clock rate, say 0.2 Hz. Connect the data input to the positive rails through a

1 kΩ resistor to pull it high. Use a push-button connected between the data input and the 0 V rail to produce a low input. The 74LS174 has only Q outputs but the others have Q and Q outputs. Investigate both clocked and direct modes.

Test a J-K flip-flop (CMOS 4027, 74LS73A, 74LS112A, 74LS379) as in Fig. 12.17. Connect J and K inputs with pull-up resistors and push-buttons, as described above.

Make a toggle flip-flop from a D-type flip-flop (Fig. 12.15) or a J-K flip-flop, and investigate its action.

Ripple counters

The clock is running at 1 MHz in the simulation pictured in Fig 12.19 so propagation delays of a new nanoseconds may be ignored. If we increase the clock rate by 10 times and also plot the outputs on a 10 times scale, propagation delays become relatively more important and we can see them in the plots (Fig. 12.20). The typical delay of the 74LS73 J-K flip-flop used in the simulation is 11 ns. The plot shows that each stage changes state 11 ms *after* the output from the previous stage has gone low. This shows up best after 800 ns, where all outputs are 1, then they all change to 0, one after the other. First the clock goes

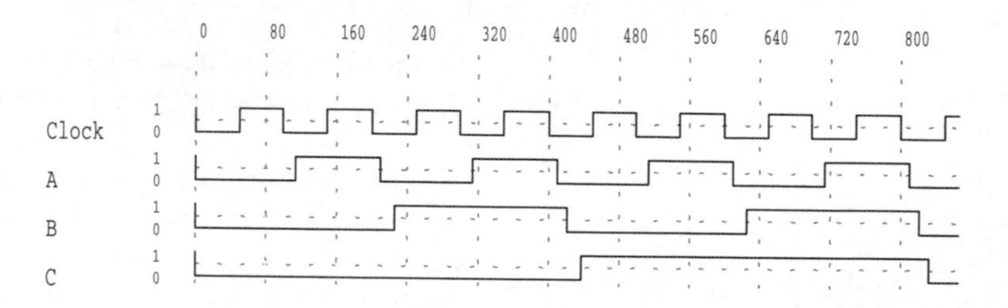

Figure 12.20 *When the clock speed is increased to 10 MHz, propagation delay becomes more noticeable and it can be seen that change of output from 1 to 0 'ripple' through the three flip-flops. This produces glitches in the counter output.*

low, then flip-flop A, then flip-flop B and finally flip-flop C. The change from 1 to 0 *ripples* along the chain. For this reason such a counter is known as a *ripple counter*. This example has shown only three flip-flops cascaded, but most counters available as ics have at least 4, and often 8 flip-flops. Some CMOS counters (for example, the 4020) have 14 stages, and can count up to $2^{14} -1$, which is equal to 16 383.

This table opposite shows the sequence of outputs in a ripple counter as the count changes from 111 to 000. The digits change from 1 to 0 in

Test your knowledge 12.8

What is the maximum count for a counter with 6 flip-flops?

Test your knowledge 12.9

How many flip-flops are cascaded to count from 0 to 31?

Maximum counts

If a counter has 4 flip-flops, the maximum count it can reach is 1111. This equals 15 in decimal. The next binary number after this is 10000, which is 16 in decimal or 2^4. So the maximum count of a 4-bit counter is $2^4 - 1 = 15$.

In general terms, a counter of *n* flip-flops can count up to

$$2^n - 1.$$

LSD: least significant digit.

order, beginning with the LSD on the right:

Digits	Decimal equivalent	
111	7	correct
110	6	error
100	4	error
000	0	correct

While changing from 7 to 0 the counter output goes very briefly through 6 and 4. This may not matter. For example, if the clock is running at 1 Hz, and the counter is driving an LED display that shows the number of seconds elapsed, we will not be able to see the incorrect '6' and '4'. The display will apparently change straight from '7' to '0'. But we might have a logic circuit connected to the output, intended to detect, say, a '6' output. This would easily be able to respond to the brief but incorrect output. This might cause it to trigger some other activity in the circuit at a time when it should not take place.

An instance of an error of this kind was shown in Fig. 11.4. In the simulation a ripple counter was used as a convenient way to provide inputs 000 to 111 to the logic network. However, when the counter input changed from 011 to 100 (at 800 ns), it went briefly through a stage 010. As can be seen from Fig. 11.4 at about 500 ns, this is an input which gives a low output from the network. Consequently, there is a low glitch on the output at 800 ns. The logic network has been fast enough to respond to the incorrect output.

Synchronous counters

Clocking all the flip-flops at the same time eliminates the ripple effect. Instead of letting each stage be clocked by the output from the previous stage, they are all clocked by the clock input signal. Fig. 12.21 shows such a counter made up from three J-K flip-flops. It is known as a *synchronous counter.* The J and K inputs of the first flip-flop are held high, as in the ripple counter, but the J and K inputs of following flip-flops are connected to the Q output of the previous stage. Fig. 12.22 shows the output of the synchronous counter, plotted with the same clock speed and time scale as Fig. 12.20. There is still an 11 ns

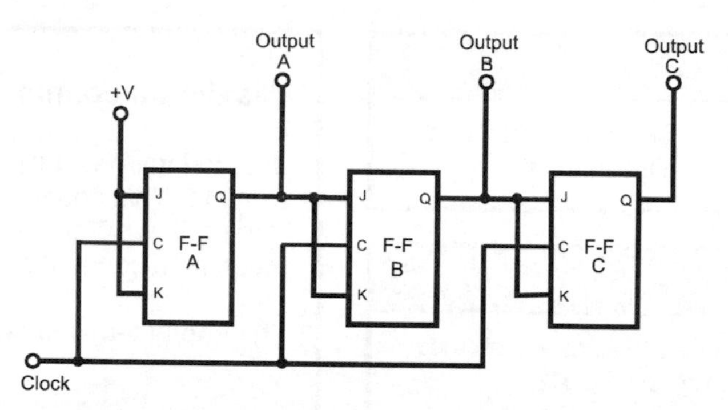

Figure 12.21 *A synchronous counter, in which all the flip-flops are clocked simultaneously, avoids the glitches that occur in ripple counters.*

delay between the clock going low and the change in the outputs of the flip-flops. But flip-flops change state at *the same time* (synchronously). This is clearly shown at 800 ns where all three change at once. The count goes directly from 111 to 000 with no intermediate stages.

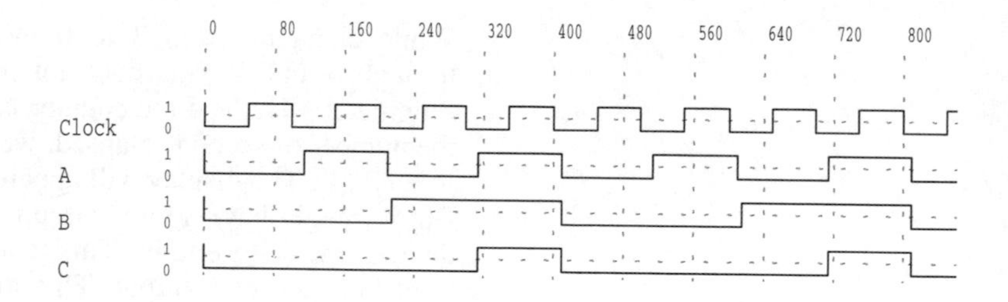

Figure 12.22 *This is the output from the synchronous counter of Fig. 12.21 when the clock is running at 10 MHz. Compare this with the output of the ripple counter in Fig. 12.20.*

Counters can also be built from D-type flip-flops, operating as toggle flip-flops.

Data registers

Another important use for flip-flops is as *data registers* or stores. A single D-type flip-flop can be used to store a single bit of data, as already described. An array of four or eight such registers, all connected to the same clock, can be used to store 4 bits or 8 bits of data. Data registers are available as MSI devices, such as the CMOS 4076 and the 74LS175 which store 4 bits.

Shift registers

Shift registers are often used to manipulate data. Fig. 12.23 shows a shift register built from four D-type flip-flops. The same clock triggers them all. At each rising edge of the clock the data present at the input

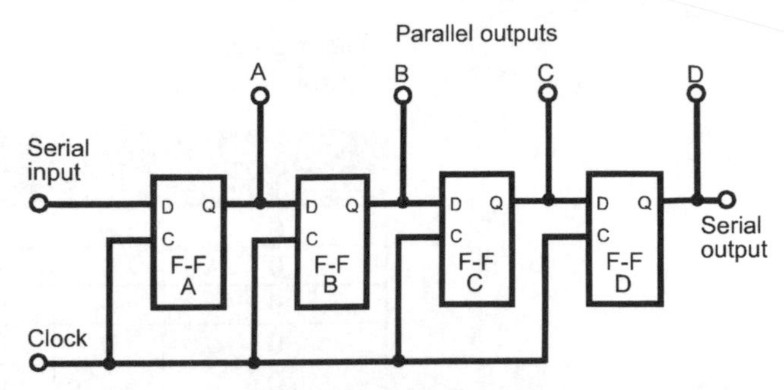

Figure 12.23 *A shift register can be built from D-type flip-flops. When they are clocked, data is shifted one step along the chain.*

is loaded into flip-flop A and appears at its Q output. At the same time, the data that was previously in A is shifted into flip-flop B. Similarly there is a shift from B to C. Data that was in D is replaced by the data that was in C, so the data that was in D is lost. If 4 bits of data are presented in turn at the input and the clock goes through 4 cycles, the registers hold all 4 bits of data. This table illustrates what happens when four bits of data (*wxyz*) are shifted serially into a register that is holding four 0's. We begin with the least significant bit, *z*:

Data at serial input	Registers			
	A	B	C	D
z	0	0	0	0
y	z	0	0	0
x	y	z	0	0
w	x	y	z	0
0	w	x	y	z

Going from one line of the table to the next represents a shift caused by the clock going high. Data is shifted into the register one bit at a time. This is called *serial input*. At any time the data in the four registers can be read from the four registers through their Q outputs. These are *parallel outputs*.

Serial and Parallel Data Transfer

Serial transfer: data is transferred, one bit at a time, along a single line.

Parallel transfer: data is transferred several bits at a time, along a set of parallel lines. The set of parallel lines is known as a *data bus*, and often consists of 8 or more lines.

The register also has a serial output. If we read the data at this output for four more clock cycles the bits are shifted out one at a time:

Data at serial input	Registers A	B	C	D	Data at serial output
0	w	x	y	z	z
0	0	w	x	y	y
0	0	0	w	x	x
0	0	0	0	w	w
0	0	0	0	0	Register empty

Shift registers are made as integrated circuits with varying features. The one described above operates in serial-in-parallel-out mode (SIPO) and also in serial-in-serial-out mode (SISO). There are also registers that operate in parallel-in-parallel-out (PIPO) and parallel-in-serial-out (PISO) modes. Registers also vary in length from 4 bits and 8 bits (the most common) up to 64 bits. Most are *right-shift* registers, like the example above, but some can shift the data in either direction.

Shift registers are used for data storage, especially where data is handled in serial form. The bits are shifted in, one at a time, held in the latches, and shifted out later when needed. PISO registers are used when data has to be converted from parallel to serial form. An example is found in a modem connecting a computer to the telephone network. The computer sends a byte of data along its data bus to the modem. Here it is parallel-loaded into a PISO register. Then the data is fed out, one bit at a time, and transmitted serially along the telephone line. The reverse happens at the receiving end. The arriving data is fed serially into a SIPO shift register. When the register holds all the bits of one byte, the data is transferred in parallel to the receiving computer's data bus.

Activities 12.3 Counters and registers

Investigate each of the circuits listed below, either on a breadboard or by using a simulator, using the suggested ics. Consult data sheets for connection details.

Build a ripple counter from three or four J-K flip-flops, as in Fig. 12.18 using 74LS73A or 74LS112A ics which are negative-edge triggered. Note that 74LS73A, has a K' input instead of K.

Investigate the action of a 4-stage MSI ripple counter such as the 74LS93 and the 7-stage CMOS 4024 counter.

Run the 4-stage MSI ripple counter at, say 10 kHz, then use an audio amplifier to 'listen' to the output from each stage.

Design and build a circuit to check the output of a 4-stage ripple counter for glitches. An example is the output '6' which occurs during the transition from '7' to '8'. Connect a set-reset flip-flop to the output of the detector circuit to register when the glitch has occurred.

By using logic gates to detect an output of 1100 (decimal 10), and reset the counter, convert the 74LS93 to a BCD (binary-coded decimal) counter with outputs running from 0000 to 1001.

Design and build a circuit to count in seven stages, from 000 to 110.

Investigate the action of a 4-stage MSI synchronous counter, such as the CMOS 4520 and the 74LS193A (up/down).

Use the circuit you devised in the previous investigation to check for output glitches in a synchronous counter.

Build a shift register from D-type flip-flops as in Fig. 12.23.

Demonstate the action of shift registers such as CMOS 4014 (PISO, SISO), CMOS 4015 (SIPO, SISO), 74LS96 (SIPO, SISO, PIPO, PISO), 74LS165A (PISO). A good way of displaying the parallel outputs is to connect each output to a transistor switch driving an LED. The shifting of data is revealed by the shifting pattern of lit and unlit LEDs.

Static RAM

For storing a large amount of data we use an array of many flip-flops on a single chip. A large array of this kind is usually called a *memory*. Each flip-flop stores a single bit of data. Each flip-flop can have data stored into it individually. The flip-flop is either set or reset, depending on whether the bit is to be a 0 or a 1. This is called *writing,* because the process of storing data has the same purpose as writing information on a piece of paper. Each flip-flop can also be *read* to find out whether the data stored there is 0 or 1. A RAM chip may have several hundreds or even thousands of flip-flops on it. To write to or read from any particular one of these, we need to know its *address*. This is a binary number describing its location within the array on the chip. This is discussed in more detail later.

Because at any time we can write to or read from any flip-flop we choose, we say this is *random access*. A memory based on this principle is a *random access memory*, or RAM. This is different from a SISO shift register, for example, from which the bits of data must be read in the same order in which they were written. Once data is stored in a flip-flop it remains there until either it is changed by writing new data in the same flip-flop, or the power supply is turned off and all data is lost. This type of RAM is known as *static RAM*, or *SRAM*.

Dynamic RAM

There is another type of RAM known as *dynamic RAM,* or *DRAM*. In this, the bits are not stored in flip-flops but as charges on the gate of MOSFET transistors. If a gate is charged, the transistor is on; if the gate has no charge, the transistor is off. Writing consists of charging or discharging the gate. Reading consists of finding out if the transistor is on or off by registering the output level (0 or 1) that it produces on the data line. The storage unit of a DRAM is smaller than a flip-flop, so more data can be stored on a chip. Also it takes less time to write or read to a DRAM, which is important in today's fast computers. The main problem with DRAMs is that the charge on the gate leaks away. This means that the data stored there must be refreshed at regular intervals. Under the control of a clock running at about 10 kHz, the charge is passed from one transistor to another, and the amount of charge is topped up to the proper level at each transfer. The data is kept 'on the move', giving this type of memory the description 'dynamic'. The need to refresh the memory regularly means that a certain amount of the computer's operating time must be set aside for this, and the RAM can not be used for reading and writing data during this time. The operating system is therefore more complicated than that needed for SRAM, which is available for use at any time.

Memory control

Different types of RAM store different numbers of bits arranged in different ways. We take the 2114 RAM as an example. This is not *typical* since it is one of the smallest memory ics commonly available. But it illustrates the principles of storing data. The 2114 stores 4096 bits (usually referred to as 4 kbits), which are writable and readable in groups of four. We say that the chip stores 1024 4-bit *words.* The chip has four data terminals (D0 to D3) through which a word may be written or read (Fig. 12.24). Since there are 1024 words, each needs its own address. 1024 is 2^{10} so we need 10 address lines, A0 to A9. Then we can input every address from 0 (binary 0 000 000 000, all inputs low) to 1023 (binary 1 111 111 111, all inputs high).

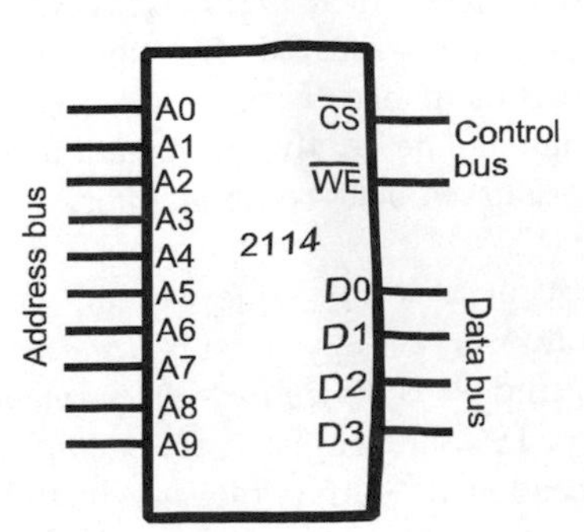

Figure 12.24 *The ten address inputs of the 2114 SRAM allow for storage of 2^{10} (= 1024) words of data, each of 4 bits.*

Bus: a set of parallel lines in a large logical circuit, each one of these carrying one bit of data, or of an address, or a control signal.

As well as the connections to the data and address busses, there are two pins connected to the control bus. One of these is $\overline{\text{CHIP SELECT}}$ ($\overline{\text{CS}}$). The bar over its name and symbol indicate that this must be made low to have the desired effect. When it is high the chip is unselected; its data terminals are in a high-impedance state so the chip is effectively disconnected from the data bus. Before we can write to or read from the RAM the $\overline{\text{CS}}$ input must be made low.

The data terminals act either as inputs or outputs, depending on whether the $\overline{\text{WE}}$ (write enable) pin is made high or low. With $\overline{\text{CS}}$ low and $\overline{\text{WE}}$ high, the chip is in read mode. Next we put a 10-bit address on the address bus. This selects one of the words and the 4-bits of data that it comprises are put on to the data bus. With $\overline{\text{CS}}$ low and $\overline{\text{WE}}$ low, the chip is in write mode. Data present on the bus is written into the addressed word in memory and stored there.

Often in computing circuits we handle data in bytes, consisting of 8-bits. Some RAMs are organised to store data in this format. For example, the 6264 RAM stores 64 kbits, organised as 8192 bytes, each of 8 bits. This needs 8 data input/outputs and 13 address lines. In most SRAMs the bits are addressed individually, not as words or bytes. The 4164 SRAM for example stores 64 kbits as single bits. The chip needs only one data input/output, but 16 address lines. On the whole SRAMs have shorter access times than DRAMs. The 2114 is relatively slow, taking 250 ns to write or read data. Most DRAMs are faster than this, averaging around 100 ns. By contrast, most SRAMs have access times of 60 ns.

Activities 12.4 RAM

Investigate the action of a SRAM such as the 2114. A circuit which can be set up on a breadboard or simulator is shown in Fig. 12.25. The CHIP SELECT' input is held low permanently, and so are all but one of the address inputs. Addressing is controlled by the flip-flop. When this is reset the address is 0 and when it is set the address is 1. The flip-flop is set or reset by briefly shorting one of its inputs to 0 V.

When S1 is not pressed the WRITE ENABLE' input is high, putting the chip in read mode. Data already stored at the selected address as a 4-bit word is displayed on the LEDs.

To write data into an address:

Select the address.

Set up the data that is to be written by temporarily connecting the data terminals to +5V (1) or 0V (0). On a breadboard use flying leads.

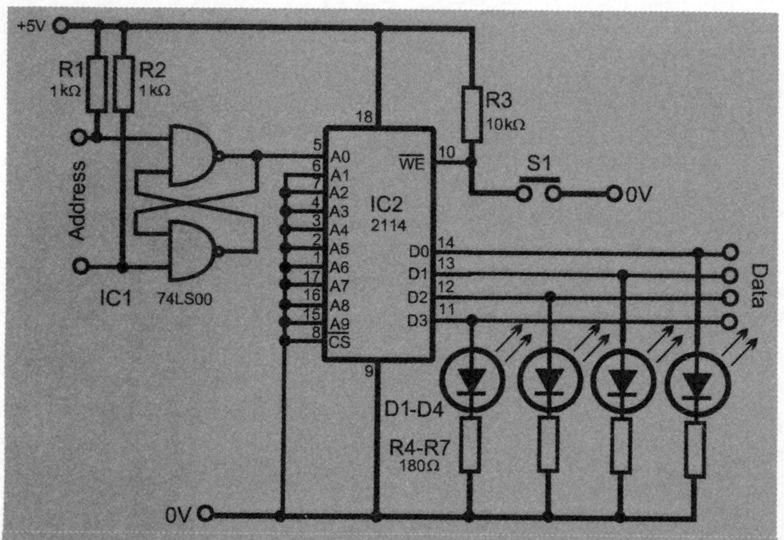

Figure 12.25 *A breadboarded circuit for investigating the action of the 2114 SRAM.*

Press and release S1. This writes the data into the RAM.

Remove the connections (flying leads) from the data terminals.

Read the newly stored data on the LEDs.

If you select the other address, then return to the same address, the stored data will still be there. Experiment with storing different values at the two addresses.

Add another flip-flop connected to A1. It is then possible to address four words (00, 01, 10 and 11). In these addresses store the data for a traffic-light sequence, assuming D0 is red, D1 is yellow and D2 is green (D3 is not used). By running through the addresses in order you can simulate a changing traffic light.

Devise a circuit to operate these traffic lights automatically.

Problems on logical sequences

1 Explain the working of a flip-flop built from two NAND gates.

2 Describe how the 555 ic is used as a pulse generator.

3 Explain how to make a pulse generator from two NAND gates, a capacitor and a resistor.

4 Design an astable circuit using the 555 ic to obtain a mark-space ratio of less than 1.

5 Describe how to make an astable using two CMOS NOT gates, a capacitor and two resistors.

6 Show how to build a data latch using logic gates and use a truth table to explain how it works.

7 What are the advantages of using clocked master-slave logic devices.

8 Explain what a J-K flip-flop does and how three such flip-flops may be connected to make a ripple counter.

9 What is the main drawback of a ripple counter? Explain how a synchronous counter overcomes this problem.

10 Draw a circuit diagram for a serial-in parallel-out shift register, using 4 D-type flip-flops. Explain its action by means of a table.

11 Explain the difference between static and dynamic RAM. What are the advantages and disadvantages of each type?

Multiple choice questions

1 A logical circuit with two stable states is NOT called a:

A monostable.
B flip-flop.
C multivibrator.
D bistable.

2 If the supply voltage is 12 V, the voltage across the capacitor of a 555 monostable as the pulse ends is:

A 12 V.
B 4 V.
C 3 V.
D 8 V.

3 A 555 monostable has a 3.3 nF capacitor and a 15 kΩ resistor. The supply voltage is 15 V. The length of the pulse is:

A 49.5 μs.
B 34.65 μs.
C 495 μs.
D 54.45 μs.

4 If the supply voltage is reduced to 7.5 V in the monostable described in question 3, the length of the pulse is:

A unchanged.
B doubled.
C halved.
D divided by 3.

5 If the symbol for an input terminal has a bar drawn over it, it means that:

A the input must always be low.
B the input does nothing when it is high.
C a negative input produces a negative output.
D the input is made low to bring about the named response.

6 Which of these statements about a toggle flip-flop is NOT true?

A The $\overline{Q}$ output is the inverse of the Q output.
B The output follows the D input.
C Clocking reverses the state of both outputs.
D The $\overline{Q}$ output is connected to the D input.

7 A J-K flip-flop has its J and K inputs connected to the 0 V line. Its Q output is high. It is triggered by positive-going edges. When will its Q output go low?

A On the next positive-going clock edge.
B When the Set input is made high.
C On the next negative-going clock edge.
D Never.

8 A RAM has 11 address inputs and 4 data inputs/outputs. The number of bits that it stores is:

A 8192.
B 2048.
C 44.
D 4096.

13 Display devices

Summary

The two most often used kinds of display are those using light-emitting diodes and the liquid crystal displays. The format for both types is usually the 7-segment display, which produces the figures 0 to 9 and certain alphabetic characters. Two other formats are the starburst and dot matrix display, both of which can produce figures and all alphabetic characters. The production of characters is controlled by special decoder ics.

Light-emitting diodes

Light-emitting diodes, or LEDs, are widely used as indicators and in displays. Although filament lamps and neon lamps are still used, they have been replaced by LEDs in very many applications. LEDs outlast filament lamps, are cheaper and require very little current. They are available in a wide range of shapes, sizes and colours.

An LED has the electrical properties of an ordinary rectifying diode. The most important practical difference (apart from their production of light) is that they are able to withstand only a small reverse voltage. For most LEDs the maximum reverse voltage is 5 V, so care must be taken when using LEDs to mount them the right way round.

LEDs are usually operated with a forward current of 20 mA, with a maximum of up to 70 mA, depending on the type. There are also low-current types that operate with 7 mA. But if power is limited, for instance in battery powered equipment, most types of LED give a reasonable light output with as little as 5 mA. The current through the LED must generally be limited by wiring a resistor in series with it (see box on p. 212).

The brightness of an LED is specified in candela (cd) or millicand (mcd). Typical LEDs are in the range 2-10 mcd. Super-bright LEDs h

Series Resistor

To calculate the value of the resistor to place in series with an LED we need to know:

- the current required.
- the supply voltage.
- the forward voltage drop across the LED.

The forward voltage drop depends on the current and the type of LED. It is typically 2 V when the LED is passing 20 mA. This is higher than for a typical diode, for which it is usually between 0.6 V and 0.7V.

The value of the series resistor is:

$$R = (\text{supply} - \text{drop})/\text{current}$$

Example

Calculate the series resistor required when the current is 20 mA, the supply voltage is 9 V, and the forward voltage drop (at 20 mA) is 2 V.

$$R = (9 - 2)/0.02 = 350\ \Omega$$

The nearest E24 value is 360 Ω, but 330 Ω or 390 Ω would usually be close enough.

Test your knowledge 4.1

Calculate the series resistor required for an LED when the curent is 15 mA, the supply voltage is 6 V and the forward voltage drop is 1.6 V.

brightnesses of tens or even a few hundred millicandela, while the brightest types are rated at several candela. Viewing angle is also important, for some types achieve high brightness by concentrating the light into a narrower beam. This may make them less conspicuous in a display if the viewer is not directly in front of the panel. Standard LEDs have a viewing angle of 120°, but the angle may be as small as 15° in some of the brighter types.

Hexadecimal: a number system based on powers of 16. The digits are 0 to 9 and A to F. Digits A to F are the equivalents of decimal numbers 10 to 15.

When a piece of equipment, such as a clock or a frequency meter, needs to display numerical values it may use an LED display in the seven segment format (Fig. 13.1). The segments are called *a* to *g*, as shown in the figure. The segments may be lit in various patterns to produce the digits 0 to 9, as in Fig. 13.2. It is also possible to produce capital letters A to F for use in displaying hexadecimal numbers.

Seven-segment displays

Seven-segment displays are made in a wide range of types. Most are red but some are green or yellow. The height of the display ranges from 7 mm to 100 mm. Most have a single digit but there are some with 2 or 4 digits. The decimal point is an eighth LED, usually round

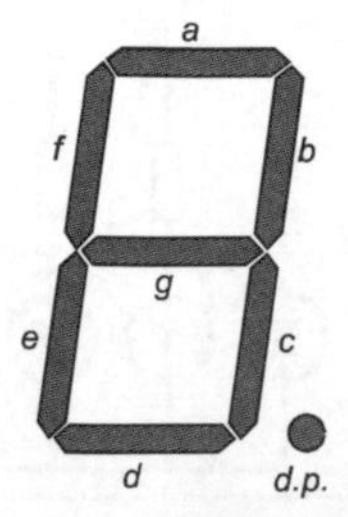

Figure 13.1 *Seven bar-shaped LEDs make up a 7-segment display for figures 0 to 9. There is also a decimal point, usually on the right, though some displays have a second one on the left.*

Figure 13.2 *This is how a seven-segment display shows the figures 0 to 9. The 6 and 9 may sometimes have no tails (segments* a *and* d *respectively).*

in shape. All displays fall into two categories, *common anode* or *common cathode*. In the common anode display the anodes of each LED are connected internally to a common terminal (Fig. 13.3). The segments are lit by drawing current through their cathodes to current *sinks*. Each segment needs a series resistor. The common cathode display has the cathodes of all segments connected internally to a common terminal (Fig. 13.4). Current *sources* are connected to the anodes, each through a series resistor.

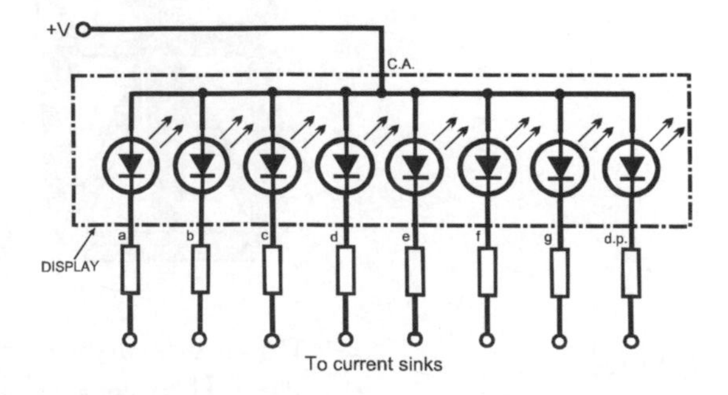

Figure 13.3 With a *common-anode display, all anodes are connected to the positive supply and current is drawn from the individual cathodes through the series resistors.*

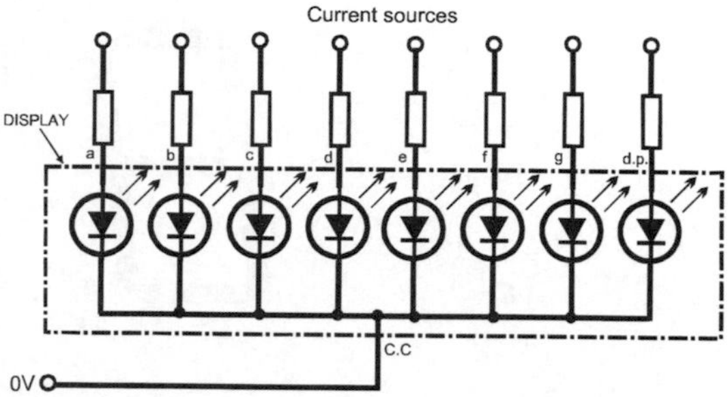

Figure 13.4 *With a common cathode display, all cathodes are connected to the 0 V rail and current is supplied to the individual anodes through the series resistors.*

Starburst display

The starburst LED display consists of 14 segments arranged as in Fig. 13.5. This is able to produce any figure or alphabetic character, and many other symbols. Another type of display that can do the same thing is the *dot matrix display.* This usually has 35 circular LEDs arranged in 5 rows and seven columns. The cathodes of all LEDs in the same column are connected to a single line and there is a separate line for each column. The anodes of all LEDs in the same row are connected to a single line and there is a separate line for each row. To light a given LED we supply current (through a series resistor) to its cathode line and make its anode line low. The LEDs forming the character are lit one at a time but in rapid succession. This action is repeated so quickly that each illuminated LED appears to glow steadily and all appear to be on at the same time.

> **Test your knowledge 13.2**
>
> Draw diagrams to show how the letters of your name can be formed on a starburst display.

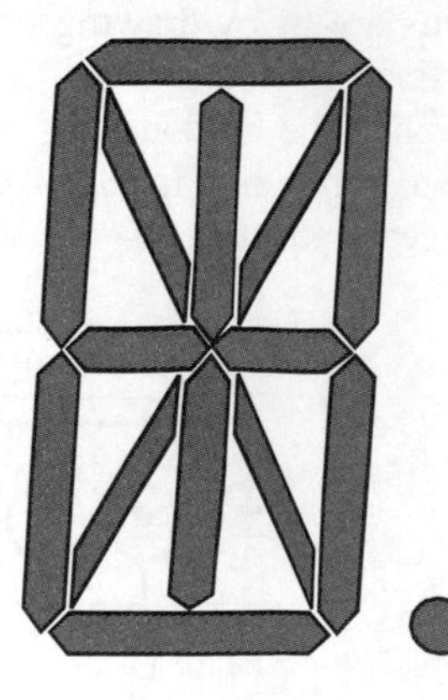

Figure 13.5 *A starburst display can produce any figure, or alphabetic character, and many symbols.*

Dot matrix display

Dot matrix displays require complicated logic circuits to drive them. There are special LSI or VLSI ics for this purpose, or the task may be performed by a microcontroller. The dot-matrix system is ideal for travelling-message displays and other effects in which the characters appear to move sideways or up and down, to flash or to change size.

Activity 13.1 LED displays

Investigate the action of various MSI devices used for driving seven-segment LED displays. Connect a display to the outputs of the ic, remembering to include series resistors of suitable value. Then see the effect of different combinations of inputs.

Devices to work with include:

Open-collector outputs: require a pull-up resistor (see Fig. 10.2).

Active-low: the output goes low if a segment is to be switched on.

74LS47: BCD decoder-driver, with open-collector outputs. The outputs are active-low so a common-anode display is required. The input is a 4-bit binary value in the range 0000 to 1001 (decimal 0 to 9). Inputs 1010 to 1111 produce 'nonsense' output on the display. This device has ripple-blanking and lamp test inputs. Try cascading two or more devices and use the ripple-blanking output and input.

CMOS 4033: decade counter/decoder. This is a synchronous decade counter (0 to 9) which has its outputs decoded to drive a 7-segment display. Outputs are active-high so a common-cathode display is used. The ic does not provide high output currents but can be used with low-current displays. It has ripple-blanking and lamp test inputs. Try cascading two or more devices and use the ripple-blanking output and input.

4511: Seven-segment decoder/latch/driver. Outputs are active-high, so use a common-cathode display. Series resistors are essential. Display goes blank for inputs 1010 to 1111. It has a STORE input to latch the display while input changes.

Liquid crystal displays

An LED display is difficult to see if exposed to bright light but has the advantage that it is highly visible in the dark. A liquid crystal display (LCD) has the opposite features. It is difficult to see in dim light but easy to read in direct sunlight.

An LCD consists of a layer of liquid crystal sandwiched between two layers of glass. Conductive electrodes are printed on the inside of the front glass, in the format of a 7-segment display. Very narrow printed conductive strips connect each segment to its terminal at the edge of the display. There is a continuous but transparent coating of conductive material on the inside of the back glass, forming the *back plane*. Normally the liquid crystal is transparent but, if an electric field exists between the back plane and one of the electrodes, the material becomes (in effect) opaque in that region. The connecting strips are too narrow to produce an image. Thus the characters appear black on a light grey ground. The display can be read by reflected light. In darkness, it may have a back light to make it visible, but the display is not as easy to read under these conditions as an LED display.

The action of the LCD depends on an *alternating* electric *field* between the back plane and the electrodes. The segment electrodes must be made alternately positive and negative of the back plane. If this is not done, the electrodes become plated and the device eventually stops working. The driving circuit requires an oscillator running at 30 Hz to 200 Hz. This can be built from two NOT gates as in Fig. 12.9. It is the *field* (or potential difference) that produces the effect, but the amount of *current* passing through the material is very small. Because of this, the power required for driving LCDs is very much less than that used by an LED display. A typical 4-digit LCD needs only 2 or 3 microamperes compared with at least 10 mA *per segment* for an LED display. LCDs are ideal for portable battery-powered equipment such as watches, clocks, calculators, kitchen scales, and multimeters.

The fact that the electrodes only have to be printed on the glass means that it is inexpensive to produce a new LCD design for a new product. Setting up the manufacture of a new design of LED display is much more difficult and expensive. As a result of this, a varied range of LCDs is obtainable. Displays are made with 2 to 6 digits, usually including a colon in the centre for timing applications. There may often be a plus and minus sign for metering applications, and short messages or symbols indicating conditions such as 'alarm on', 'snooze', 'flat battery', and 'overflow'.

LCDs are also made in the dot matrix format. Some of these are able to display messages of up to 4 lines of up to 20 characters. In calculators and meters the LCD may be used for displaying graphs. Such complicated displays are usually controlled by a microcontroller.

Activity 13.2 Liquid crystal displays

Investigate how LCDs are driven by MSI and LSI devices. Among the MSI devices are:

CMOS 4056: This has a 'Display frequency in' input at pin 6. This is supplied from the astable clock (such as Fig. 12.9) which also goes to the back plane. The outputs driving the LCD segments which are to be displayed are *180° out of phase* with the clock signal. This causes an alternating voltage between the segments and the back plane. The segments appear black. The outputs driving the other segments are *in phase* with the clock signal. There is zero volts between these segments and the display. The segments are invisible. This ic can also be used for LED displays.

If pin 6 is connected to 0 V, outputs are active-high and are used to drive a common-cathode display. With pin 6 connected to the positive display, outputs are active low, and suitable for a common-anode display. This ic requires three power supply lines, +5 V, 0 V and –5V.

CMOS 4543: This is similar to the 4056, but does not require the –5 V supply.

Problems on display devices

1 Make a systematic list of the different types of displays and indicators used in equipment, including domestic, business, industrial, leisure, and sporting equipment

2 What is a light-emitting diode? Outline its characteristics.

3 Draw a diagram of a seven-segment display and explain how LEDs arranged in this format are used to display the figures 0 to 9.

4 Describe the action of an ic used for controlling a seven-segment LED display.

5 Compare the features of LED displays and liquid crystal displays.

6 Draw the format of a 7 × 5 dot-matrix display, showing how it displays a letter *d*.

Multiple choice questions

1 Outputs *c, d, e, f,* and *g* of a 7-segment decoder ic are at logic high, the rest are at logic low. If a common-cathode display is connected to the ic, the figure displayed is:

A 6.
B 2.
C 0.
D 9.

2 Outputs *e* and *f* of a 7-segment decoder ic are at logic high, the rest are at logic low. If a common-anode display is connected to the ic, the figure displayed is:

A 4.
B 3.
C 1.
D 0.

3 LCDs are the best kind of display for a digital watch because they:

A are easy to read in the dark.
B are resistant to damage.
C take very little current.
D can be made very small.

14 Conversion devices

Summary

A flash converter is the fastest way of converting an analogue signal to its digital equivalent. For higher precision, but with longer conversion time, the successive approximation technique may be used. Digital signals may be converted to analogue form using an operational amplifier adder circuit with weighted inputs. Greater precision is obtainable with an R-2R ladder converter.

Analogue-to-digital converters

Flash converter

For many applications, such as converting analogue audio signals into digital signals in real time, a high-speed converter is essential. A flash converter consists of a number of comparators connected at their (+) inputs to the input voltage v_{IN}, which is the voltage to be converted (Fig. 14.1). Their (–) inputs are connected to a potential divider chain consisting of 8 resistors. One end of this is at 0 V and the other is connected to a reference voltage v_{REF}. The voltage at successive points in the chain increases in steps from 0 V to v_{REF}.

There are three cases:

- If v_{IN} is 0 V or close to 0 V, the (+) inputs of all comparators are at a lower voltage than the (–) inputs. As a result, the outputs of all comparators are low (equivalent to logic low, or 0).
- At the other extreme, if v_{IN} is close to v_{REF}, the (+) inputs of all comparators are at a higher voltage than the (–) inputs. All outputs are high (equivalent to logic high, or 1).
- At intermediate input voltages there are comparators at the lower end of the chain, which have high outputs, and comparators at the top end which have low outputs (Fig. 14.2).

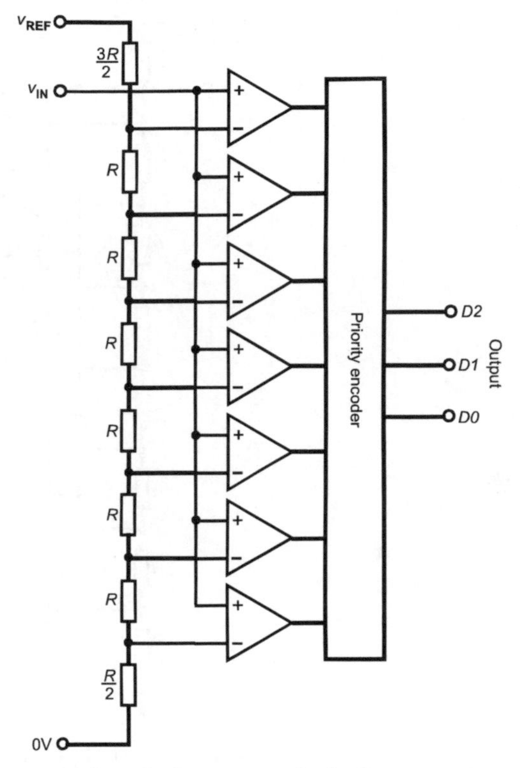

Figure 14.1 *A flash converter is the fastest analogue-to-digital converter, but it needs several hundred comparators to give high precision.*

The result is that, as v_{IN} is increased from 0 V to v_{REF} the outputs change like this:

00000000 when $v_{IN} = 0$ V
00000001
00000011
00000111
00001111
00011111
00111111
01111111
11111111 when $v_{IN} = v_{REF}$

The outputs are passed to a priority encoder. This has the same action as the circuit of Fig. 11.17, but has 8 inputs and 3 outputs. The outputs indicate the highest-ranking high input. As v_{IN} is increased from 0 V to v_{REF} the binary output increases from 000 to 111 (decimal 0 to 7). The output from the encoder indicates the voltage range that includes v_{IN}.

In Fig 14.2 the reference voltage is 8 V to make calculating the ranges easier to understand, but it could be any other reasonable voltage.

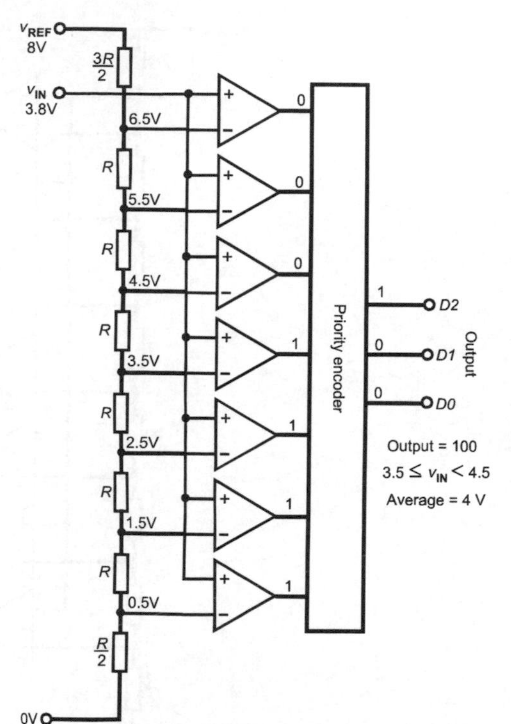

Figure 14.2 *The comparators at the lower end of the chain (least significant bits) have high output, and those at the upper end (most significant bits) have low output. Here, with v_{IN} = 3.8 V, the lower four have high outputs and the upper three have low outputs. The output of the encoder is 100 (decimal 4), which is equal to the input voltage, to the nearest volt.*

The value of R (within reason) does not affect the working of the circuit. The essential point is that all resistors must be *precisely* R, or R/2, or 3R/2. This is usually easy to arrange if all the resistors are made on the same chip.

Voltage ranges

The resistors in the chain all equal R, except for those at the top and the bottom. Using the formula for potential dividers and numbering the resistors from 0 to 7, starting at the 0 V end, the voltage at a point between the *n*th resistor and the (*n*+1)th resistor is:

$$v_n = v_{REF} \times \frac{nR + R/2}{6R + R/2 + 3R/2} = v_{REF} \times \frac{2n+1}{16}$$

This formula gives the voltages marked beside the (–) inputs of the comparators in Fig 14.2. Because the lowest resistor is R/2, the voltage steps up from 0.5 V in 1 V steps. Thus the ranges run from 0 V to 0.5 V (output 0), 0.5 V to 1.5 V (output 1), 1.5 V to 2.5 V (output 2) and so on up the chain. At all stages, except the last, the output equals the *average* voltage over the range. Or, in other words, it indicates the voltage *to the nearest volt.*

A flash converter takes only the time required for the comparators to settle, plus the propagation delay in the gates of the encoder. Typically, a flash converter produces its output in 10 ns to 2 μs, depending mainly on the number of bits. It can convert signals in the megahertz ranges and does not need the signal to be held in a sample-and-hold circuit while it does the conversion. The main disadvantage of the 3-bit circuit in Fig. 14.1 is that it has only 8 possible output values, 0 to 7. Yet the analogue input varies smoothly over a range of several volts, with an almost infinite range of values. The ADC in Fig. 14.1 has 7 converters to produce a 3-bit output. The rule is that it takes $2^n - 1$ comparators to produce an n-bit output.

Example

A flash ADC with 8-bit output requires $2^8 - 1 = 255$ comparators. These give 256 possible output values, from 00000000 to 11111111 (0 to 255 in decimal).

Even this does not give really high precision. Flash ADCs are made for 4-bit, 6-bit and 8-bit conversion. Some of the 8-bit converters and all of those with more bits (the largest have 12 bits) employ a technique known as *half-flash*. This is a compromise that requires fewer comparators but works in two stages and therefore takes longer.

Successive approximation converter

A successive approximation ADC operates on an entirely different principle. Fig. 14.3 is a block diagram of its main sections. The heart of the DAC is the *successive approximation register* which holds a binary value, usually with 8 or 16 bits. The control logic is driven by the clock and sets the register to a series of values in a systematic way. It gradually 'homes' on the value that is the digital equivalent of v_{IN}.

When conversion begins, the register is set to its 'half-way' value. With 4 bits, this is 1000. If the maximum input voltage is 2 V (approximately, see later) this is equivalent to 1 V. As well as going to the terminal pins of the ic, the output from the register goes to a digital-to-analogue converter (DAC), which is also present on the chip. This converts the digital value to analogue and produces an output of 1 V. This goes to a comparator, where it is compared with v_{IN}. The result of this comparison tells the logic whether to increase or decrease the value in the register, so as make it closer to v_{IN}. This is what is meant by successive approximation.

Example

Given that v_{IN} = 4.8 V, the successive approximations are as shown in the table overleaf.

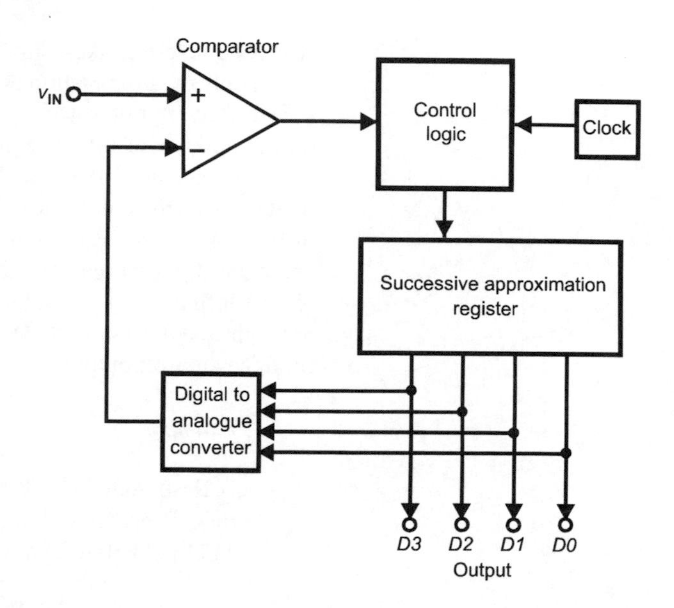

Figure 14.3 *A successive approximation ADC systematically adjusts the content of the register until the output of the DAC is as close as possible to the input voltage v_{IN}.*

Clock cycle	Register set to	DAC output (V)	v_{IN} lower or higher?	Next step?	Next register setting
1	1000	4	higher	increase register	1100
2	1100	6	lower	decrease register	1010
3	1010	5	lower	decrease register	1001
4	1001	4.5	higher	increase register	1010

The control logic takes each bit of the register, one bit per clock cycle. It starts with the most significant bit. To increase the register, the next bit to the right is set to '1'. To decrease the register, the current bit is reset to '0' and the next bit to the right is set to '1'. In this example, the output at the end of 4 clock cycles is 1010, equivalent to 5 V.

The result is to the nearest bit. In this example 1000 is equivalent to 4 V, so the least significant bit is equivalent to 0.5 V. The output is read as 5 V, which means that the input may actually have any value between 4.5 V and 5.5 V.

Example

Given that v_{IN} = 3.8 V, the table below shows the stages of the conversion:

Clock cycle	Register set to	DAC output (V)	v_{IN} lower or higher?	Next step?	Next register setting
1	1000	4	lower	decrease register	0100
2	0100	2	higher	increase register	0110
3	0110	3	higher	increase register	0111
4	0111	3.5	higher	increase register	1000

The first approximation takes the DAC output down too far, but it gradually returns to 4 V, which is the closest to the input of 3.8 V.

As with the flash converter, precision is low with a small number of bits. But increasing the bit number is only a matter of lengthening the register. ADCs are made with up to 18 bits and a conversion time of 20 μs. Some ADCs of this type are not as fast as this, even those with fewer bits, and may take up to 100 μs to convert.

The searching routine of the successive approximation method is unsuited to a rapidly changing input. For example, if the input is 3.8 V as in the example tabled above, the first clock cycle decreases the DAC output from 4 V to 2 V. Suppose that v_{IN} rises to 6 V before the second clock cycle. This could happen because the signal is changing rapidly or perhaps because of an interference 'spike' on the signal. The first digit is now set at '0' so the DAC output can never rise higher than 4 V, even with carry-over. The output error is 2 V. It is not possible for the searching routine to home on a moving target. For this reason, it is usual to employ a sample-and-hold circuit (Fig. 7.3) to hold v_{IN} while it is being converted.

Activities 14.1 Analogue-to-digital converters

On a breadboard or simulator, set up a 2-bit flash converter, complete with priority encoder. Measure the input voltage ranges needed to produce the output states 00, 01, 10 and 11.

Investigate the action of a 4-bit, 6-bit or 8-bit ADC using the manufacturers' data sheet as a guide. Supply the input from a variable resistor connected across the supply lines to act as a potential divider. Use a multimeter to measure the input voltage and monitor the digital outputs using a probe.

Using the manufacterers' data sheets, investigate the action of a flash converter ic. The CA3304E is a 4-bit flash ADC, and the CA3306E converts to 6 bits. They include a zener diode which can be used to provide a reference voltage.

The ACD0804LCN is an 8-bit ADC using the successive approximation technique. The digital output is held in latches which makes this device suitable for connection to the data bus of a microcontroller system. Investigate its action on a breadboard.

Digital-to-analogue converters

Operational amplifier adder

The operational amplifier adder (Fig. 6.12) is a simple way of converting digital signals to their analogue equivalents. In this application (Fig 14.4) the data inputs are logical inputs, so their voltage is either 0 V (logical 0) or a fixed high voltage v_H (logical 1). The resistors are *weighted* on a binary scale so that:

$$R_3 = 2R_F \qquad R_1 = 2R_2 = 8R_F$$
$$R_2 = 2R_3 = 4R_F \qquad R_0 = 2R_1 = 16R_F$$

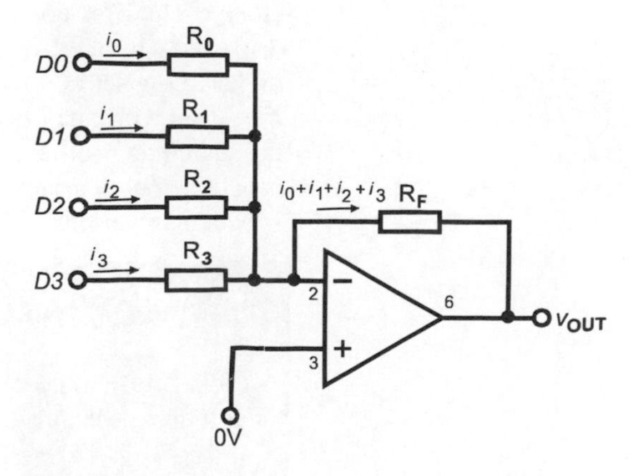

Figure 14.4 *An operational amplifier adder with binary weighted inputs makes a digital-to-analogue converter.*

Example

If $R_F = 10$ kΩ, then $R_3 = 20$ kΩ, $R_2 = 40$ kΩ, $R_1 = 80$ kΩ, and R_0 = 160 kΩ.

For any given data input, the current flowing through the input resistor is v_H divided by the resistance. If resistances are doubled at each stage, as above, currents are halved. For a high data input, the current through R_2 is half that through R_3, the current through R_1 is half that through R_2 and so on.

Example

If $R_F = 10$ kΩ and $v_H = 5$ V, then:

$$i_3 = v_H/R_3 = v_H/2R_F = 5/20000 = 250\ \mu A$$

Similarly, because of the doubling of resistances, the currents corresponding to other high-level data inputs are:

$$i_2 = 125\ \mu A$$
$$i_1 = 62.5\ \mu A$$
$$i_0 = 31.25\ \mu A$$

When a data input is low, the current through the resistor is zero. When more than one data input is high, the current through R_F is the total of the input currents. The output voltage of the converter is therefore proportional to the sum of the inputs, the inputs being weighted on the binary scale.

Example

Given a data input 1010, calculate the output, assuming other conditions are as set out above.

The total current is the sum of i_3 and i_1. Total current = 250 + 62.5 = 312.5 μA. $R_F = 10$ kΩ. The circuit is inverting, so output is −312.5 × 10 [μA × kΩ] = −3.125 V.

Test your knowledge 14.1

Given the same converter details, calculate the output voltage when the data input is 1011.

The output represents the binary input on a scale in which v_H is equivalent to binary 10000 (decimal 16). We therefore have to multiply the output voltage by $-16/v_H$.

Example

In the example, the multiplier is −16/5 = −3.2.
Multiplying the output voltage:

$$-3.125 \times -3.2 = 10 \text{ (exactly)}.$$

This is the same value as binary 1010.

Another example

Using the same circuit with data input 0111.

Total current = $i_2 + i_1 + {}_{i0}$ = 125 + 62.5 + 31.25 = 218.75 μA
Output voltage = 218.75 × 10 [μA × kΩ] = –2.1875 V
Multiplying the output voltage:

$$-2.1875 \times -3.2 = 7 \text{ (exactly)}$$

This is the same value as binary 0111.

Integer: whole number.

This method of conversion is almost instantaneous, taking only the time required for the amplifier to settle. The main limitation is that the circuit needs an input resistor for each binary digit. With four resistors, as in the figure, the maximum value that can be converted is 1111. Only the integer values from 0 to 7 can be converted. To convert any integer in the range 0 to 255 requires 8 data inputs. When there are many data inputs, the difficulty is obtaining a set of resistors that is sufficiently precise in value. In the 4-bit converter a 1% error in R_3 will make the current through it too large or too small by 25 μA. This is almost as great as the current through R_0, leading to an error of almost ±1 bit.

R-2R ladder converter

Most DACs are based on the R-2R 'ladder'. The ladder is fabricated as part of the ic and consists of resistors of value R and 2R only. The *exact* value of R does not matter provided that all R resistors have the *same* value and all 2R resistors are exactly twice that value. This is much easier to do than to provide a set of precisely weighted resistors as required for the adder converter described above. Usually R is in the region of 10 kΩ and 2R is 20 kΩ.

Fig. 14.5 illustrates the circuit of a 4-bit converter, though most DACs of this type have 8, 12 or even as many as 18 bits. The switches shown in the circuit are actually CMOS switching devices under logical control. The switches may be switched either to the 0 V rail (representing a data input of 0), or to the inverting input of the op amp (representing a data input of 1). Because the inverting input of the op amp is a virtual earth, it makes no difference to the flow of current in the resistor network whether the switches are to 0 V or to the inverting input. Whichever way the switch is set, the current flowing through the switch is the same. In Fig. 14.5 the positions of the switches correspond to a data input of 1010.

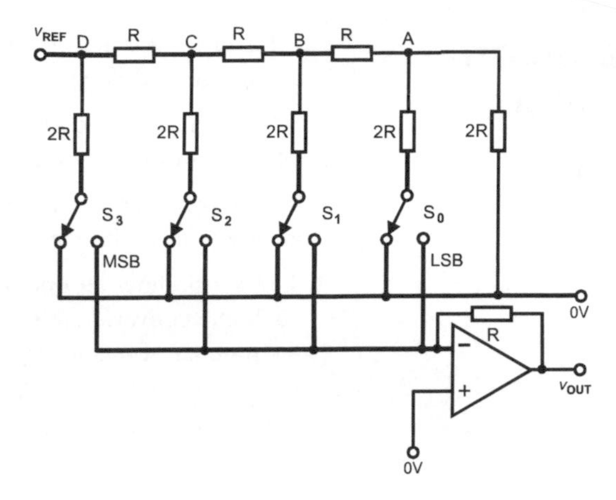

Figure 14.5 *An R-2R 'ladder' is the basis of this DAC.*

Test your knowledge 14.2

If in the converter of Fig. 14.5 the reference voltage is 5 V, what is the voltage output when the data input is 1001?

Because of the resistor values the current flowing through switch S2 is half that flowing through S3. Similarly, the current flowing through S1 is half that through S2, and so on. We have binary weighted currents (as in the previous DAC) and the op amp sums these. This makes the negative voltage output of the op amp proportional to the data input. Its value does not depend on the value of R. As before, there is a scaling multiplier, which is $v_{REF}/16$ in a 4-bit converter. The output voltage is multiplied by this to obtain the value of the digital input.

Example

If $v_{REF} = 2.5$ V, the voltage output corresponding to a digital input of 1010 (decimal 10) is $10 \times 2.5/16 = -1.5625$ V.

Activities 14.2 Digital-to-analogue converters

On a breadboard set up a 4-bit operational amplifier adder as in Fig. 14.4. Resistances could be $R_F = 10$kΩ, $R_3 = 20$ kΩ, $R_2 = 40$ kΩ (39 kΩ and 1 kΩ in series), $R_1 = 80$ kΩ (68 kΩ and 12 kΩ in series) and $R_0 = 160$ kΩ. Use resistors of 1% tolerance. Connect inputs to the positive supply line or to 0 V to obtain high- and low-level inputs. Measure the output with a multimeter. This circuit could also be modelled on a simulator.

Set up a demonstration R-2R ladder converter on a breadboard or simulator, using Fig. 14.5 as a guide. Measure voltages at various points, including points A to D.

The DAC0801LVCN is an 8-bit digital-to-analogue converter. Investigate this on a breadboard, referring to the manufacturers' data sheets.

Problems on conversion devices

1 Describe the action of an 8-bit flash converter. What is the main advantage of this type of converter?

2 Explain the principle of the successive approximation technique of analogue-to-digital conversion.

3 Compare the flash and successive approximation techniques of analogue-to-digital conversion.

4 Describe how an operational amplifier may be used as a digital-to-analogue converter. What is the difficulty of using this technique when the number of bits is large?

5 Describe the action of a digital-to-analogue converter based on the R-2R ladder technique.

Multiple choice questions

1 The number of comparators in a 6-bit flash converter is:

A 6.
B 31.
C 63.
D 64.

2 The number of clock cycles taken by an 8-bit successive approximation converter to produce its output is:

A 8.
B 4.
C 7.
D None of these.

3 An op amp adder is used as a digital-to-analogue converter. The input resistor for D0 is 15 kΩ. The input resistor for D2 is:

A 30 kΩ.
B 15 kΩ.
C 10 kΩ.
D 7.5 kΩ.

4 The fastest analogue-to-digital converter is:

A an R-2R ladder.
B an operational amplifier.
C a successive approximation converter.
D a flash converter.

15 Telecommunication systems

Summary

Telecommunications systems share the same basic structure, whether they are based on cables, optical fibre, or radio. For various reasons it is usual to modulate the signal on to a carrier. The systems employed for modulation by analogue signals include amplitude modulation, frequency modulation, phase modulation, and pulse modulation. Pulse modulation includes pulse width modulation, pulse position modulation, and pulse code modulation, the latter also being used for transmission of binary data. Frequency shift keying is another technique used for transmitting binary data. Some types of modulation reduce the unwanted effects of noise. Large numbers of signals may be transmitted simultaneously by modulating them on to groups of carrier waves with frequencies spaced a certain distance apart.

Tele-communications

The job of a communication system is to convey information from one place to another place that is some distance away. In this book we are concerned with communication systems which employ electronic devices. These include communication by electrical cable (telegraph and telephone systems), optical fibre and radio. These systems are collectively known as telecommunication systems, to distinguish them from other systems of communication such as smoke signals, semaphore and homing pigeons.

Communications systems have many features in common, as illustrated in Fig. 15.1. This is a big simplification because there are hundreds of different communications systems in use. Some do not make use of all the stages shown in the figure and listed in the table. In others, the functions of different stages may be combined into one stage so that they can not be separated. Specialist systems may have additional stages that are not shown here.

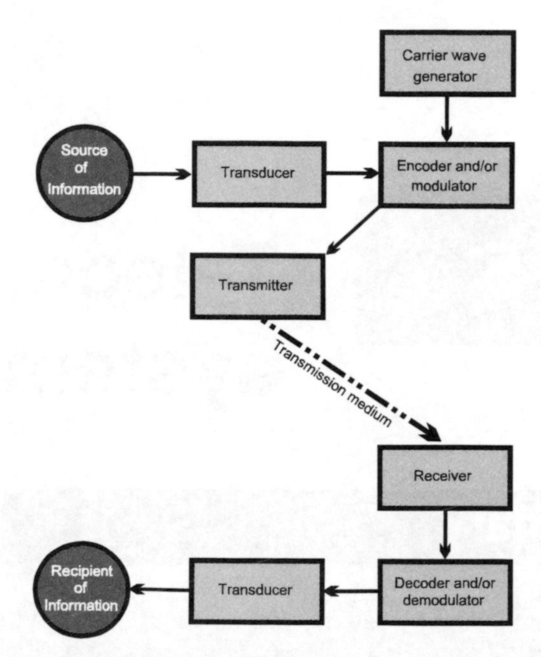

Figure 15.1 *A telecommunications system usually has the stages illustrated in this diagram. Information passes through the system from the source to the recipent, taking various forms on its journey.*

There are some terms in the table on p. 231 that you may not understand. Use the index to find where in the book these terms are explained. The term 'transducer' is used loosely in the table. Not all of the devices mentioned are true transducers. Many of the features of telecommunications systems listed in the table are fully described in Chapters 16 to 19.

Except for sending audio signals as a varying voltage on a local telephone line, analogue information is sent more often by superimposing it on a signal that has a considerably higher frequency. This process is called *modulation* and the signal of higher frequency is known as the *carrier*. The amplitude, frequency, or phase of the carrier is modified in proportion to the amplitude of the analogue signal.

Carriers and modulation

Amplitude modulation (AM) is illustrated in Fig. 15.2. The signal is a sinusoid (a). There is a short period at the beginning (on the left in the diagram) when there is no signal. The carrier is also a sinusoid but with a higher frequency, and shows its unmodulated amplitude on the left of curve (b). As soon as the signal starts, the amplitude of the carrier varies according to the signal superimposed on it. This is the *modulated carrier*.

In this example, the signal causes the amplitude of the carrier to vary

Protocol: an agreement, often international, covering the way in which something is done. Example: HTTP, hypertext transport protocol, used in the World Wide Web.

Stage	Function	Cable	Optical fibre	Radio
Source of information	Stores or originates information	Stored: computer memory, computer disk, compact disc, human brain, video recording, book. Original: temperature, colour, pH, weight, sound, a new idea in a human mind.		
Transducer	Converts information into electrical signal	Microphone, keyboard, photocell, thermistor, electromagnetic head of a disk drive. This stage may include an amplifier.		
Encoder	Makes signal suitable for transmission	Companding analogue signals, converting analogue to digital, compressing digital, parity bits, formatting to various protocols (ASCII, e-mail, teletext), conversion to pulse codes. Making data into packets. No encoding in telephone connection to local exchange.		
Modulator	Modulates a carrier wave with the signal	Analogue signals by amplitude, frequency, phase or pulse modulation. Digital signals by frequency shift keying, pulse code modulation. In local telephone circuits: not used for speech, but modems used for fax and Internet communication.		
Carrier wave generator	Produces sinusoid of fixed frequency and amplitude	Radio-frequency oscillator.	Laser or light-emitting diode usually in the infrared band.	Radio-frequency oscillator.
Transmitter	Transfers signal to transmission medium	Line driver.	Modulated laser or LED.	Radio-frequency amplifier and antenna.
Transmission medium	Carries signal to receiver	Twisted pair of wires or coaxial cable carries current.	Narrow glass or plastic fibre carries light.	Free space allows propagation of electro-magnetic waves.
Receiver	Takes in signal and converts it to an electrical signal	Line receiver.	Photo-diode.	Antenna and radio frequency amplifier.
Demodulator	Extracts signal from modulated carrier	Reverses the modulation process.		
Decoder	Converts signal to its original electrical form.	Reverses most of the encoding process.		
Transducer	Converts signal to its original physical form.	Loudspeaker, electron gun in TV tube, electric motor in remote-controlled robot.		
Recipient of information	Receives and possibly responds to information.	Data recorder, TV set, computer memory, human mind, printer, fax machine, pager, computer monitor.		

by as much as ±50%. The *modulation depth* (or *modulation factor*) is 50%. In the figure, the carrier frequency is 20 times that of the signal. It is limited to that amount because it would not be easy to draw a diagram with the frequency any higher. In practice, the ratio of frequencies is much higher.

Example

An AM radio transmitter on the Medium waveband has a frequency of 800 kHz. It is modulated by an audio signal of 1 kHz. With these typical values, the carrier frequency is 800 times the signal frequency.

AM bandwidth

An unmodulated carrier is of a single frequency, f_c. When it is modulated, the composite signal (carrier plus modulator) consists of many frequencies. These comprise f_c itself and, for each modulating frequency f_m, the sum and difference frequencies $f_c + f_m$ and $f_c - f_m$. As there is usually many different modulating frequencies present, their sum and difference frequencies occupy two bands above and below the carrier frequency (Fig. 15.3). These are the *sidebands*, referred to as the *upper sideband* and *lower sideband*. The amplitude of signals in the sidebands is usually less than that of the carrier, as in the figure. Therefore most of the power in the signal is in the carrier and less in the sidebands. But the carrier itself provides no information, so AM is wasteful of power.

If the highest modulating frequency is f_{max}, the modulated signal occupies a band of frequencies extending from $f_c - f_{max}$ to $f_c + f_{max}$. The bandwidth of the signal is $2f_{max}$.

Both sidebands contain identical information so there is no point in transmitting them both. In single-sideband (SSB) transmission, the carrier and one of the sidebands is suppressed before transmission. The power of the transmitter is devoted to transmitting the single remaining sideband. This is much more efficient. Also in SSB, the bandwidth of the signal is reduced to f_{max}. This means that transmitters may be spaced more closely on the radio spectrum.

In a telephone system transmitting speech, the highest frequency present is 3.4 kHz. The bandwidth of the modulated signal is 6.8 kHz, but is reduced to 3.4 kHz by SSB transmission.

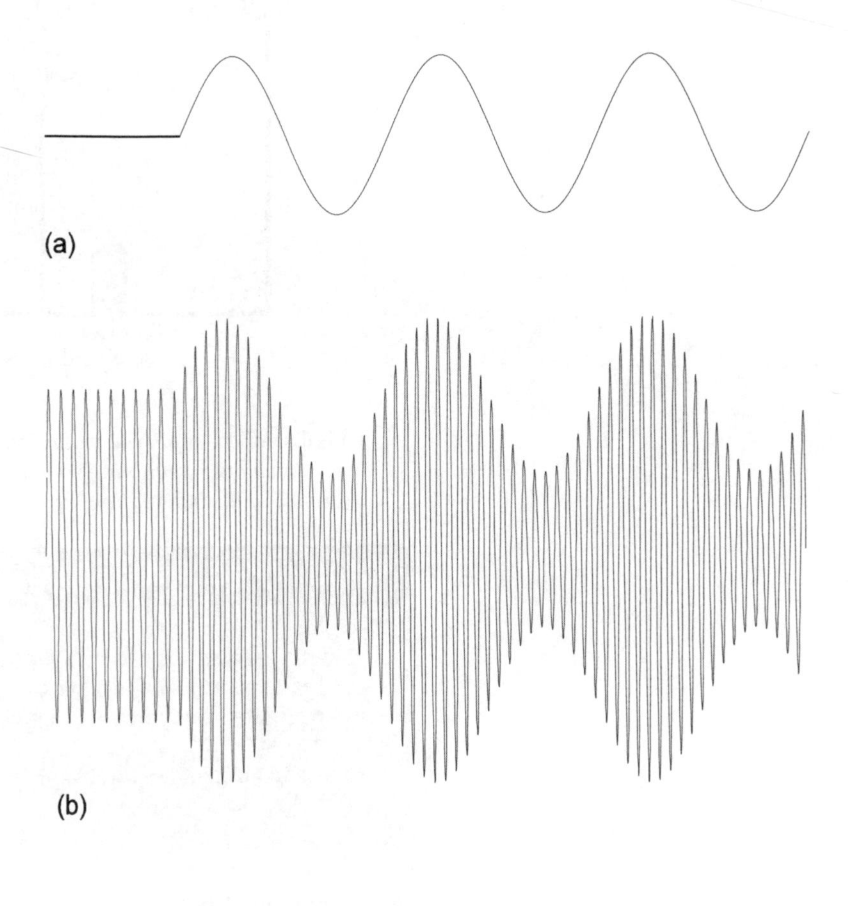

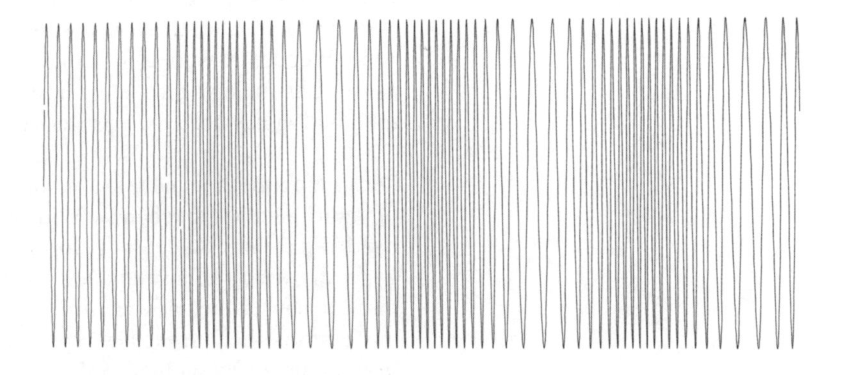

Figure 15.2 *Signal (a) is used to modulate the amplitude of a carrier, producing the modulated waveform (b), which has constant frequency. In (c) the signal modulates the frequency and amplitude remains constant.*

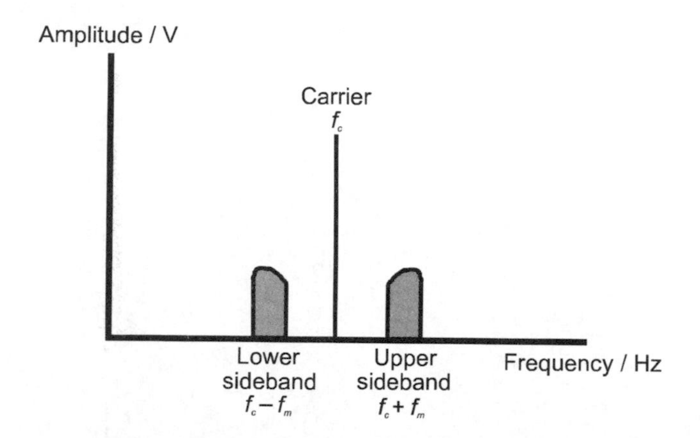

Figure 15.3 *Amplitude modulation of the carrier generates two sidebands containing frequencies that are the sum and difference of the modulating frequencies and the carrier.*

Activity 15.1 Amplitude modulation

Use an oscilloscope to examine the signal from a radio set tuned to an AM transmission. Look for indications of interference caused by local electromagnetic sources.

Practical circuits for AM and FM are described in Chapter 19.

Frequency modulation (FM) is illustrated in Fig. 15.2c. Amplitude remains constant but frequency increases and decreases with the signal voltage. On the left there is no signal and the unmodulated frequency is seen. Frequency increases as the signal rises above zero and decreases as it falls below zero. As in the previous figure, the amount of modulation has been exaggerated to make the diagram clearer.

Depending on the amount of modulation, an FM signal may have one or more pairs of sidebands. With more sidebands it requires greater bandwidth than AM. A high-frequency carrier (VHF or UHF) is generally used because this allows room for more sidebands between the carrier frequencies of different transmitters. With this type of modulation, the carrier itself contains information about the signal, so FM is more efficient than AM.

A further advantage of FM is that it is not affected by changes in the amplitude of the signal. If reception conditions are poor, an AM signal varies in amplitude, being received strongly at times but fading at other times. This does not happen with FM, since changes in amplitude have no effect on frequency.

Phase modulation (PM) has many similarities to FM. PM signals are receivable on an FM receiver and the other way about, but the relative amplitudes of high and low audio frequencies are altered.

Analogue information may also be transmitted by various systems of pulse modulation:

Pulse modulation

Pulse width modulation (PWM), or *pulse duration modulation* (PDM) is the commonest system in use and is illustrated in Fig. 15.4. The analogue signal is sampled at regular intervals, as indicated by the points marked on the analogue curve. The PWM signal consists of a series of pulses with their leading edges occurring at the sampling times. The width of each pulse is proportional to the amplitude of the analogue signal at the sample point. The scaling is such that an instantaneous amplitude of 0 V results in a pulse of half width. The pulse is wider when the signal voltage is positive. The pulse is narrower when the voltage is negative.

At the receiving end the original signal is recovered by sending the pulsed signal through a lowpass filter. A rough description of how this works is to take the two extreme conditions. A succession of pulses wider than average results in an increase in voltage output from the filter, so reconstructing the original positive-going input voltage. Conversely, a succession of narrow pulses, corresponding to a negative input voltage, results in a falling output from the filter.

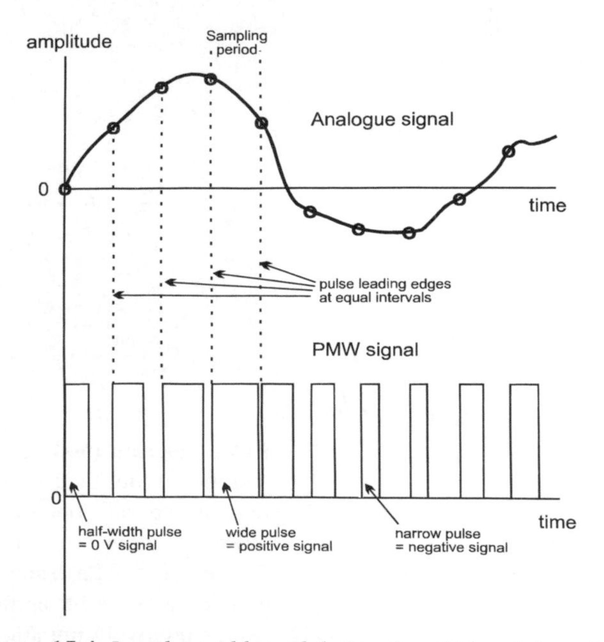

Figure 15.4 *In pulse width modulation the widths of the pulses are in proportion to the instantaneous value of the analogue signal at each sampling.*

The circuits for producing PWM pulses and for recovering the original signal are relatively simple. A further advantage is that spikes and noise on the transmission system affect pulse amplitude but have little effect on pulse length. A disadvantage of PWM is that the transmitter has to be powerful enough to generate full-length pulses yet, on average, it is producing only half-length pulses. This makes inefficient use of the transmitter. In effect, the transmitter operates at only half the power it could otherwise generate.

Activity 15.2 Pulse width modulation

This is a simple illustration of the principles of PWM. Demonstrate it by using a breadboarded circuit, or by a simulator.

Set up the circuit shown in Fig. 15.5. The input signal may come from a sine-wave generator, a microphone, a radio receiver or CD player. The op amp compares the input signal with the signal from a triangle-wave generator. The output of the op amp is a PWM signal.

Filter the output with a basic RC lowpass filter to recover the original signal.

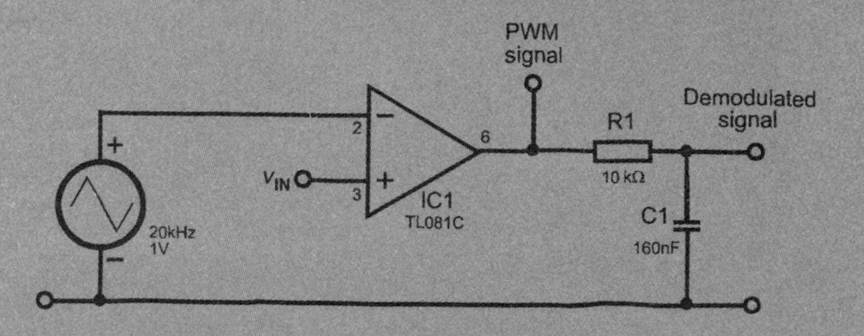

Figure 15.5 *This circuit demonstrates pulse width modulation and how the signal may be demodulated by a simple lowpass filter.*

Pulse position modulation (PPM) uses pulses of constant amplitude and length, but their timing varies. With no signal, the pulses occur at regular intervals, one for each sample. If a signal sample is positive the pulse is delayed by an amount proportional to the signal voltage. Conversely, if the sample is negative the pulse occurs earlier. The pulses make use of the full power of the transmitter, and the transmission is relatively immune from noise. Drawbacks of this technique are that circuits are relatively complicated and that the transmitter and receiver must be synchronised, which is not required for PWM.

Pulse code modulation (PCM) begins with the analogue signal being converted to digital form, using an analogue-to-digital converter (Chapter 14). Each sample is converted 8-bit binary value. This can have any value in the range 0 to 255. The sample values are automatically formatted into larger groups for transmission, extra bits being added to aid the checking of errors on reception and to synchronise the receiver with the transmitter. Data from several sources may be combined, transmitted simultaneously, and separated again at the receiving end. Very high transmission rates may be used, the fastest being over 500 megabits per second with the capability of transmitting 7680 channels simultaneously.

PCM may also be used for signals that are digital in origin, such as signals sent from one computer to another. Whether the signal originates in analogue or digital form, the values transmitted are either 0's or 1's. It is easier for the receiver to distinguish between 0's and 1's than it is to respond to analogue voltage levels, so digital systems are much less subject to error than analogue systems. Pulses become distorted on a long journey but their shape can be restored at relay stations spaced along the route. These receive pulses which are distorted, but which are still distinguishable as 0's and 1's. The station re-transmits the signal, but with correctly shaped pulses. Digital relay stations can be much more widely spaced than the amplifying stations used on an analogue line. Noise and spikes on the signal lines have much less effect on a digital signal. For these reasons, PCM is widely use in telecommunications. It is also used in the recording and playback of compact discs.

Frequency shift keying

Frequency shift keying is one of several related systems for transmitting digital information. FSK uses two frequencies, one to represent 0 and the other to represent 1. In one version of FSK, a 0 is represented by a burst of 4 cycles at 1200 Hz, and a 1 is represented by a burst of 8 cycles at 2400 Hz. The two frequencies are easily distinguishable at the receiving end and the original sequence of 0's and 1's is reconstructed. This is a relatively slow system used by modems operating over ordinary telephone lines. Many types of fax machine also use FSK with audio frequencies when establishing contact with another fax machine over the telephone line.

Pulse modulation

Pulses can be transmitted directly as pulses, but more often they are used to modulate a high-frequency carrier signal before transmission.

Modulation of any kind makes it possible to send a large number of different signals simultaneously on *one* channel of transmission. For example, as explained in the box, a voice signal requires a bandwidth of 3.4 kHz to reproduce adequately understandable speech. If such a signal is SSB modulated on to a 64 kHz carrier, the lower sideband extends down to 64 – 3.4 = 60.7 kHz. The frequency space between

Bit rate and Baud rate

Both of these are used to express the rate of transmission of data. They are not the same thing.

Bit rate is the number of binary bits (0's or 1's) transmitted in 1 second.

Baud rate is the number of 'signal events' transmitted per second. If one signal event (for example, a pulse of given length or amplitude, or a burst of one of the FSK frequencies) represents one binary digit, the Baud rate and the bit rate are equal. But in many systems a signal event represents more than one bit.

Example

In QPSK, the phase of the carrier is shifted by four different amounts to code the binary groups 00, 01, 10 and 11. Each signal event (in this case a shift of phase) is equivalent to two bits of data. A Baud rate of 19200 is equivalent to 38400 bits per second.

60.7 kHz and 64 kHz must be reserved for this transmission. But there is plenty of frequency space above 64 kHz, which can be used for transmitting other signals. The next higher carrier could be at 68 kHz which, when modulated, occupies from 64.7 kHz up to 68 kHz. This does not overlap the band occupied by the signal on the 64 kHz carrier. In this way we can stack a group of voice signals on carriers spaced 4 kHz apart. A *group* of 12 speech signals modulated on to carriers spaced 4 kHz apart occupies a 48 kHz bandwidth.

Modulation can be taken a step further. *Groups* of signals can themselves be modulated on to carriers of even higher frequency. If a number of these carriers are spaced 48 kHz apart, they produce a *super-group.* This process of stacking signals by modulating them on to groups and supergroups (and even higher groupings) of frequencies makes it possible to transmit hundreds of signals simultaneously along a single transmission channel such as a coaxial cable. At the receiving end, tuned circuits and demodulators are used to separate out the super-groups, then the groups, then the original modulated signals, and finally the individual analogue signals. These are then routed to appropriate destinations.

The process of transmitting many signals simultaneously by dividing the available bandwidth between them is known as *frequency division multiplexing.* The same principle is used with signals of other kinds.

The bandwidth needed depends on the nature of the original signal. Musical signals, for instance, require a greater bandwidth than speech. Typically a musical signals need 15 kHz bandwidth for a single transmission, so the carriers must be spaced more widely apart. For the highest quality the bandwidth is increased to 20 kHz. Bandwidths for video transmissions are rated in tens of megahertz.

Making best use of channel width

Channel width is the range of frequencies at which signals can be transmitted on a channel. It is the bandwidth of the channel itself. Channel width is usually much wider that the bandwidth of individual signals. This allows the channel to be used for transmitting many signals and many groups of signals simultaneously.

Frequency division multiplexing divides the available channel width between a number of simultaneous signals. An example is given on p. 238.

Time division multiplexing is another approach to conveying many signals on a single channel. Signals from several sources are allocated to short time slots (in the region of 1 s) and transmitted one after another in quick succession. At the receiving end the signals are sorted out and routed to their proper receivers. This system is often used in telephone exchanges and in radio satellite communications.

Pulses

In theory a pulse waveform rises instantly to its full amplitude, stays there for a while then falls instantly to zero (Fig. 15.6a). This never happens in practice because a voltage can only increase when current flows, and current takes *time* to carry charge from one place to another. For this reason, a pulse rises and decays (falls) in a finite time, giving rise to a graph like Fig. 15.6b. By definition, the *rise time* is the time taken for the voltage to rise from 10% to 90% of its final value. Conversely, the decay time is the time taken to fall from 90% to 10% of its original value. The *pulse width* is the length of time the voltage is 90% or more of its maximum value.

In a telecommunications system the shape of the pulse may be modified by the filtering effect of any circuit through which it passes. There is always a certain amount of resistance in a conductor. There is capacitance between the conductor and nearby conductors, including those connected to ground. This resistance and capacitance have the same effect as the resistor and capacitor on the left of Fig. 7.16. The act as a lowpass filter. Fig. 15.7 shows what happens to a 10 kHz pu signal after it has been through a lowpass filter with –3 dB poir

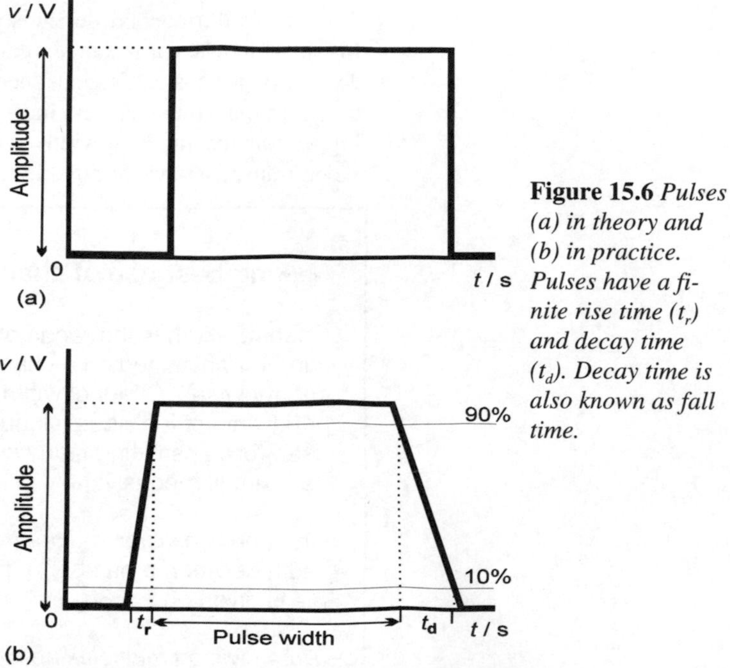

Figure 15.6 *Pulses (a) in theory and (b) in practice. Pulses have a finite rise time (t_r) and decay time (t_d). Decay time is also known as fall time.*

33 kHz. The pulse has lost its square shape. Both the rise time and the decay time of the signal have been extended. The shape of the pulse may be squared up by passing it through a Schmitt trigger circuit (Fig. 7.8). If this is not done there could be problems when feeding this signal to a logic circuit. Some logical devices require that the rise time should be less than a specified minimum, so the device might not operate reliably when presented with a signal such as that in Fig. 15.7.

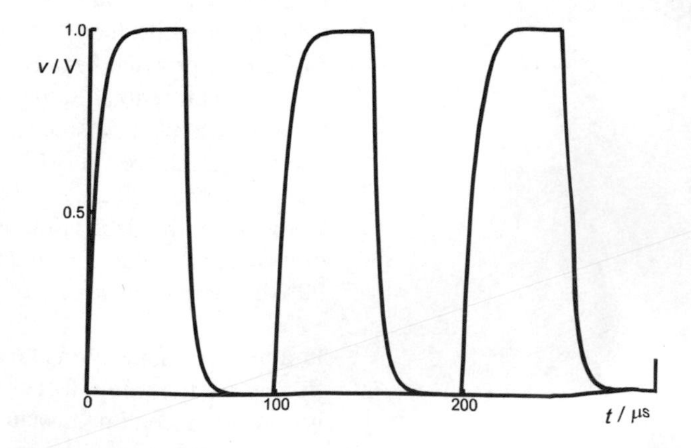

Figure 15.7 *When a pulsed signal passes though a circuit, its shape becomes distorted, mainly by the capacitances in the circuit.*

A further effect that sometimes occurs is illustrated in Fig. 15.8. Here a train of pulses, amplitude 1 V, repetition rate 20 kHz, is sent along 100 m of cable. Because of the capacitance between the two conductors and because of their inductance, the signal that emerges at the other end is far from a square pulse in shape. It shows *overshoot* and *ringing*. Overshoot is the term used to describe how the voltage level rises or falls beyond the final level. Here the overshoot is approximately 0.2 V. Ringing is the way in which the voltage oscillates until it finally settles to the final level. This is due to the system tending to oscillate at a particular frequency, because of the capacitance and inductance in it. Overshoot and ringing are not necessarily a serious problem. If the overshoot and ringing are not too great, the original well-squared form of the pulses may be regained by passing the signal through a Schmitt trigger circuit

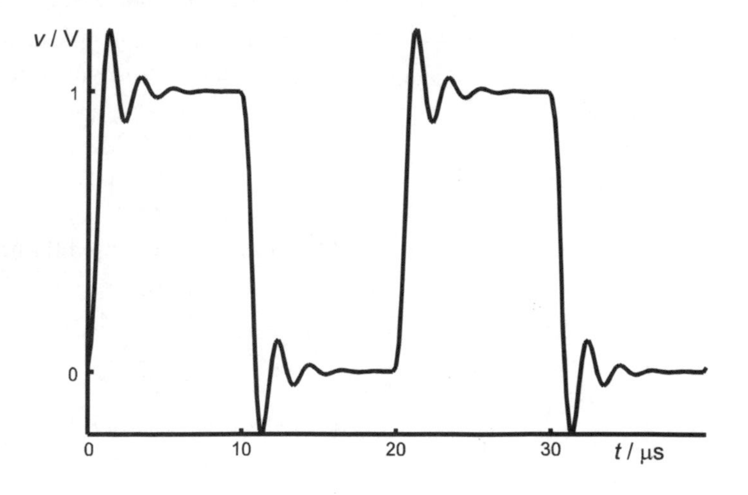

Figure 15.8 *Another type of distortion occurs when a pulse is passed through a circuit or cable which has appreciable inductance as well as capacitance.*

Problems on telecommunications systems

1 Match the stages of Fig. 15.1 against the stages of transmission of information by: a smoke signal, semaphore, a video surveillance system in a supermarket, and any other named system of your choice.

2 Name some sources of information that make up the first stage of a telecommunications system. In what form(s) does the information exist?

3 Explain, giving examples, the functions of transducers in telecommunications systems.

4 What is modulation? Explain the difference between amplitude modulation and frequency modulation.

5 What is a sideband? In what way does the production of sidebands affect the bandwidth of a modulated carrier?

6 Describe how an analogue signal is transmitted and received using pulse width modulation.

7 Compare pulse width modulation with pulse position modulation.

8 Describe how a binary signal is sent using frequency shift keying.

9 Explain what is meant by the amplitude, rise time and decay time of a pulse signal.

Multiple choice questions

1 An AM signal has a modulation depth of 20%. This means that:

A the signal frequency may vary by as much as 20%.
B the modulated carrier amplitude varies by as much as 20%.
C the signal frequency is 20% of the carrier frequency.
D the amplitude of the signal is 20% of the carrier amplitude.

2 A modulation system in which there are pulses of constant amplitude and length but with varying timing is known as:

A PPM.
B PWM.
C PCM.
D PDM.

3 One of the systems used for transmitting binary data is:

A PCM.
B PWM.
C PDM.
D PWM.

4 Of the systems listed below, the one most subject to interference from spikes and noise is:

A FSK.
B AM.
C FM.
D PPM.

5 When a pulse signal oscillates before settling to its new level, the action is known as:

A overshoot.
B the time constant.
C ringing.
D oscillations.

16 Transmission media

Summary

The three media used in telecommunications are cable, optical fibre and (for radio) free space. These have very different characteristics, which are examined in this chapter. As well as studying the propagation of signals in the medium, we look at the way in which signals are transmitted into the medium and the way in which they are recovered from it.

Cable

Telecommunication by cable is an extension of the passing of signals along the conductors within the circuit in which the signal originates. In the simpler systems the wires of the transmitting circuit are directly connected to the cable and this in turn is wired directly to the receiving circuit. In a wired intercom system, for example, the sound signal is carried along a pair of wires from the master station to one of the remote stations. When the master station is transmitting to the remote station, the remote station is in essence just a loudspeaker on the end of a long lead. The signal is unmodulated. The distance between transmitter and receiver is limited to a few hundred metres. The behaviour of the external circuit can be analysed by applying the usual rules of circuit analysis, such as Ohm's Law, the Principle of Superposition and others. Communication between a telephone in the home and the local exchange falls into the same category.

Most cable telecommunications differ from the above in one or both of the following ways:

- the distance between transmitter and receiver is greater.
- the signal is modulated at frequencies much higher than audio frequencies.

These factors have important effects on transmission. To understand these we first look at the *types of cable* used and also examine the properties of *transmission lines*.

Types of cable

A wide range of cables is available but the two types most commonly used are the *twisted pair* and *coaxial cable* (Fig. 16.1). The twisted pair consists of two insulated wires twisted around each other. Because the wires are unshielded (that is, they have no metallic sheath around them) there is capacitance between the two wires and between each of the wires and ground or any other nearby conductors. Because they are twisted around each other, the capacitances of each wire to ground or to external conductors is balanced. Any capacitative effects are common mode and therefore have little or no effect on the signal. A cable of this kind is described as *balanced*.

Common mode: Affecting both wires equally.

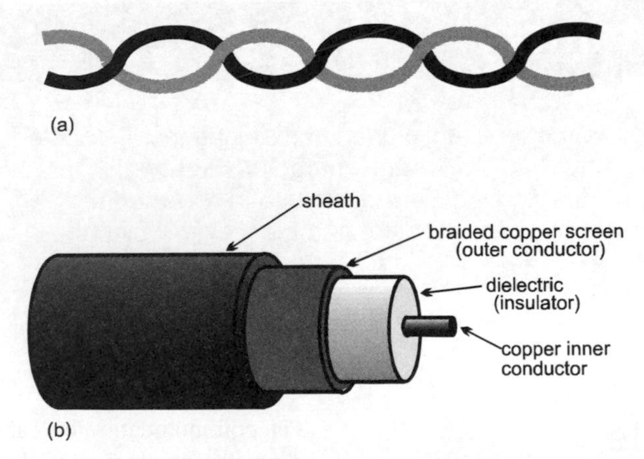

Figure 16.1 *The two types of cable most often used in telecommunications are (a) the twisted pair, and (b) coaxial cable.*

The induction of signals produced by external magnetic fields is minimised by twisting. In any circuit which contains one or more loops, an external magnetic field passing through each loop induces an e.m.f. in the loop. But in a twisted pair, the direction of induced e.m.f.s in the wires is reversed in adjacent 'loops' (Fig. 16.2) so they cancel out.

Coaxial cable is widely used for carrying radio-frequency signals. The signal is carried in the inner conductor which may be a single copper wire or may consist of a number of narrower copper strands. Stranded wire is used when flexibility is important. The inner conductor is insulated from the outer conductor by the dielectric, a layer of polythene or polyethylene. The output conductor is made of copper and it usually braided to give flexibility. This is connected to ground so it serves to shield the inner conductor from external magnetic fields. It also prevents magnetic fields produced in the inner conductor from spreading to other nearby circuits. Because there is no capacitance

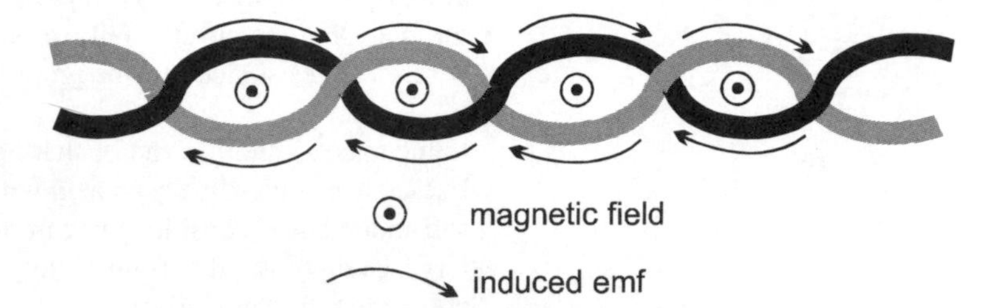

Figure 16.2 *The loops of the twisted pair are threaded by a magnetic field increasing in a direction upward through the paper of the page. By Fleming's Left-hand Rule, this induces e.m.f.s in a clockwise direction in each loop. Because of the twisting, these e.m.f.s cancel out in adjacent sections of the wires. Changes in the intensity or direction of the field affect all loops equally.*

between the inner conductor and external conductors, and because the inner conductor is shielded from external magnetic fields, a coaxial cable is described as *unbalanced*.

Transmission lines

When a twisted pair or coaxial cable is being used for transmitting signals over a large distance, or the signal has high frequency, or both circumstances apply, we have to consider the properties of the medium in more detail. We regard it not as a mere length of conductor but as a *transmission line*. There is no sharp distinction between a simple lead, such as a loudspeaker lead, and a transmission line such as a submarine telephone cable. It is just that as length increases, or as transmission frequency increases, the distinctive properties of the line become more and more important and have to be taken into account.

Not all transmission lines extend for kilometres under the ocean. In microwave transmitters and receivers the frequency is so high (up to tens of gigahertz) that the wavelength of the signal in the conductor is only a few tens of millimetres. Then, even conductors joining one part of the circuit to another part need to be treated as transmission lines. At the very highest frequencies the mechanism of propagation of the signal becomes rather different. We use *waveguides* (box-like metal channelling or metal strips) instead of ordinary cable. There is not enough space here to discuss this topic further.

A transmission line may be considered to be an electronic circuit in its own right. The circuit in Fig. 16.3 represents a balanced transmission line by its equivalent electronic components. The resistors R represent the resistance of the copper conductors. In reality the resistance is spread evenly along the whole length of the wires, but here we represent it by individual resistors. The resistors in Fig. 16.3 are the *lumped equivalents* of the evenly distributed resistance of the wire.

If the cable is a short one carrying audio signals, resistance is the only property of the line that is important. Ohm's Law is all that is needed for analysing the circuit. With short runs of a suitable gauge of copper wire, even resistance may be ignored.

Inductance is another factor that adds to the impedance of the cable. Inductance is usually associated with coils, but even a straight wire has self-inductance (consider the wire to be part of a one-turn coil of very large radius). In the figure, this too is represented by its lumped equivalent, the two inductors.

Test your knowledge 16.1

In what way does the capacitance of a capacitor change if the plates are made closer together?

Capacitance between the conductors is represented by the capacitors C, joining the wires at regular intervals. The size of the capacitance depends on the spacing between the conductors and the material (dielectric) between them. Usually the dielectric is plastic, often PVC in the case of a twisted pair, or polythene in a coaxial cable.

Finally, there is always a small amount of leakage of current through the insulation or dielectric, represented in lumped form in Fig. 16.3 by the conductances G.

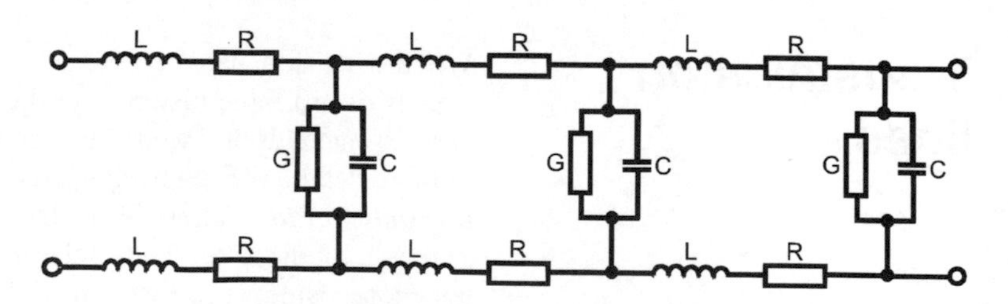

Figure 16.3 *The lumped equivalent of a transmission line. In theory the line consists of an infinite number of units like these three, spaced equally along the line.*

Using lumped equivalents reduces the transmission line to a relatively simple 'circuit' which can be analysed by the usual techniques. For instance, we know that the reactance of capacitors becomes very low at high frequencies, and it is easy to see from Fig. 16.3 that capacitance then becomes much more important than conductance. Conversely, the reactance of the inductors becomes very high at high frequencies, so inductance becomes much more important than resistance. As a result, conductance and resistance may be ignored at high frequencies. The lumped equivalent circuit consists only of inductance and capacitance, as in Fig. 16.4. This arrangement of inductance and capacitance has the properties of a lowpass filter. Signals passing along the line lose relatively more of their high-frequency components. This has the effect of making pulses more rounded and spread out. If this occurs, adjacent pulses may partly merge, causing errors in the signal at the far end of a long line.

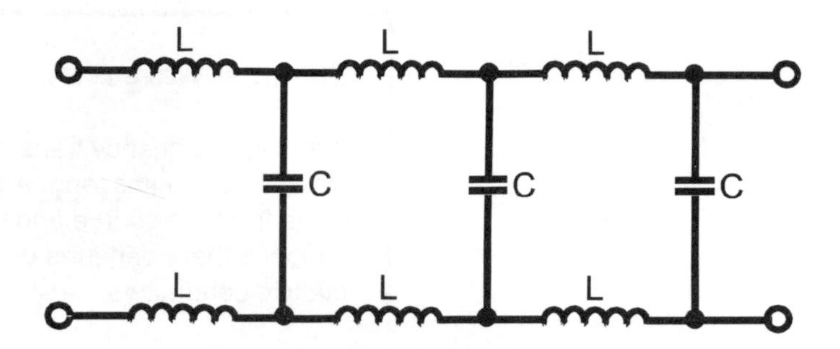

Figure 16.4 *At high frequencies, resistance and conductance may be ignored, leaving only inductance and capacitance.*

Test your knowledge 16.2

Explain why the circuit of Fig. 16.4 acts as a lowpass filter.

Inductance and capacitance store and release energy but they do not absorb it. In *theory*, there is no loss of energy or power along a transmission line with a high-frequency signal. In *practice,* some of the energy escapes as electromagnetic radiation (in twisted pairs, but less in coaxial cable), or as heat generated in the conductors when current passes through them. Consequently, there is a fall in signal power along the line. This can be ignored for relatively short lines, which we then consider to be *lossless.*

Characteristic impedance

The most important feature of a transmission line is its *characteristic impedance*, Z_0. This is defined as the ratio of the voltage and current (v/i) at the input of a line of infinite length. If the impedance of the signal source is equal to Z_0, all of the power generated by the source is transferred to the line. Similarly, if there is an impedance Z_0 at the far end of the line, all the power of the signal is transferred to this. It follows that a lossless line transfers all of the power from the source to the receiver. But for this to happen the source and receiver impedances must be *matched* (equal) to the line impedance. If the impedance at the end of the line is *not* equal to Z_0 (including when it is an open circuit, or when it is a short-circuit) part or all of the energy of the signal is reflected back along the line (see box on p. 248). This causes standing waves to form in the conductors, and less power is transferred to the receiver. For efficient transmission of signals, this is to be avoided by careful matching.

The characteristic impedance of a cable at high frequencies depends on the inductance and capacitance per unit length of cable. It is independent of frequency, provided that frequency is high. In other words, the cable behaves like a pure resistance; that is, it remains fixed whatever the frequency. In practice, cables are made with standard characteristic impedances and transmission equipment has input or output impedances that match these standards. The most common value of Z_0 for coaxial cable is 50 Ω, another common value being 75 Ω. Twisted pairs have standard impedances of 300 Ω or 600 Ω.

Standing Waves

With high-frequency transmissions the cable is several or many wavelengths long. A succession of waves enters the cable from the source and is carried to the far end. What happens there depends on what is connected there between the two conductors:

A matched load circuit (having an impedance equal to Z_0): All of the power of the signal is transferred to the load.

A short circuit: No power is dissipated in this so the power is turned to the cable in the form of waves travelling back toward the source. Where the two conductors are short-circuited, they must be at equal voltage. The amplitude of the wave is zero at this point. Interference between the forward and returning waves sets up a pattern of standing waves, as shown in Fig. 16.5a. There are alternating nodes (zero amplitude) and antinodes (maximum amplitude). The nodes are spaced $\lambda/2$ apart, with an antinode at the source end (representing the alternating voltage supplied by the source) and a node at the far end (because of the short-circuit).

Open circuit: As with a short circuit, no power is dissipated at the far end, so the waves are reflected back along the cable (Fig. 16.5b). The far end is a point of maximum voltage change so an antinode occurs at this point. As with a short circuit, there is an antinode at the source end.

An unmatched receiving circuit (having an impedance that is *not* equal to Z_0): Some of the power is transferred to the load, but some is reflected back along the cable. This produces standing waves of smaller amplitude.

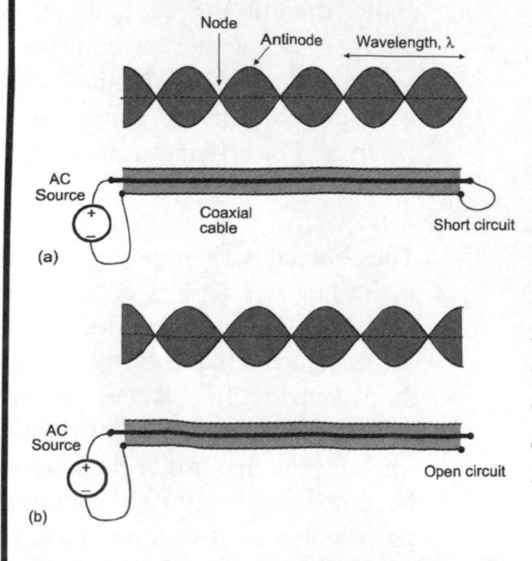

Figure 16.5 *What happens when a signal is sent along a transmission line depends on the loading at the far end. If this is (a) a short circuit or (b) an open circuit, a pattern of standing waves is set up along the line.*

Line driving

For audio signals, a pair of centre-tapped transformers (Fig. 16.6) may couple the transmitter and receiver to the line. The taps are grounded so that, when one line of the twisted pair is positive of ground, the other line is negative by an exactly equal amount. The line is *balanced.* This minimises electromagnetic interference because any such interference will affect both lines equally, making one more positive and the other less negative. The interference is *common-mode* and has no effect at the receiver. If the source circuit has a high output impedance a buffer is needed between it and the transformer to match the source to the line.

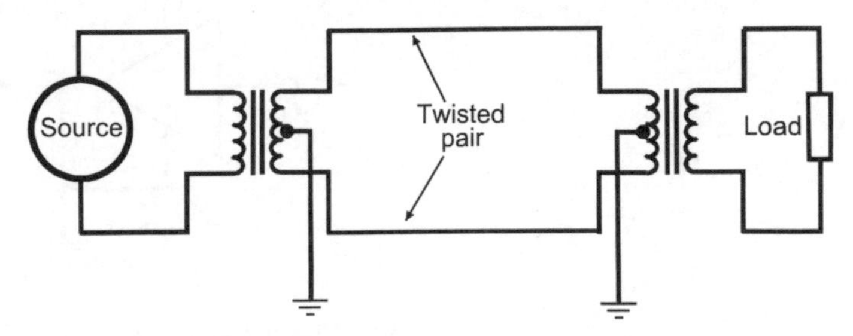

Figure 16.6 *Source and load may be transformer-coupled to a balanced transmission line to minimise EMI.*

Bandwidth: In the case of drivers, the maximum frequency at which the driver still has useful gain.

Antiphase: One is the inverse of the other.

Audio signals may be fed directly to the line using a line driver. A wide range of these is available as integrated circuits, which usually have several drivers and receivers in the same package. Each driver is a differential amplifier with low output impedance to match the impedance of the line and with fast slew rate and high bandwidth. It has two outputs (Fig. 16.7) which are in antiphase for connection to the two lines of a balanced transmission line. The receiver is a differential amplifier, and the two wires of the transmission line are connected to its (+) and (−) inputs. It has fast slew rate, high bandwidth and high common-mode rejection. Although most types of driver and receiver are intended for use with balanced lines, they are also able to operate with unbalanced lines, in which one path is grounded and the signal passes along the other.

Test your knowledge 16.3

What is the meaning of *slew rate*?

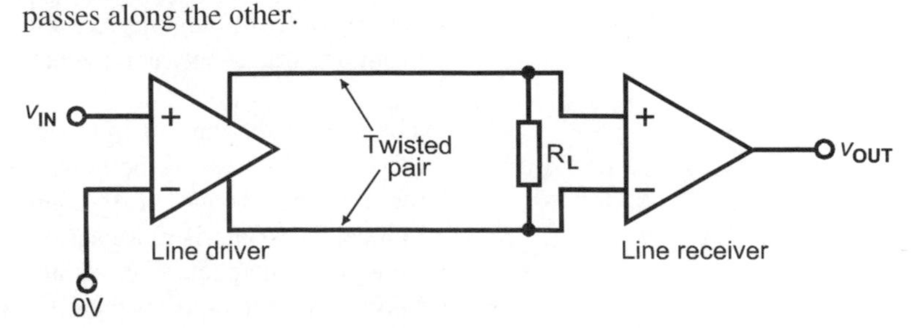

Figure 16.7 *The essentials of using a line driver and receiver to send data along a twisted pair, or any other form of balanced line. The line is terminated in a load resistor equal to the characteristic impedance of the line.*

For digital or pulsed signals the line driver is usually a logic gate, often INVERT or NAND. TTL buffer gates can be used (Fig. 16.8), though special line driver gates are preferred, because these have higher slew rate and higher bandwidth. CMOS drivers are used in low-power circuits. The receiver is also a logic gate, sometimes an INVERT gate. It may have Schmitt trigger inputs to 'square up' the distorted incoming pulse and to reduce the effects of noise on the line.

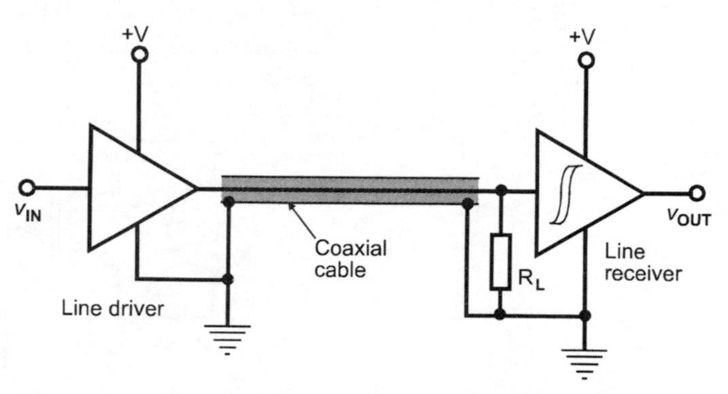

Figure 16.8 *A basic circuit for using a line driver and receiver to send data along an unbalanced line. The line is terminated in a load resistor equal to the characteristic impedance of the line, usually 50 Ω or 75Ω in the case of coaxial cable.*

Crosstalk

When several lines are close together, the current passing along one line may cause current to flow in one or more adjacent lines. This may be the result of capacitance between the lines or the effect of inductance. The effect is known as *crosstalk,* and often occurs in multicored cables and in the ribbon cables used for connecting computers to peripherals. It may be eliminated in multicored cables by having each of the cores enclosed in a separate earthed shield, though this makes the cable considerably more expensive. Connecting alternate lines to ground may reduce crosstalk in ribbon cables. Crosstalk also occurs between adjacent tracks on circuit boards, particularly if the tracks run side-by-side for an appreciable distance. It is avoided by running grounded tracks between the active ones.

Peripheral: Equipment such as a monitor, keyboard, printer, modem, or CD drive which is connected to a computer.

The size of the current induced by a magnetic field is proportional to the *rate of change* of the magnetic flux. This in turn is proportional to the rate of change of the current generating the field. Therefore induced crosstalk is more serious at high frequencies. The effect is the same for capacitative crosstalk because capacitance decreases with frequency and high-frequency signals pass more readily across the capacitative barrier.

In data signalling at high rates, the pulses have very rapid rise-times, and these give rise to the most serious crosstalk problems. Spikes several volts high may be generated on an adjacent line at the begining and end of each pulse. This effect can completely disrupt the signal on the adjacent line. With many types of line driver the rise-time can be adjusted so that it is no more rapid than necessary, so reducing the crosstalk.

Activity 16.1 Transmission line

Set up a circuit in which a sinusoidal voltage source feeds a signal through a source resistor into a transmission line, and there is a load resistor at the far end. The transmission line is a 100 m length of twisted pair or coaxial cable. The resistors values are equal to the characteristic impedance of the line, which is usually given in the retailer's catalogue. This circuit may be breadboarded, or set up with simulation software (Fig. 16.9). Use an oscilloscope or the corresponding software feature to examine voltage signals at the input and output of the line. Measure the signal voltage amplitudes and calculate their maximum power, using the equation $P = V^2/R$. Run the test with various load resistors in the range 10Ω to 1 kΩ to find which value resistor obtains the greatest amount

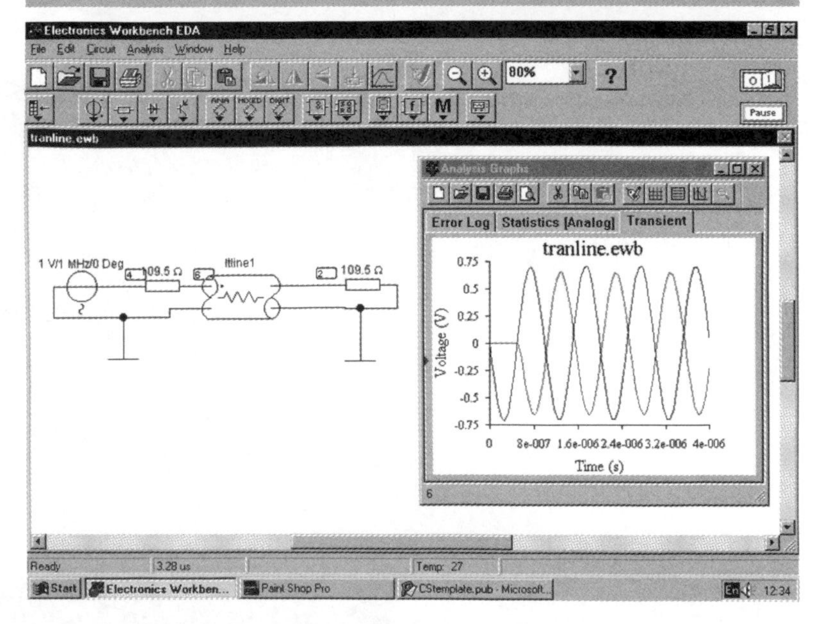

Figure 16.9 *A simulation, using* Electronics Workbench ®, *measures the voltages at the source and load of a lossy transmission line 100 m long. The voltage at the load remains zero for a short time after the run begins because of the time taken for the signal to travel along the line. Source and load impedances are matched to that of the line, so source and load voltage signals have approximately equal amplitude but with some loss due to the resistance of the line.*

of power from the line. Try this for sinusoidal signals for a range of frequencies from 1 MHz up to 20 MHz.

Here are some hints to help with setting up and interpreting the tests. At high frequency, $Z_0 = \sqrt{(L/C)}$, where L and C are the inductance and capacitance per metre. Typical values for a twisted pair are $L = 600$ nH and $C = 50$ pF, which gives $Z_0 =$ 109.5 Ω, independent of the length of the line. If the simulator requires a value for R, a typical value is 0.17 Ω per metre. Obtain values for various other types of line from manufacturer's data sheets. To calculate the delay time taken for the signal to pass along the line, begin with the velocity of the signal in free space, which is 300×10^6 m/s. The velocity is less than this in a conductor, usually between 0.6 and 0.85 of the free space velocity. A typical delay for a 100 m line is 475 ns.

Replace the sinewave generator with a square-wave generator and investigate the effect of the line on the shape of the pulses at various frequencies, with matched and unmatched loads.

Activity 16.2 Transformer coupling

Set up the circuit of Fig. 16.6, using 1:1 line-isolating transformers. Apply an audio-frequency sinusoid at one end (through a source resistor) and use an oscilloscope to examine the signal across the load resistor. Calculate the power of the input and output signals for various combinations of source and load resistors in the range 50 Ω to 10 kΩ, and at various frequencies.

This circuit can also be run on a simulator.

Activities 16.3 Line drivers and receivers

Using data and circuits from the manufacturer's data sheets, set up a 100 m balanced transmission line. For audio signals (sinusoids in the audio frequency range) use drivers such as the SSM2142 and receivers such as the SSM2143. Use an oscilloscope to investigate the transmission of signals along the line.

Set up a 100 m unbalanced coaxial transmission line and use the 75172 line driver and 75173 receiver to transmit digital data along the line. Examine pulse shape at the source and the load. Alternatively, use a CMOS driver/receiver ic such as the ADM205.

Recommended Standards

Standard interfaces between transmission lines and transmitters or receivers are recommended by the EIA (Electronic Industries Association). The Recommended Standards (RS) cover voltage levels, types of connector, pin allocations and many other features. This means that drivers and receivers built by different companies but conforming to these standards will be able to communicate with each other without error.

The most widely used standard is RS-232, which allows for one driver at one end of the cable and one receiver at the other. The cable may be up to 15 m long. The driver output signal is between ±5 V and ± 15V. The maximum data rate is 20 kbits per second. New standards have been adopted to cater for the increasing amount of data transmission. The recent RS-485 standard allows up to 32 drivers and 32 receivers to be connected to one line, which may be up to 1.2 km long. The maximum data rate is 10 Mbits / sec.

Optical fibre

When a ray of light meets the boundary between a medium (such as air) and another transparent medium (such as glass), which is optically denser, it may be partly *reflected* and partly *refracted*. Ignoring the part that is reflected, Fig. 16.10 shows what happens when the ray passes from glass to air. The ray is bent or refracted *toward* the normal. Fig. 16.11 shows what happens when the ray passes from the medium with higher n (glass) to the medium with lower n (air). All depends on the angle θ_1 at which the ray strikes the boundary. If it strikes it perpendicularly, it passes straight out of the glass into the air. If it strikes at a small angle to the normal, part of it passes out and is refracted *away from* the normal. Part is reflected (not shown in the figure). As the angle θ_1 is increased, a point is reached at which the refracted ray *just* skims along the surface ($\theta_2 = 90°$). If θ_1 is increased a little more, the ray can not escape from the glass. The *whole* of it is reflected. This is called *total reflection*. The value of θ_1 at which this happens is the *critical angle, c*. For a typical glass-air surface, c is about 42°. Total reflection occurs whenever θ_1 is equal to or greater than c.

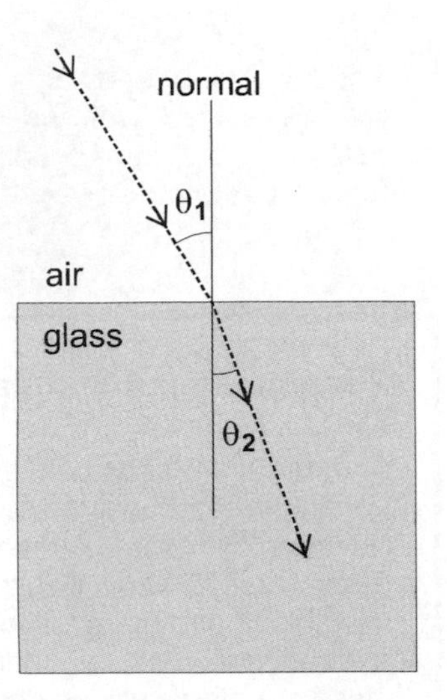

Figure 16.10 *When a ray of light passes from one medium to another it is refracted. Note that in this and later figures the grey tone represents the fact that the glass has a higher refractive index than the air,* not *that the glass is tinted. In fact, the glass used in optical fibres is exceptionally clear.*

Normal: A line perpendicular to the boundary at the point where the ray strikes it (Fig. 16.10).

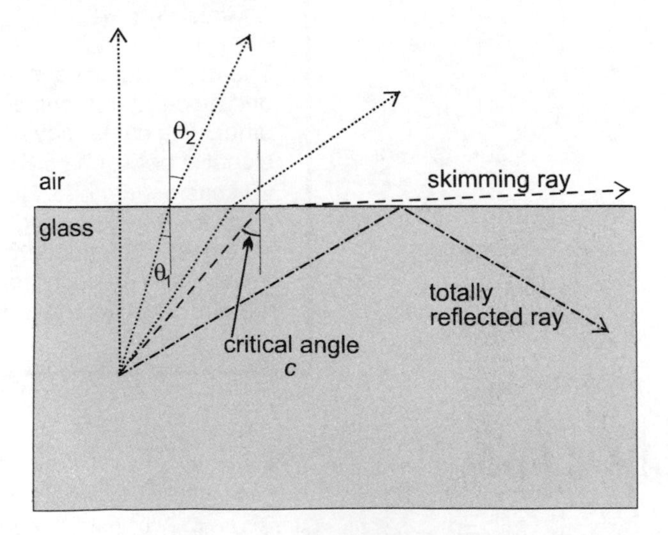

Figure 16.11 *If a ray of light is travelling in glass and meets the glass-air boundary at an angle equal to or greater than the critical angle, the ray is totally reflected back into the glass.*

The angles at which these effects occur depend on the wavelength (colour) of the light. This causes a beam of ordinary white light to spread into its component colours. The diagrams assume that the light is monochromatic so the ray remains narrow.

Monochromatic: light of a single wavelength.

In Fig. 16.12, light is passing along a narrow cylinder of glass. When it strikes the glass-air surface, it is not able to emerge into the air. This is because θ_1 is greater than c. The ray is totally reflected. It passes along inside the cylinder and may be totally reflected many more times

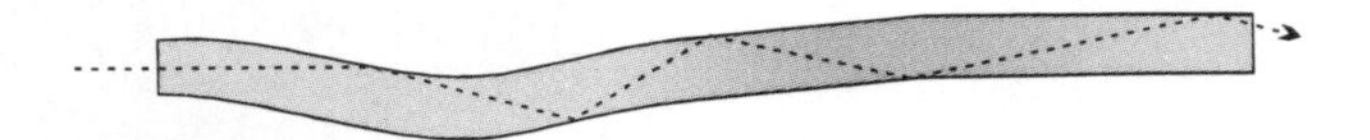

Figure 16.12 *A ray of light travelling in a glass fibre is totally reflected when it strikes the surface of the glass. If the fibre has slight bends in it, the ray is conducted along the fibre without loss.*

before it emerges from the other end. This is the basis of the *optical fibre*. Even if the fibre is curved, as in the figure, the light travels from one end to the other without emerging.

It may sometimes happen that a fibre is too sharply curved. Then θ_1 is less than c and the light emerges. Light also emerges if there are scratches in the surface of the glass, producing local regions where θ_1 is less than c.

A practical fibre optic cable consists of a narrow fibre of glass, the *core*, surrounded by a layer of glass of lower optical density, the *cladding*. Fig. 16.13 illustrated this type, known as *step index* cable. The light passes along the core and is totally reflected when it strikes the boundary between the core and cladding. Note that the reflecting surface is the inner surface of the cladding, so it protected from becoming scratched. In *graded index* fibre the refractive index decreases gradually from the centre of the core outward. The ray is bent gradually as it enters the outer regions and is eventually returned to the central region. Its path along the cable is smoothly curved instead of being a sharp zigzag. This type of cable produces less distortion of light pulses and is best for long-distance communication.

The glass used for fibre optic cables is of very high purity and clarity prepared in a clean environment. The loss of signal due to absorption in the glass is about 0.15 dB (3.5%) per kilometre. Cheaper cables

Refractive index

The amount by which a ray is refracted (Fig. 16.10) depends on the relative refractive index n of the two media. If Medium 1 is a vacuum or air ($n = 1$), the index for Medium 2 may be calculated from:

$$n = \sin \theta_1 / \sin \theta_2$$

For any given pair of media, n is a constant. In the figure, $\theta_1 = 31°$ and $\theta_2 = 20°$, making $n = 1.5$. This is typical for an air-to-glass index.

Test your knowledge 16.4

A ray of light travelling in air strikes an air-glass surface at an angle $\theta_1 = 35°$. The angle of the refracted wave $\theta_2 = 23°$. What is the relative refractive index of the glass?

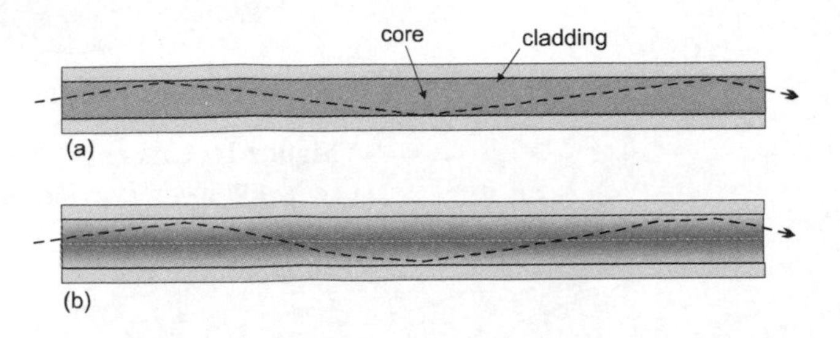

Figure 16.13 *Optical fibre usually has (a) a core of high refractive index surrounded by a cladding of low refractive index, or (b) the refractive index of the core decreases gradually from the centre out. In (a) the ray is totally reflected every time it strikes the boundary. In (b) it is gradually redirected back to the centre of the core whenever it deviates into the outer region.*

made from polymer are used for short-distance runs (up to 100 m). They have a higher absorption rate. Fibres of either type are enclosed in a tough plastic sheath for protection.

Light sources

The light signal is provided by a light-emitting diode or a solid-state laser, both of which produce monochromatic light. This in enclosed in a metal housing which clamps the LED firmly against the highly polished end of the fibre. LEDs are inexpensive and reliable, but their power is low. The light output is proportional to the current. The intensity (amplitude) of the light signal is easily modulated and this is the most commonly used modulation system.

Solid-state lasers are more expensive than LEDs and require more complicated circuits to drive them, but their power is higher. Their output may be modulated in intensity, frequency or phase, and also by the angle of polarisation. Like the LEDs, they are mounted in a metal housing.

At present, most of the LEDs and lasers operate in the infrared region but, with further research, operation is being extended into the visible light spectrum. Because of the very high frequency of light waves it is possible to modulate the light signal at very high rates, measured in terahertz (1 THz = 10^{12} Hz). Consequently, the bandwidth of optical channels is around 100 times wider than that of radio channels. Such a wide frequency range provides enormous bandwidth for stacking simultaneous transmissions on a single channel. There is room to carry up to 100 000 voice channels simultaneously over a single optical fibre channel. This is one reason why optical fibre is becoming more and more widely used for telecommunications.

Light receivers

The devices used for detecting the signal and converting it to an electrical signal need to have very high sensitivity and the ability to operate at high frequencies. The most commonly used receivers are avalanche photodiodes, PIN photodiodes and phototransistors. Avalanche photodiodes (APDs) are operated with reverse bias. When stimulated by light, a few electron-hole pairs are generated. These move in the electric field, striking other atoms in the lattice and generating more electron-hole pairs. The effect is similar to that in an avalanche Zener diode. The result is a rapid increase in the number of electron-hole pairs and a rapid increase in the current passing through the diode. The response time is about 200 ps, so the device is suited to high-frequency operation.

Intrinsic material: pure, undoped semiconductor, with low conductivity.

PIN diodes are so-called because they have a layer of intrinsic (i -type) material between the n-type and p-type material. This widens the gap between the p-type and n-type so that the capacitance between them is reduced. With reduced capacitance, PIN diodes are more suited for operation at high frequencies. The wider gap also increases the absorption of light at the junction, making the diode more sensitive. The diodes are reverse-biased in operation, and light causes the generation of electron-hole pairs, which increases the leakage current. In a circuit such as Fig. 16.14 the increased current through the resistor produces an increased voltage, which is then amplified.

Test your knowledge 16.5

What charge carriers are present in the depletion region of a reverse-biased diode?

A phototransistor is enclosed in a transparent package, or in a metal can with a transparent cap. It is connected in circuit in the same way as an ordinary npn junction transistor, except that it is not necessary to make any connection to the base. When light falls on the transistor it generates electron-hole pairs in the base layer. A current flows from base to emitter, which is equivalent to the base current in an ordinary junction transistor. Because of transistor action, a much larger current flows from collector to emitter. Their amplifying action makes phototransistors more sensitive than photodiodes, but less able to operate at very high frequencies.

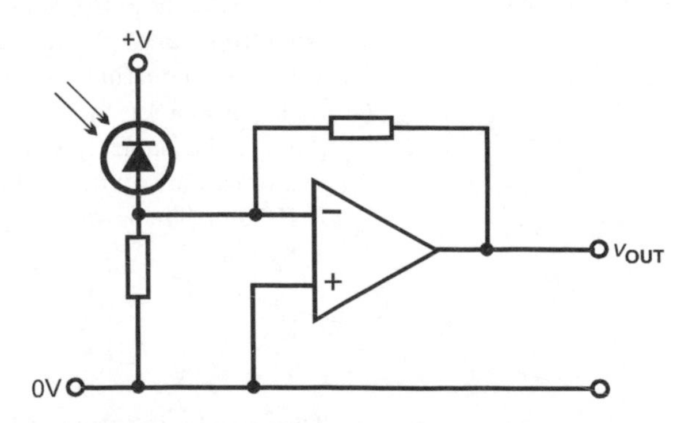

Figure 16.14 *A basic circuit for converting an AM optical signal into an AM electrical signal.*

Advantages of optical fibre

The main advantages of optical fibre are:

- **Bandwidth:** optical systems have a higher bandwidth than any other system of telecommunications. Many more signals can be frequency multiplexed or time multiplexed on to a single channel.
- **Cable cost:** the cost of optical fibre is greater than that of copper, but it is falling. Offset against this is the fact that, on long distances, repeater stations can be up to 160 km apart. Often they are much closer than this because the power required for the repeater unit is supplied along a pair of copper wires running in the cable alongside the fibres. This limits the distance between repeaters to about 10 km. Local power supplies are used if the distance between repeaters is greater than this. Even so, optical fibres give longer runs compared with copper cables, which need repeaters every few kilometres.
- **Immunity from EMI:** unlike copper cable, optical cable is immune from the effects of external electromagnetic fields. It is particularly suited to industrial applications where electrical machinery and heavy switching generate strong interference. There is no crosstalk between fibres bundled in the same cable.
- **Safety:** faults in copper cabling may generate excess heat, which may lead to fire. This is much less likely to occur with optical fibre.
- **Security:** surveillance equipment can be used to read messages being transmitted on copper cables. This is not possible with optical fibre.
- **Corrosion:** unlike copper cabling, optical cable does not corrode.

Radio transmission

When an electron is accelerated, it radiates energy in the form of electromagnetic waves. We can think of the waves spreading through free space like ripples on a pond, but in three dimensions. The wavefronts are spherical and continually expanding. Electromagnetic waves include (in order of increasing wavelength), gamma rays, X-rays, ultraviolet, visible light, infrared, microwaves and radio waves. All are the same type of radiation and have similar properties. For example, they all travel through free space with the same velocity, 300×10^6 m/s. Their wavelength (l) is related to their frequency (f) by the equation:

$$\lambda = \frac{3 \times 10^8}{f}$$

They can be reflected, refracted, and diffracted, for example, and they can be absorbed by the medium they are travelling through. But there are differences; for example, gamma rays and X-rays can penetrate dense matter such as metals but radiation of longer wavelength can not. These radiations and also ultra violet are ionising radiations. They cause free electrons to be knocked from the atoms of materials as they pass through. They can be very damaging, possibly lethal, to living tissues. They also produce the ionised layers of the upper atmosphere which are so important to radio transmission.

Radio waves have the longest wavelengths of all types of electromagnetic radiation. The shortest wavelengths (10 mm to 100 mm) are found in the Super-high Frequency band (30 GHz to 3 GHz). The longest wavelengths (10 km to 10000 km) are in the VLF band. All are used for telecommunications of various kinds, with interest nowadays being centred on the highest frequencies.

VLF: Very Low Frequency radio, 30 Hz to 30 kHz.

Many radio transmitters have an antenna that radiates the waves in all directions with more-or-less equal strength. It is *broadcast* to everyone within range. This is a feature of radio communication that other forms of telecommunication do not normally have. A radio signal spreads through free space. We can communicate with many people at once, as when a politician or a disc jockey speaks to their audience by radio. This is a facility which, until recently, only radio could provide. Now a few countries are installing cable networks, which are the equivalent of radio broadcasting. The costs limit this to the most densely populated areas of the more wealthy countries. Another benefit of radio is that we can broadcast messages to particular individuals without having to know exactly where they are. Ship-to-shore radio, mobile telephones and many paging systems rely on radio to convey messages to receivers that are not in a known place.

Test your knowledge 16.6

Calculate the longest wavelength of the VLF band.

Ground waves

A signal that is broadcast in all directions can travel from the transmitter to the receiver in two ways. If the frequency is less than 500 kHz, it spreads close to the surface of the Earth as a *ground wave.* It is not able to pass out of the atmosphere because of the ionised layers referred to earlier. These have a lower refractive index than the atmosphere, so the radio waves of lower frequency (below 100 kHz) are reflected back toward the Earth's surface. Ground waves of lower frequency are more affected by the nature of the surface, particularly when it consists of moist soil or water. The wave fronts nearest the surface are slightly held back so that, instead of being vertical, they are tilted forward (Fig. 16.15). The result of this is that the waves follow round the curvature of the earth. VLF transmissions can travel for thousands of kilometres in this way and can reach almost any part of the world.

Unfortunately, the surface of the Earth is not completely smooth. Ranges of hills can block the path of ground waves so that the signal does not reach areas just beyond the hills. This limits the range of transmission of ground waves in many terrains. However, if the rang

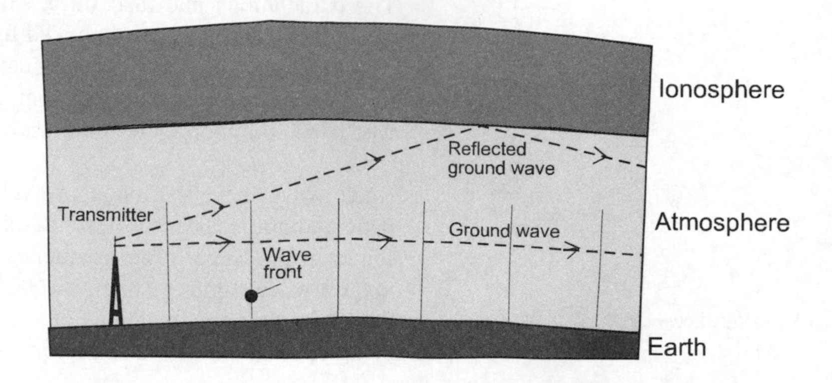

Figure 16.15 *Ground waves travel in a direction more or less parallel to the Earth's surface. Although low-frequency ground waves can encircle the Earth, those of higher frequency have relatively short range and are frequently blocked by high ground. This figure is not to scale.*

has a sharp edge, the waves may be diffracted and then stations immediately behind the hills are able to receive the signal.

Sky waves

As frequency increases above 500 kHz the radio waves are able to penetrate the lower ionised layers. As they pass upward through the layers the refractive index is decreasing. The effect is similar to that of light passing along a graded index fibre. The waves are refracted along a curved path and are eventually directed back to Earth again. These are known as *sky waves* (Fig. 16.16). These reach the earth at a

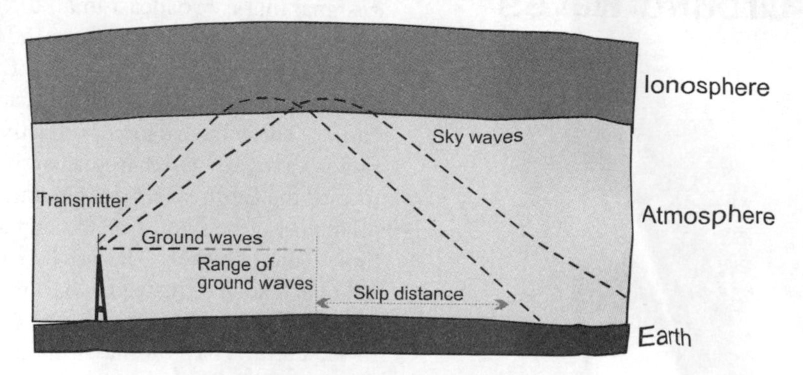

Figure 16.16 *Radio waves with frequencies between about 500 kHz and 25 MHz are refracted back to Earth in the ionosphere, because of the lower refractive index of the ionised layers. They then become sky waves, which are important for long-distance communication. This figure is not to scale.*

Ionised layers

Ultra violet radiation from the Sun causes molecules in the upper atmosphere to become ionised. Several ionised regions form at various heights above the surface. Together, these are known as the *ionosphere*. They occur from about 50 km above the surface up to about 750 km. The regions are not permanent, varying with time of day, the season, or the number of sunspots. This is why long-distance radio transmission is not entirely reliable.

considerable distance from the transmitter and are important for long-distance communication. The main drawback is that reception of sky waves varies with the state of solar activity, with regional weather, and with the time of day. At a certain distance from the transmitter, there may be a point that is too far away for the ground waves to reach, but is too near for the sky waves to reach. The signal is not received at this distance, which is known as the *skip distance*.

Because of the curvature of the earth and the ionised layers, some of the sky waves skim past the Earth's surface and once more head up toward the ionised layers (Fig. 16.17). Once again they are refracted back toward the surface. In this way the signal from a sufficiently powerful transmitter may *hop* around the world.

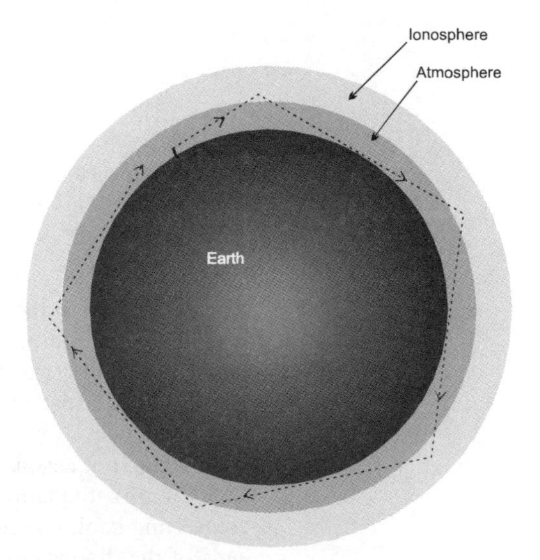

Figure 16.17 *Sky waves may be repeatedly refracted in the ionosphere and circle round the Earth several times in good conditions. This produces annoying echoes, repeated at intervals of about 1/7 second. This figure is not to scale.*

Above about 25 MHz, radio waves are able to penetrate the ionosphere and pass out into space. These frequencies are therefore suitable for communication with spacecraft on interplanetary missions. On the Earth's surface the transmission path must be a straight line from transmitter to receiver, without any reflections on the way. These are known as *line-of sight transmissions,* and are widely used in communications links. Repeater stations are located on high ground to relay the transmission to areas out of sight of the main transmitter.

Antennas

There are many types of antenna, but one of the most often used in the dipole (Fig. 16.18). It consists of a vertical metal rod or wire, slightly less than half a wavelength long, and divided into two sections. It is not practicable to build a tower to support a dipole for low-frequency transmitters so other types of antenna are used. Dipoles can also be mounted horizontally.

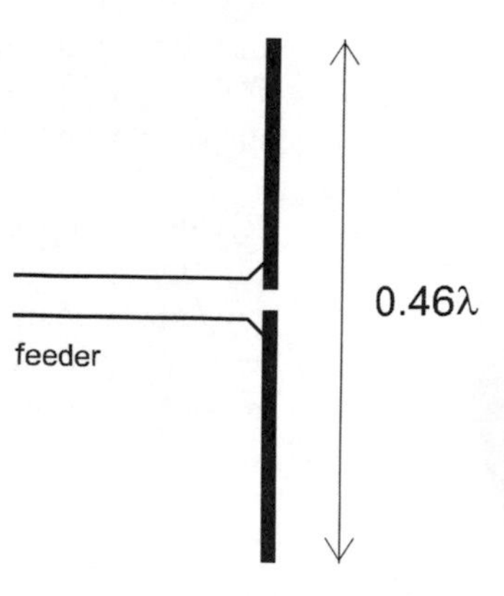

Figure 16.18 *The motion of the electrons in the standing waves of a half-wave dipole antenna generates radio waves.*

The two halves of the dipole are fed from the output of the transmitter, usually by coaxial cable (75 Ω). An alternative is a special *feeder cable*, which has two stranded copper conductors running parallel with each other and spaced about 9 mm apart. Typically it has a characteristic impedance of 300 Ω. The halves of the dipole are insulated from each other so it acts as an open circuit at the end of a transmission line (Fig. 16.5b). Standing waves appear in the dipole (and also in the connecting cable). A node of the voltage waves is located at the outer ends of the dipole and an antinode at its centre. This assumes that the total length of the two dipole elements is equal to half a wavelength. In practice, the velocity of electromagnetic waves in the conductor is less than it would be in free space, so the element is made 0.46λ long instead of 0.5λ.

The dipole is an example of a *standing wave antenna*, in contrast to some other types, which are *travelling wave* antennas. The formation of standing waves means that the dipole is resonating at the frequency of the carrier wave, and radiating part of the energy fed to it as electromagnetic radiation. At a receiver, a dipole antenna intercepts arriving electromagnetic radiation, which induces currents in it. If the frequency is correct, the antenna resonates and standing waves are formed in it. The antenna receives signals of its resonant frequency much more strongly than those of other frequencies. This helps to make a receiver more selective. Although a dipole transmits or receives best at its resonant frequency, it is also able to operate reasonably well at frequencies that are within ±10% of this.

Directional transmissions

The vertical dipole radiates equally strongly in all horizontal directions so is suitable for broadcasting. But it is wasteful of power to broadcast a signal in all directions when it is intended only for one particular receiver. The dipole may be made more directional by adding further elements to it. Fig. 16.19 shows a dipole with two *parasitic elements,* known as the *director* and the *reflector*. These have about the same length as the dipole elements. In an antenna being used on a transmitter the parasitic elements pick up the signal from the dipole, resonate at the signal frequency, and re-radiate the signal. They act as two signal sources spaced certain distances apart so that they reinforce the signal being radiated from the dipole. The signal is stronger than that radiated from the simple dipole, but is concentrated in one direction. The array may have further directors added to it to increase its gain and make it even more directional. Multiple directional dipoles are also known as *Yagi arrays*.

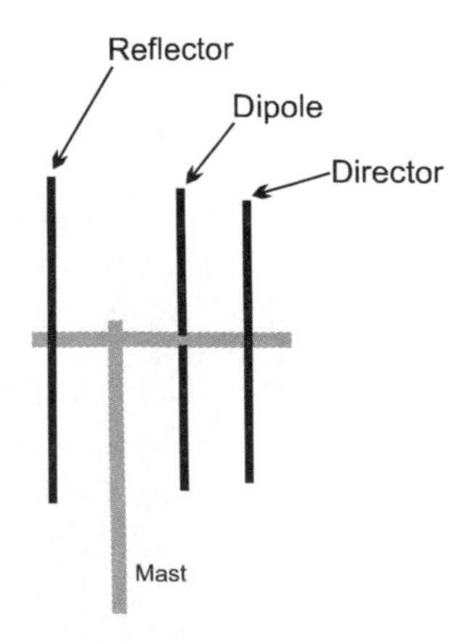

Figure 16.19 *Addition of a reflector and director makes the dipole antenna more directional and increases its gain.*

As a receiving antenna, a dipole with parasitic elements is used for picking up signals from weak or distant stations. A common example is the antenna used in fringe TV reception areas. The electromagnetic field causes electrons in the elements to oscillate to and fro. If the frequency of the incoming transmission is correct, the elements all resonate to the incoming signal and re-radiate it. At the dipole, their radiations are added to the incoming signal, so increasing the signal passing to the receiving circuit. Standing waves are set up in the dipole and the parasitic elements. Some of the energy of these waves passes to the receiving circuits. Arrays with up to 18 parasitic elements are used for reception of TV signals in the most difficult areas.

The use of directional dipoles makes it possible to set up microwave links with repeater stations at intervals. Theoretically the stations at either end of a link should be able to 'see' each other, for this is line of sight transmission. The range depends partly on terrain. Repeater stations are generally located on hill-tops or on mountain-tops when available. Other local conditions may extend the range slightly above the line-of sight range. In practice, the maximum range is about 50 km, though it may be as far as 100 km in some localities. The number of microwave repeater stations required on a long route is far fewer than for cable or optical fibre links.

The use of a resonant (or standing wave) antenna is important in many types of radio communication. It allows reception of weak signals from distant transmitters or, conversely, allows us to use transmitters of lower power. In some applications, antennas half a wavelength long are too long for convenience. For example, for radio pagers in the VHF band the wavelength is at least 1 m, so a half-wave dipole would have to be at least 50 cm long. This is not convenient for a device that is to be carried in a jacket pocket. A much shorter antenna is used in such equipment, but tuning components are added to the receiver circuit to produce the equivalent of a resonating antenna.

Radio waves may be reflected by a metallic surface, but the surface does not have to be a continuous sheet. Perforated metal or a wire mesh may be used for cheapness and light weight. For reflection to occur, the reflector or *dish* must have dimensions that are several times greater than the wavelength. Transmissions are most directional when the antenna is mounted in front of a reflector which is parabolic in section (Fig. 16.20). If the antenna is placed at the focus of the reflector, all radiation incident on the reflector, is reflected as a parallel-sided beam. Conversely, all rays arriving from a distant transmitter, being parallel with each other, are concentrated on an antenna placed at the focal point. Radio-telescopes have very large dishes and similar dishes are found at the ground stations of satellite communications systems. Smaller reflectors are used in line-of-sight communications links and for receiving satellite television at home.

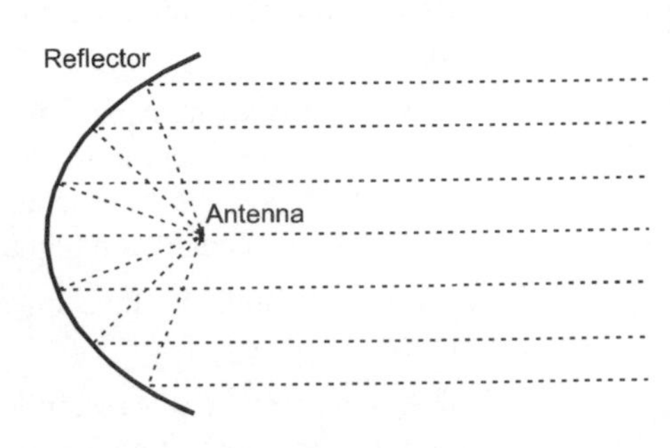

Figure 16.20 *For frequencies in the megahertz bands and above, a parabolic dish concentrates the energy from the antenna into a narrow beam. Conversely, the dish may concentrate a weak signal on to the antenna to enable a distant transmitter to be heard.*

Problems on transmission media

1 Compare the structure and properties of a twisted pair with coaxial cable.

2 Explain what is meant by the lumped equivalent of a transmission line. Draw the lumped equivalent for a line carrying a high-frequency signal.

3 What is meant by the *characteristic impedance* of a transmission line? Why is it important for the impedances of source and load to be matched to that of the line?

4 Explain why, in theory, there is no loss of signal along a transmission line.

5 Describe how a signal source and a load may be transformer-coupled to a balanced transmission line.

6 What is meant by *crosstalk* and how can it be reduced or avoided?

7 Describe what happens when a ray of light travelling in air meets an air-to-glass surface.

8 Describe what is meant by *critical angle*.

9 What is step index optical fibre and how does it work?

10 Describe how an LED is used as the light source for optical fibre.

11 Describe the action of an avalanche photo-diode as an optical fibre receiver.

12 What are advantages of optical fibre as a medium for telecommunications?

13 Describe the propagation of (a) ground waves and (b) sky waves.

14 Give an example of (a) reflection, (b) refraction, and (c) diffraction of radio waves.

15 Describe the structure and action of a half-wave dipole antenna.

16 Give examples of the use of radio for (a) broadcasting and (b) person-to-person communication.

17 Describe how a parabolic reflector is used for directional radio transmission.

Multiple choice questions

1 At high frequencies the most important properties of a transmission line are:

A resistance and capacitance.
B capacitance and inductance.
C resistance and conductance.
D inductance and conductance.

2 When we say that the load on a transmission line is matched, we mean that:

A it has the same capacitance as the line.
B the end of the line is short-circuited.
C it has the same impedance as the line.
D standing waves are formed along the line.

3 A balanced transmission line is one in which:

A one conductor is grounded.
B both conductors carry the same signal.
C the line is shielded against electromagnetic interference.
D the signals on the conductors are the inverse of each other.

4 Total reflection occurs when a ray of light:

A passes from air to glass.
B is refracted away from the normal.
C strikes the surface at an angle less than the critical angle.
D strikes the surface at an angle greater than the critical angle.

5 It important not to bend an optical fibre too sharply because:

A it may break.
B it alters the refractive index.
C light may escape at the bend.
D the sheathing plastic may split.

6 The electromagnetic waves of longest wavelength are:

A infra red.
B radio.
C gamma rays.
D ultraviolet.

7 Sky waves are refracted in the:

A ionosphere.
B atmosphere.
C stratosphere.
D free space.

8 The frequency of ground waves is:

A up to 500 kHz.
B over 500 kHz.
C above 25 MHz.
D below 30 kHz.

17 Noise

Summary

Noise is any unwanted signal in an electronic circuit. It may be produced by electromagnetic interference within the circuit or from nearby equipment, by mechanical effects, by the nature of electrical conduction, or as a result of signal processing. The extent to which a signal is degraded by noise is expressed as the signal–to–noise ratio. Noise is reduced by various methods depending on its origin. These include shielding and filtering, or circuit designs intended to minimise noise. Noise is an important factor in telecommunications, making it necessary to use repeaters and regenerators on long transmission paths.

Sources of noise

Noise is any unwanted signal and every kind of circuit is subject to it. When a radio receiver is turned up to full volume but is not tuned to a station; there is a steady hiss from the speaker. This noise is *noise* in the electronic sense. It may have originated in the circuit of the radio set, in the surrounding district, or in space. But noise does not have to be audible noise. A TV set that is not tuned to a station shows 'snow' on its screen. This too is noise. In digital data transmission, noise is the erratic reception of 1s for 0s or 0s for 1s. Obviously, noise is something that must be reduced or preferably eliminated.

EMI

There are many sources of *electromagnetic interference*:

- **Electrical machines:** These include everything from a domestic refrigerator to heavy industrial machinery. Whenever switch contacts open there is a momentary arcing. If the load is inductive, as it often is (for example, switching an electric motor) the e.m.f generated between the contacts is high, perhaps hundreds of volt The arc is a high current in which electrons are being rapic accelerated. This generates an electromagnetic field, a radio w which spreads for an appreciable distance from the sou

It produces a series of pulses in any conductor that it meets. Motor vehicles come into this category for their ignition systems generate their sparks by inductive circuits, producing electromagnetic radiation at the same time. Electromagnetic noise can be picked up by a circuit, and possibly be amplified by the circuit. Or it may be picked up on mains wiring and enter equipment through the power supply circuit.

- **Mains hum:** In mains-powered equipment the 50 Hz (or 60 Hz in USA) mains frequency may pass through the power supply circuits and appear as *ripple* on the output. In may also enter the circuit by induction in an earth loop (Fig. 17.1). Another source of noise arising in the power supply is high-frequency *switching noise,* which is generated in switch-mode power supply units.
- **Thunderstorms:** Flashes of lightning are similar to the arcing in electrical machines, though on a larger scale. They produce intense electromagnetic fields that can be heard as noise on a radio receiver many kilometres away.
- **Cosmic electromagnetic action:** Cosmic radiation arriving from Outer Space produces electromagnetic waves when it interacts with the ionosphere, and these are picked up by sensitive electrical equipment.

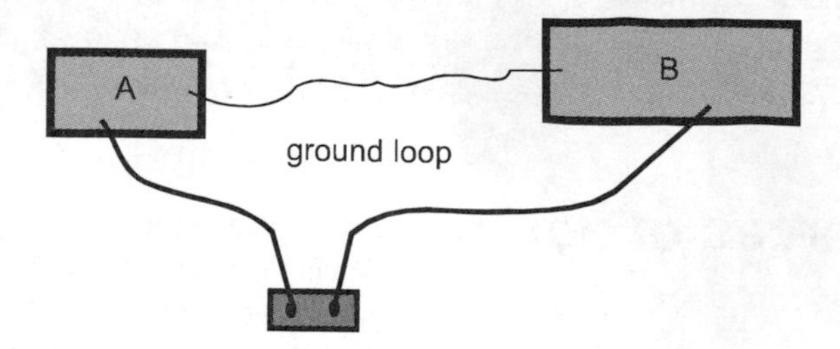

Figure 17.1 *Two pieces of equipment are plugged into mains sockets; both are connected to the mains earth line. A shielded cable joins them, and the shield is earthed at both ends. This completes a loop. Any electromagnetic fields passing through this loop generates a current in it, causing interference in the equipment. To prevent interference, break the loop by earthing the connecting cable at only one end.*

Mechanical effects

In an amplifier based on thermionic valves, any vibration of the equipment may lead to vibrations of the electrodes within the valves. This makes the characteristics of the valve fluctuate, causing a noise signal to be generated. The effect is known as *microphony*. It is not a common problem nowadays, when valves are seldom used, but a similar effect is found in cables. If the cables are vibrated, the capacitance between the conductors varies, producing a noise signal.

The noise sources described above are more often classified as *interference*, rather than as noise. They can often be reduced or even eliminated by shielding, by filtering, or by good design and layout. They are not discussed any further here.

Electrical conduction

The types of noise listed next all arise through the nature of electrical conduction. They occur in all electronic circuits and are difficult to reduce:

- **Thermal noise:** This is also known as *Johnson noise*. Electrical conductors all contain charge carriers, most often electrons. Although there is a *mass motion* of carriers when a current is flowing, there is always a *random motion* superimposed on this. Their motion is made random by forces between the carriers and the molecules in the lattice as the carriers pass by. Random motion produces random changes in *voltage*, called *thermal noise* (Fig. 17.2). This is what causes the hissing in the output of an audio circuit. It is similar to the noise made by other random processes such as steam escaping from a steam valve, or water running through shingle on the seashore. The amount of thermal noise depends on the amount of random motion of the charge carriers and this depends on temperature.

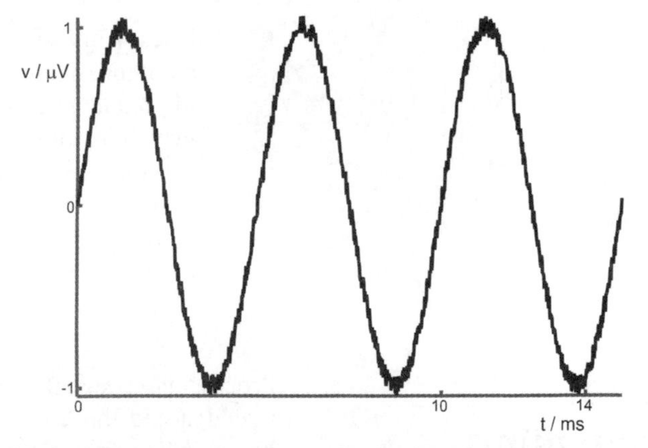

Figure 17.2 *A 200 Hz sinusoid, amplitude 1 µV, has picked up a 60 nV noise signal, mainly due to thermal noise. The sinusoid now has a randomly spiky form and, if this noisy signal is sent to a speaker, we hear a hissing sound as well as the 200 Hz tone.*

At higher temperatures, the carriers gain extra energy from vibrations of the crystal lattice of the conductor, and so random motion becomes more important. The noise level depends on the absolute temperature, so a conductor needs to be cooled to absolute zero to eliminate this type of noise. Thermal noise also depends on resistance and on bandwidth. Being completely random, thermal noise has no particular frequency. Or we can take the other viewpoint and say that it contains *all* frequencies at equal amplitudes. Because it contains all frequencies, it is often called *white noise*. If a circuit has a wide bandwidth, a wider range of noise frequencies can exist in it. The total amount of noise in the system is greater. Because thermal noise depends on resistance, it is produced by all components in a circuit, but particularly with those that have high resistance.

- **Shot noise:** With a steady current the *average* number of charge carriers passing a given point in a circuit is constant. But the number passing *at any instant* varies randomly. The random variation of *current* produces another form of white noise called *shot noise*. Shot noise is a variation of current but the current passes through resistances in the circuit and so becomes converted to variations in voltage. So the effect of shot noise is added to thermal noise. Like thermal noise, shot noise includes all frequencies, so the amount of it in a circuit increases with bandwidth. If currents are large (A or mA) the random variations in current are relatively small and shot noise is low. But for currents in the picoamp range, shot noise is *relatively* greater, and may have an amplitude that is 10% or more of the signal amplitude. Shot noise occurs most in semiconductors.
- **Flicker noise:** This is also known as *1/f noise* because its amplitude varies inversely with frequency (Fig. 17.3). It is only important when signal frequency is less than about 100 Hz, and may be ignored at frequencies greater than 1 kHz. The cause of it is unknown. It occurs in semiconductors, possibly because of imperfections in the material. It also occurs in valves, where it is possibly an effect of irregularities in the surface of the cathode.

Signal processing

Processing a signal does not necessarily produce the exact result intended and this can give rise to noise. An example is *quantisation noise*. When an analogue signal is being converted to digital form, a smoothly varying analogue voltage is converted to a stepped digital voltage. The steps introduce an irregularity into the signal which has the same effect as noise. If the conversion has a resolution of 12 bits or more, the stepping usually can be ignored. One bit in 2^{12}, is 1 in 4096, which is an error of only 0.024%.

Signal–to–noise ratio

There are several ways of expressing the amount of noise present in a signal. It is not usually the absolute power of the noise that is important, but rather the power of the noise relative to the power of the signal itself. For this reason, one of the most often used ways of expressing

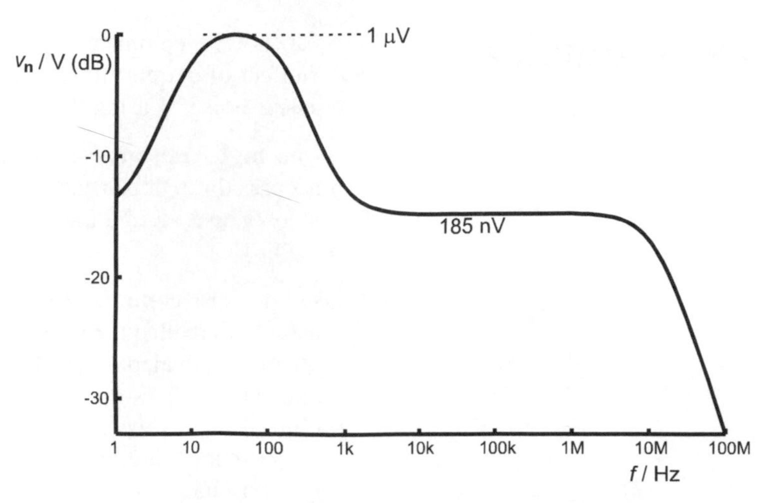

Figure 17.3 *This graph is an analysis of the common-emitter amplifier of Fig. 2.5. It plots the amplitude of the noise signal at the output of the amplifier for frequencies in the range 1 Hz to 100 MHz. Over most of the audio range the noise is 185 nV, due to thermal noise and shot noise. The noise level drops at both very low and very high frequencies because these frequencies are outside the bandwidth of the amplifier, so they are not amplified.*

rms: root mean square value, approximately 0.7 of the amplitude, if the signal is sinusoidal.

the amount of noise is the *signal–to–noise power ratio.* A ratio of two powers is most conveniently stated on the decibel scale. Because power is proportional to voltage squared, the signal-to-noise ratio is defined as:

$$\mathrm{SNR} = 10 \log_{10}(v_s^2/v_n^2)\ \mathrm{dB}$$

In this equation, v_s is the rms signal voltage and v_n is the rms noise voltage. The equation has the standard form used for expressing a ratio of powers in decibels.

Test your knowledge 17.1

If the rms noise level is 15 nV and the rms signal level is 350 mV, how much is the SNR?

Example

A signal has an rms value of $v_s = 2.4$ V. The rms noise level is $v_n = 7$ nV. Calculate the signal-to-noise ratio.

$$\mathrm{SNR} = 10 \log_{10}(2.4^2/(7 \times 10^{-9})^2) = 171\ \mathrm{dB}.$$

The SNR is used to calculate the degree to which a circuit, such as an amplifier, introduces additional noise into an already noisy signal. If $\mathrm{SNR_{IN}}$ and $\mathrm{SNR_{OUT}}$ are the SNRs of the signal at the input and output, the *noise figure* of the amplifier is given by:

$$\mathrm{NF} = \mathrm{SNR_{IN}}/\mathrm{SNR_{OUT}}$$

Noise reduction

Noise can not be eliminated from a circuit because noise is an inevitable effect of current flowing through components. But it is possible to reduce noise by using the techniques listed below:

- **Avoid high resistances:** Thermal noise is generated when currents pass through resistance and is proportional to the resistance. For low noise, design the circuit so that resistors are as small as possible.
- **Use low-noise components:** Some components are specially manufactured to reduce their noise levels. The BC109 transistor is the low-noise equivalent of the BC107 and BC108 because it has low flicker noise. FETs have no gate current, so no thermal noise is produced. JFETs have less shot noise and flicker noise than BJTs because there is only a small leakage current through their gates. MOSFETs have no shot noise. Many of them have low resistance, so thermal noise is low too. Using FETs in a circuit before the signal is amplified avoids introducing noise that will be amplified by later stages. Often a two-stage amplifier has a MOSFET for the first stage, followed by a BJT in the second stage. Low-noise versions of op amps are another way of avoiding noise.
- **Use low currents:** These result in low thermal noise so it is best to keep current low in the early stages of an amplifier or similar circuit. A collector current of 1 mA is suitably low for a BJT amplifier stage.
- **Restrict bandwidth:** Fig. 17.3 shows that the noise levels in the amplifier fall off at low and high frequencies because they are outside the bandwidth of the amplifier. Restricting the bandwidth can further reduce noise. In this particular amplifier the upper −3dB point is at 11 MHz. For many applications there is no need to pass signals of such high frequency. In these cases, reducing the high-frequency response will eliminate much of the noise, without affecting harming its performance in other ways. The bandwidth of radio frequency amplifiers needs to cover only the carrier wave frequency and the sidebands which occur about 15 kHz to either side of this. Restricting the bandwidth to such a narrow range means that noise in very much reduced. In general, noise may be reduced by restricting the bandwidth of a circuit to just that range of frequencies it needs to do its job.

Noise in telecommunications

In all three modes of telecommunications the signal gets progressively weaker as the distance from the transmitter decreases. We say that it is *attenuated*. Attenuation is usually expressed in decibels. The amount of attenuation depends on the transmission medium. In transmission lines, signals of higher frequency are attenuated more than those of low frequency. For example, in RG58 coaxial cable attenuation is 0.33 dB per 10 m of cable at 1 MHz, rising to 7.65 dB at 1 GHz. This limits the rate at which data can be transmitted. Optical fibre has much smaller attenuation, typically 0.15 dB per *kilo*metre. Radio transmission in free space obeys the inverse square, so that doubling the distance reduces the signal to quarter strength (–2.5 dB). So much depends on other factors such as absorption in the atmosphere and refraction in the ionosphere that it is not possible to produce reliable attenuation figures. Under favourable conditions radio signals can be directed to and received from spacecraft and can circle the world, so the exact rate of attenuation is not easy to specify.

Test your knowledge 17.2

The attenuation of 100 m of a certain cable is 4.7 dB. What is its attenuation at 200 m?

Whatever the medium, the further it travels the weaker a signal becomes. Yet the longer the communications link the more chance of picking up interference and noise. Interference can be minimised by shielding and filtering. It is least in fibre-optic systems.

True noise is introduced at the transmitter. In many systems the transducer is resistive and generates noise on its output signal. Further noise is introduced by the amplifying and transmitting circuits. The amplifiers at the receiver also introduce noise. Noise may be kept to a minimum by using some of the general electronic techniques listed above. There are also some measures specially applicable to telecommunications systems:

- **Companding:** This term is a combination of the words *compressing* and *expanding*. The principle is to amplify the signal using a special amplifier that amplifies the lower-power signals more than the higher-power signals. In terms of audio, a lower-power signal is the sound of a subdued instrument such as a triangle, and the sound of all instruments in their quietest passages. Also the higher harmonics of all instruments are of low power. In a compander, all of these are boosted relative to other more powerful signals. The loudest instruments remain the loudest and the quietest remain the quietest but the range of power from quietest to loudest is compressed. This takes the lowest powers of the wanted signal up above the general noise level. Such selective amplification produces a distorted signal overall, but the balance is restored at the receiving end by a circuit which has the reverse action. There the lower-power signals are amplified less than the high-power signals. The range of signal powers is expanded and the original balance of the signal is restored.
- **Frequency modulation:** Noise and interference are voltage signals. If they are superimposed on an amplitude-modulated signal, they affect the amplitude and are still there after the signal has been demodulated. This does not happen with frequency modulation or phase modulation. Once the carrier has been frequency modulated, random increases or decreases of amplitude have no effect on the *frequency* so any noise acquired after modulation is lost on demodulation.

- **Repeaters and regenerators:** The signal–to–noise ratio decreases with increased transmission distance so it is essential to restore the signal at intervals along the transmission path. For analogue signals there are *repeater* stations at regular intervals along the path. These receive the signal, filter it to remove noise and interference as much as possible, amplify it, and finally re-transmit it. For digital signals there are *regenerator stations*. These receive the signal and have the relatively easy task of distinguishing between the 0's and 1's. Even if the signal is noisy and the pulses are badly distorted, the original sequence of binary digits can usually be recovered. The repeater station then transmits a new stream of perfectly formed pulses exactly representing the original signal. This is one reason why digital data is preferred for long-distance telecommunications.

Problems on noise

1 Name six source of interference in electronic circuits and state how they may be avoided or reduced.

2 Name three sources of noise in electronic circuits and state their cause.

3 What is quantisation noise?

4 Define the signal–to–noise power ratio. If the rms signal voltage is 120 mA and the rms noise voltage is 2.7 μV, what is the SNR?

5 Describe how to avoid noise in a BJT amplifier.

6 Describe how to avoid noise in a radio-frequency amplifier.

7 Explain the difference between a repeater and a re-generator on a long transmission line.

8 What is companding and how does it help reduce noise?

Multiple choice questions

1 The noise produced by the random motion of charge carriers is:

A thermal noise.
B shot noise.
C flicker noise.
D $1/f$ noise.

2 The noise produced by the varying numbers of charge carriers in a very small current is:

A thermal noise.
B shot noise.
C flicker noise.
D $1/f$ noise.

3 The main sources of shot noise in a circuit are:

A semiconductors.
B resistors.
C any form of resistance.
D capacitors.

4 One way to avoid noise in a two-stage amplifier is to:

A restrict its bandwidth.
B use a BJT in the first stage.
C set the collector current to about 10 mA.
D use biasing resistors of high value.

5 An FM transmission is less subject to noise than an AM transmission because:

A its carrier has a higher frequency.
B its amplitude is greater than that of the noise.
C noise consists of all frequencies.
D noise does not alter its frequency.

18 Radio communication

Summary

This chapter is an extension of Chapters 15 and 16. It describes radio communication systems in more detail, including the main stages of transmitters and receivers and the supersonic-heterodyne receiver.

Transmitters

The main stages of a typical radio transmitter are illustrated in Fig. 18.1. It comprises the following stages, though some may be absent from particular types of transmitter:

- **Input:** This is the source of the information that is to be transmitted. Examples are listed under 'Source of information' in the table on p. 231.
- **Transducer:** Examples are listed under this heading on p. 231.
- **Amplifier:** The signal from the transducer may have insufficient

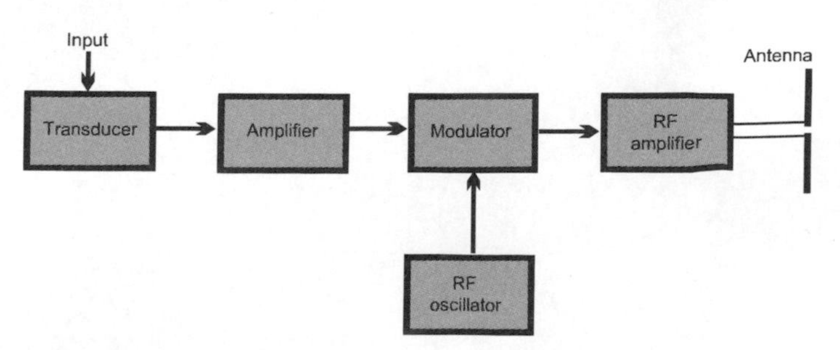

Figure 18.1 *A block diagram of a radio transmitter. The modulator mixes the signal with the oscillator output to produce a modulated carrier wave.*

amplitude, so it needs amplification at this stage. Usually this stage is an audio frequency amplifier, but an amplifier with wider bandwidth is needed for video transmissions and for high-speed digital data links. The diagram does not show an encoder, but there might be one at this stage.

- **Modulator:** this takes the amplified signal and modulates it on to the output of the radio frequency oscillator. In an amplitude-modulated (AM) transmitter, the modulator often consists of a special multiplying ic. This has two inputs. One input (the carrier) is multiplied by a varying amount determined by the instantaneous value at the other input (the signal). The output of the ic is the amplitude-modulated carrier. In a simpler system of amplitude modulation there is no modulator as such (Fig. 18.2). The amplified signal is used as the supply voltage of the RF oscillator. This makes the amplitude of the carrier increase or decrease as the signal. There are several types of modulator for frequency modulated (FM) transmitters. A simple technique for frequency modulation is to include a varactor in the tuned network of the RF oscillator. The signal is applied to the varactor so as to vary the capacitance of the network. This makes the resonant frequency of the network vary according to the signal. The output from the network is the frequency-modulated carrier.

Varactor: a diode-like device of which the capacitance depends on the voltage across it. Also known as a *varicap diode*.

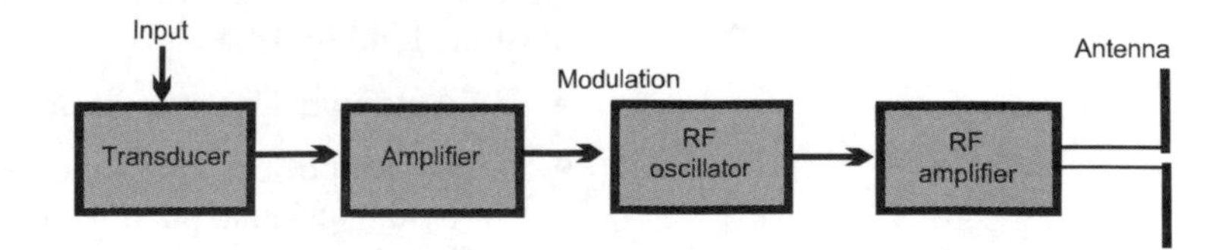

Figure 18.2 *In simpler designs of transmitter the signal from the amplifier acts directly on the oscillator circuit to modulate its output.*

Test your knowledge 18.1

Calculate the impedance of a 1 nF capacitor at 10 MHz.

Test your knowledge 18.2

Calculate the impedance of a 3 mH inductance at 10 MHz.

- **Oscillator:** Sinewave oscillators are described in Chapter 5.
- **RF amplifier:** An RF amplifier is used to increase the power of the transmitter. Fig. 18.3 shows a common-emitter BJT amplifier (from Fig. 2.5) modified as an RF amplifier. The transistor specified has good response at high frequencies. The input and output capacitors are only 1 nF so that they readily pass high frequencies. There is an inductor instead of the usual collector resistor. This has high impedance at high frequencies so a large voltage is developed across it, giving increased gain. The RF amplifier has only a narrow bandwidth, centred on the carrier frequency and extending on one or both sides to include the sidebands (Fig. 15.3).
- **Antenna:** This stage is described in Chapter 16.

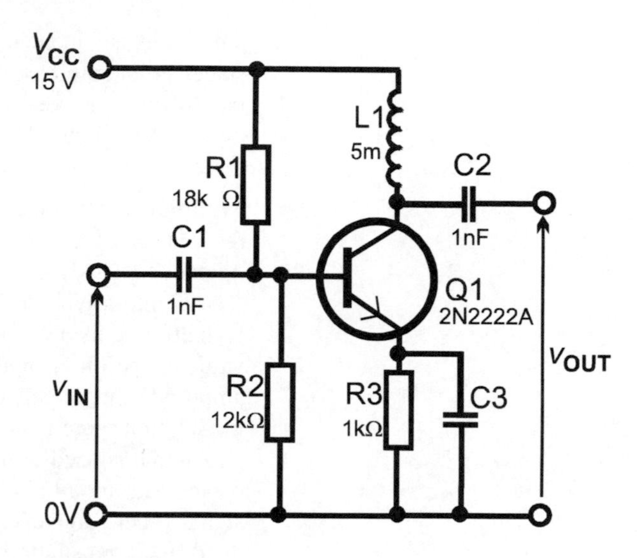

Figure 18.3 *Compare this RF amplifier with the common-emitter amplifier of Fig. 2.5.*

Receivers

In many ways a receiver is like a transmitter in reverse, but there are differences. The stages of a TRF (tuned radio frequency) receiver are illustrated in Fig. 18.4:

- **Antenna:** This stage is described in Chapter 16.
- **RF amplifier:** This not only amplifies the weak signal received by the antenna, but also plays a part in picking out one particular signal from dozens of other signals being received at the same time. The RF amplifier is a *tuned amplifier* which includes a resonant sub-circuit consisting of a capacitor and inductor. It amplifies the frequency to which it is tuned and the sidebands of the transmission, but not lower or higher frequencies. Usually the signal passes through 2 or 3 such tuned amplifiers as this gives a better separation between the wanted signals (carrier plus sidebands) and the unwanted signals (of lower and higher frequencies).

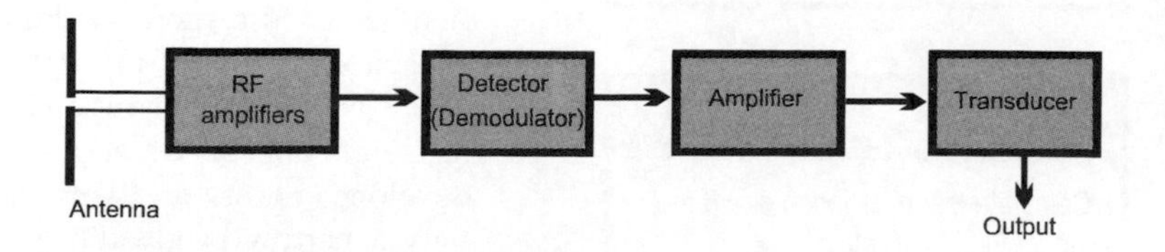

Figure 18.4 *A block diagram of a TRF radio receiver. The RF amplifier is tunable to pick up a transmission of a particular carrier frequency.*

- **Detector:** this stage demodulates the signal, but in radio receivers it is usually given the name detector. In an AM system the detector is essentially a rectifier diode, followed by a lowpass filter. Fig. 18.5 illustrates a simple detector circuit and Fig. 18.6 is a demonstration of its action. The modulated carrier (grey) is rectified by the diode. Only the positive-going half of the signal can pass through it. C1 and R1 then filter this. The output of the circuit (black) shows that the low-frequency component (the modulating signal) passes through the filter, but the high-frequency component (the carrier) is practically eliminated.
- **Amplifier:** The demodulated signal may need amplification. The bandwidth of this stage is suited to the frequency range of the signal. An audio amplifier is used for receiving a transmission of speech and music.

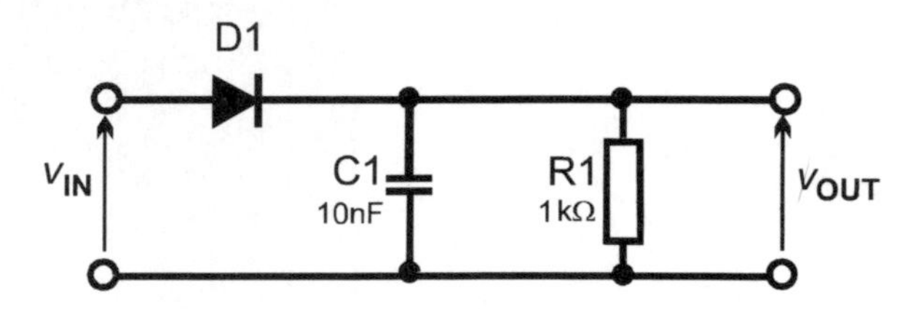

Figure 18.5 *A simple diode detector circuit demodulates AM transmissions. Because of its lower forward voltage, a germanium diode is generally used in this circuit.*

Crystal set

Semiconductor diodes had not been made in the early days of radio. Among the various detection devices that were used, the most popular was the 'crystal'. This is a piece of the mineral *galena*, which consists mainly of lead sulphide. The tip of a fine wire, the cat's-whisker, was pressed gently against the surface of the galena. The wire was usually made from gold, silver or copper. Rectification occurred at the junction between the wire and the crystal.

A basic crystal set consisted of a tunable capacitor/inductor loop, the crystal rectifier, the demodulating filter and a headset. No power supply was needed for listening to local stations. Part of the fun (or gamble?) of early radio was trying to find a 'good' spot on the surface of the crystal to touch with the cat's-whisker.

- **Transducer:** Examples are listed under this heading on p. 231.
- **Output:** This is the recipient of the information that is received. Examples are listed under 'Recipient of information' on p. 231.

The block diagrams in this chapter do not include coders and decoders, which might form part of certain radio communications systems.

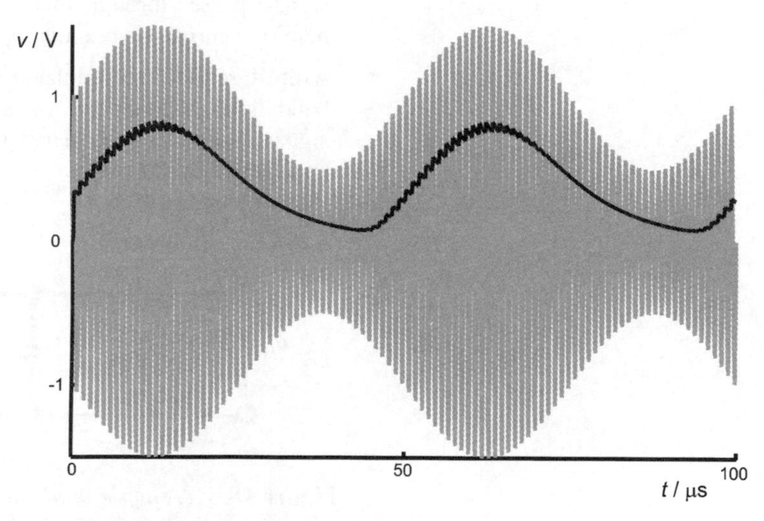

Figure 18.6 *In this simulation of the action of the detector shown in Fig. 18.5, the signal from the RF amplifier (grey) is a 1 MHz carrier, amplitude modulated at 20 kHz. The demodulated signal is in black.*

Superhet receivers

TRF: tuned radio-frequency.

Selectivity: the ability to pick out a transmission on one particular frequency from a number of other transmissions on *slightly* different frequencies.

The *supersonic-heterodyne* (or '*superhet*') receiver has a performance that is superior to that of the TRF receiver. One of the problems with the TRF receiver is that two or three tuned RF amplifier stages are needed to provide good selectivity. It is difficult to design and control a series of RF amplifiers that will all tune to precisely the same frequency within the range of frequencies that the set can receive. This problem does not arise in the superhet because there is only one RF tuned amplifier stage (Fig. 18.7). The operation of the receiver is as follows:

- **Antenna:** This stage is described in Chapter 16.
- **RF amplifier:** A single tuned RF amplifier. Its output is the modulated carrier wave.
- **Local oscillator:** This is a tunable RF oscillator. The RF amplifier and the oscillator are tuned at the same time, so that the frequency of the local oscillator is always a *fixed* amount lower than that of

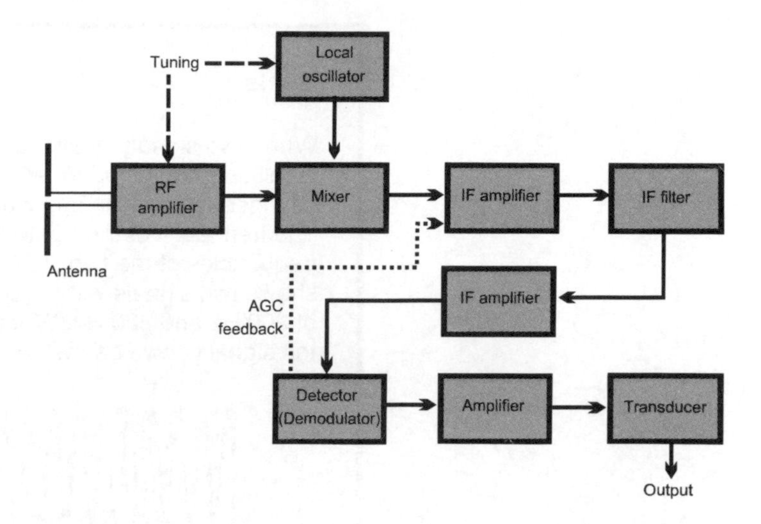

Figure 18.7 *The superhet receiver converts the carrier to an intermediate frequency so that it can be processed by stages designed to operate best at that one frequency.*

Heterodyne: producing a lower frequency from the combination of two almost-equal higher frequencies.

the amplifier. In many superhets the difference is 455 kHz.

- **Mixer:** This combines the signals from the amplifier and oscillator. They produce beats (see box and Fig. 18.8) at their frequency difference. This is known as the *intermediate frequency* (IF) and is often 455 kHz. This becomes the new carrier frequency, and has the signal amplitude modulated on it, just as it was on the signal from the RF amplifier.
- **IF amplifier:** This amplifies the signal. Because the intermediate frequency is 455 kHz, whatever the frequency of the original transmission, this amplifier is designed to give its best performance (maximum gain, minimum distortion) at 455 kHz.
- **IF filter:** This is a bandpass filter with a very narrow pass-band centred on 455 kHz. The filter is designed very exactly with sharp cut-off on either side of 455 kHz, so as to cut out signals on nearby frequencies. The narrowness of the operating band also reduces noise levels in the signal.
- **IF amplifier:** Further amplification with an amplifier designed to work best at 455 kHz.
- **Detector, amplifier, transducer and output:** As for the TRF receiver, but there is negative feedback from the detector to the first IF amplifier. If the detected signal is weak, the feedback increases the gain of the IF amplifier. This *automatic gain control* compensates for variations in signal strength, either when receiving from a weak station or when the signal from a strong station fades temporarily.

Test your knowledge 18.3

If a superhet is receiving a station at 25 MHz, and the IF is 455 kHz, what is the frequency of (a) the local oscillator, (b) the signal passing through the IF filter?

Beats

When two signals of almost equal frequency and amplitude are mixed together, the resulting signal changes regularly in amplitude. We say that it *beats*. The frequency of the beats equals the difference in frequencies of the two signals. The top part of Fig. 18.8 shows two signals with equal amplitude and frequencies of 200 Hz and 220 Hz. When these are mixed we obtain the signal shown below, which has beats at 20 Hz.

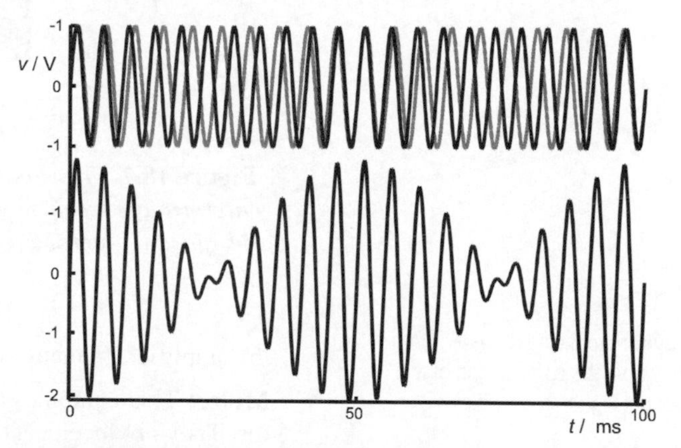

Figure 18.8 *Demonstrating the production of beats when signals of two close frequencies are mixed.*

Activity 18.1 Radio receivers

Use an oscilloscope to trace the signal through commercially-built battery-powered AM and FM radio receivers. Identify the demodulation stage and examine its input and output.

Construct a simple crystal set and investigate how it works.

Activity 18.2 Beats

Feed the output from two audio frequency generators into a mixer circuit. if you do not have a ready-made mixer, build one from an op amp, as in Fig. 6.12. Set the generators to produce sinusoidal signals of equal amplitude but slightly different frequencies. Use an oscilloscope to observe the output of the mixer and measure the beat frequency.

Problems on radio communication

1 List and describe the action of the main stages in a radio transmitter. Illustrate your answer with a block diagram.

2 Describe how a diode detector works.

3 List and describe the action of the main stages of a superhet receiver. Illustrate your answer with a block diagram.

Multiple choice questions

1 The present-day equivalent of the radio crystal and cat's-whisker is:

A a transistor.
B a capacitor.
C a diode.
D none of these.

2 The demodulation stage of a radio receiver is often called:

A the IF amplifier.
B a lowpass filter.
C the detector.
D a rectifier.

3 A varactor is a:

A variable-capacitance diode.
B variable capacitor.
C variable-frequency tuned amplifier.
D a type of triac.

4 The local oscillator in a superhet runs at:

A a frequency different from that of the carrier.
B the intermediate frequency.
C the same frequency as the carrier.
D twice the frequency of the carrier.

5 Beats are produced by mixing:

A two signals of equal amplitude.
B two signals of almost equal amplitude and frequency.
C the carrier frequency with the intermediate frequency.
D signals of equal frequency.

19 Control systems

Summary

A number of temperature control systems are discussed to illustrate some of the main features of electronic control systems. The features explained include on-off control, proportional control, open and closed loops, servo systems, set point, error signal, feedback, hysteresis, response time and overshoot. Systems for the control of rotational speed are used to illustrate some of the above features and also the use of stepper motors. Position control, both linear and angular, includes more on servomechanisms and stepper motors and introduces sequential control.

Temperature control

Temperature control often means controlling an electric heater. Sometimes it means controlling a cooling device, such as a refrigerator or an electric fan but such operations are mostly the inverse of heating applications and are not discussed here.

Most of the features of temperature control are illustrated by looking at ways of controlling the temperature of a room heated by an electric heater:

On-off control: sometimes called *bang-bang* control.

- **Manual control:** The heater has a switch that is turned on and off by hand. This is an example of *on-off control*. The heater is either fully on or fully off. If the room is too cold, the heater is turned on. The heater is turned off when the room becomes warm enough. This is not an electronic control method, but illustrates the meaning of on-off control and, in Fig. 19.1, the meaning of an open loop.
- **Thermostat, on-off control:** This is a fully automatic system used to control the temperature of an oven, a refrigerator, an incubator, a room or a building. Fig. 19.3 illustrates a typical electronic thermostat system. A very small proportion of the output from the

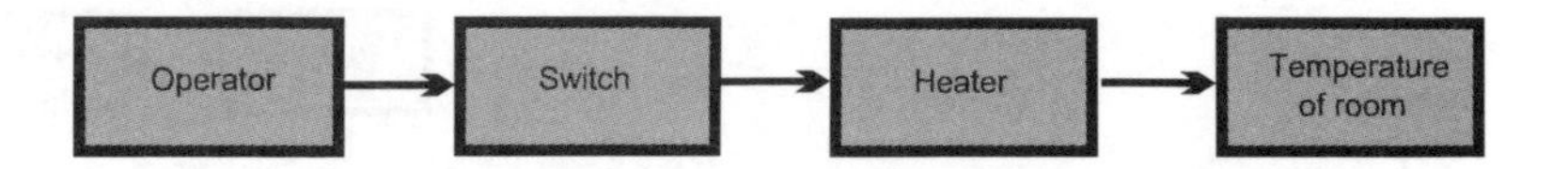

Figure 19.1 *In an open-loop system the room temperature increases until the loss of heat through open doors, open windows, walls, floor and ceiling equals that produced by the heater. The temperature of the room is determined by* outside *conditions.*

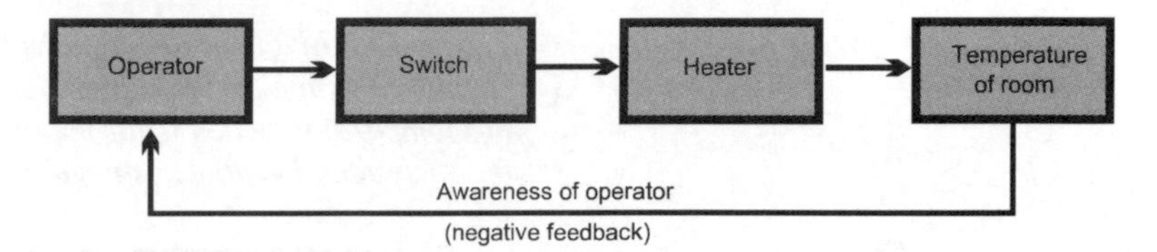

Figure 19.2 *The loop is closed when there is an operator whose task it is to keep the room at a fixed temperature.*

heater is diverted to a sensor. Thermal sensors are discussed in more detail below but, for this outline of the system, consider that the sensor produces a signal t_o that is proportional to the output temperature. The signal is usually a voltage. The signal goes to a comparator where it is compared with another signal (voltage), the *set point.* The set point (t_s) can be adjusted by the operator according to the temperature at which the room is to be held. The comparator is a differential amplifier. It amplifies the *difference* between ts and t_o. Its output is $A(t_s - t_o)$, where A is the gain of the amplifier. In a simple on-off thermostat, A is very large so that the amplifier is saturated and its output swings strongly toward the positive or the negative supply voltages. The output controls the switch, which is turned on when the output is high and is turned off when the output is low. Fig. 19.4 shows a practical thermostat circuit using a thermistor as the sensor.

Control loops

The simplest but least effective form of control is the *open loop* (Fig. 19.1). In the case of room heating, the room temperature rises steadily until the operator intervenes (if ever) or equilibrium with outside temperature is reached.

With *closed loop* control the operator is aware of room temperature and takes action to control it. The loop is closed by this action and there is *negative feedback.* When the room is too *cold*, the operator turns the heater *on.* When the room is too *hot*, the operator turns the heater *off.* Most automatic systems have a closed-loop, as can be seen in Fig. 19.2.

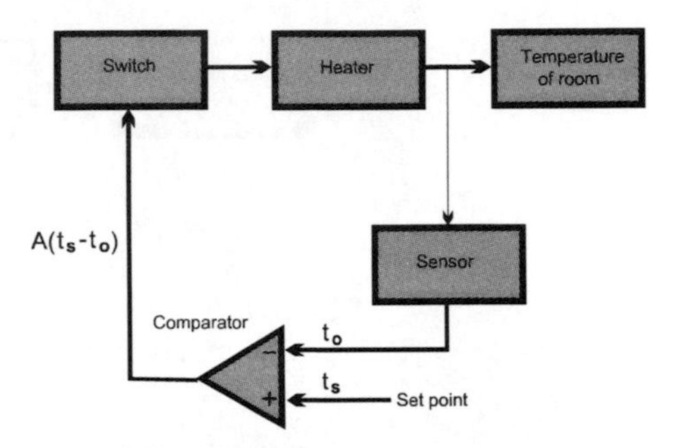

Figure 19.3 *In a thermostat the voltage output from the sensor is compared with a preset voltage level and the resulting output from the comparator switch is turned on or off. This is a closed-loop on-off system. Feedback is negative.*

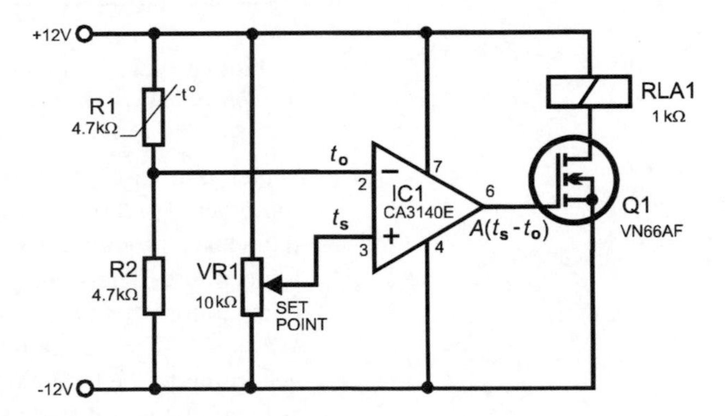

Figure 19.4 *In a practical thermostat circuit, the output voltage* t_o *from the thermistor/resistor chain is compared with the set point voltage* t_s *from VR1. Resistance of R1 increases with temperature, making* t_o *increase too. If* $t_o < t_s$, *the output of the comparator (actually an op amp wired as a comparator) is high, Q1 is turned on and the relay coil is energised. This allows current to flow to the heater element. When* $t_o > t_s$, *the comparator output is low and the heater is switched off.*

- **Thermostat, proportional control:** The output of the heater is made proportional to the difference between the actual and the required temperatures. The room is warmed up quickly by strong heating if it is cold, and after that is kept near the required temperature by moderate heating. The example in the box illustrates the principle of proportional control but gives only coarse control, as there are only three possible levels of heating. Another technique is to use a single heater and to switch it on automatically for greater or lesser proportions of the time. One way of doing this is to replace VR1 in Fig. 19.4 with a ramp generator. A UJT oscillator like that in Fig. 9.16 could be used, as in Fig 19.5, with the output taken from the junction of R1 and C1. The output rises slowly as C1 charges and falls rapidly as it

Proportional control

This is easy to understand from a manual example. The operator has an electric room heater with three separately switchable heating elements. If the room is not quite warm enough, the operator switches on one element until the room reaches the required temperature. The room can be kept at this temperature by switching this single element on and off from time to time.

If the room starts by being cold, the operator switches on two elements at first, then turns one of these off as the temperature approaches the required level. To keep both elements on at this stage would risk seriously overheating the room.

In the very coldest weather the operator might switch on all three elements at first, then reduce the number to two, and finally have only one switched on. At each stage, the number of elements switched on is *proportional* to the difference between the actual and required room temperatures.

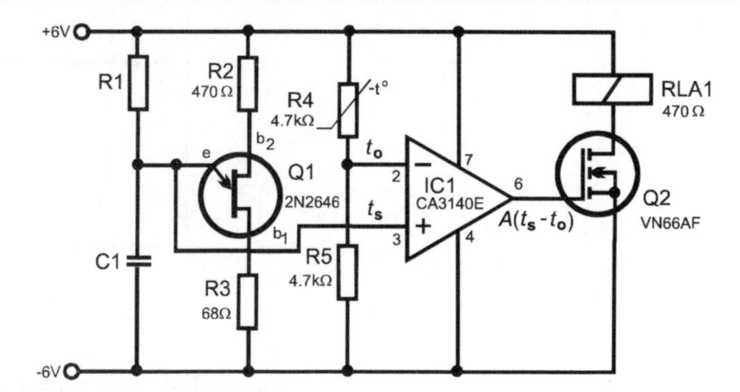

Figure 19.5 *A UJT ramp generator converts the on-off control circuit of Fig. 19.4 into a proportional control circuit.*

discharges. Fig. 19.6 shows how this is used to provide proportional control. In Fig. 19.6a the temperature is 15°C and the output from the thermistor/resistor network is $t_o = 9.7$ V. The ramp is set to oscillate at 11±3.6 V. The highest point, 14.6 V is the set point. The output of the comparator (grey line) goes high whenever the ramp voltage is greater than t_o. It is high for about 60% of the time so the heater is running at 60% of its maximum capacity. When the room has heated to 35°C (Fig. 19.6b), $t_o = 13.7$ V. The ramp voltage is greater than t_o for only about 10% of the time. The heater is working at only 10% capacity as the temperature approaches 37°C, the set point when it will equal 14.6 V. The heater is off for as long as the temperature is 37°C (or more). It comes on for short periods if the temperature falls slightly below 37°C and for longer periods if it falls drastically. This circuit and setting would be suitable for an incubator for bacterial cultures that are being grown at human body temperature. The proportional control helps to ensure that the temperature will never rise much beyond 37°C, which event could be lethal to the cultures.

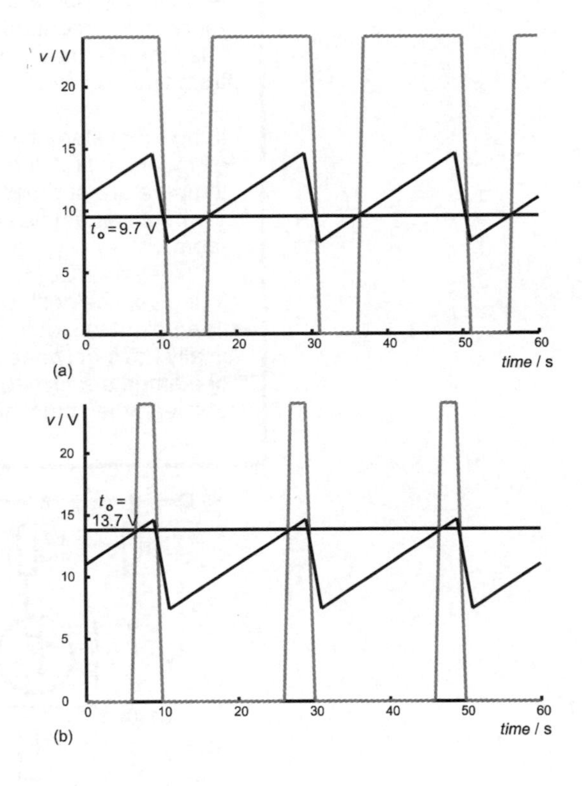

Figure 19.6 *By comparing* t_o *with a ramp voltage the proportion of time that a heater is switched on is proportional to the difference between the set temperature and the actual temperature. The horizontal line plots* t_o *when the actual temperature is (a) 15°C and (b) 35°C. The grey line shows the output of the comparator which switches on the heater.*

Feedback in a closed-loop system depends on the physical nature of the different parts of the system. Important factors include:

- **The heating element:** a thin wire heats up quickly, so the system responds quickly. We say that it has a rapid *response time*. By contrast a heavy-duty element enclosed in a thick metal sheath heats up slowly. It also continues to give off heat for some time *after* it has been switched off. We say that there is *overshoot*.

- **The way heat travels from the heater** to the place to be heated: Parts of a room may be a long way from the heater and are heated only by convection currents. Response time is long. A fan heater has a shorter response time.
- **The response time of the sensor:** a thermistor may be a relatively large disc with long response time, or it may be a minute bead with short response time. The sensor may be in contact with the air or enclosed in a glass bulb.

It is usually better if a control system has a certain amount of hysteresis built into it. In a heating system, the heater comes on when the temperature falls below a given level t_L, the *lower threshold*. It goes off when the temperature reaches a given higher level t_U, the upper threshold. With the heater off, the temperature falls but the heater does not come on again until the temperature falls below the *lower* threshold. The result of this is that the temperature rises and falls over a restricted range. Most temperature control systems have the hysteresis built into them already. The heater, the space or object to be heated, and the sensor are to a certain extent thermally isolated from each other. When the heater comes on there is lag while heat is transferred to the space or object, and to the sensor. By the time the sensor detects the increased temperature, and the heater is turned off, there is already an amount of heat stored in it. This continues to raise the temperature of the space or object. The lag provides the hysteresis.

Rotational speed control

There are several types of electric motor, differing in their design and operating characteristics. This section discusses small DC motors which have the coils wound on an armature turning between the poles of a permanent magnet. Varying the current through the coils controls the speed of a DC electric motor. There are several ways of doing this:

- **Manual control:** A variable resistor is wired in series with the coils.
- **Transistor control:** In Fig. 19.7 the speed of the motor is controlled by varying the base current. Like the manual control technique, this is open-loop control. It can not guarantee that the speed of the motor is related in any certain way to the setting of the variable resistor or to the size of the base current. The reason for this is that the mechanical load placed on it affects motor speed. This may vary for several reasons, for example: the motor is driving a vehicle which starts to go up a slope, the motor is lifting a load which is suddenly decreased, the motor is affected by uneven friction in the mechanism. A slowly rotating motor may stall completely if its load is suddenly increased.
- **Op amp closed loop control:** Fig. 19.8 has negative feedback. The op amp monitors the voltage across the motor and acts to keep it constant, equal to the control voltage, v_c. This ensures constant

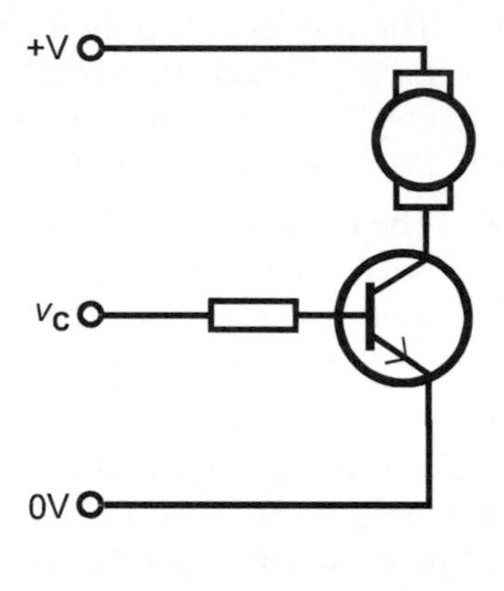

Figure 19.7 *Changing the control voltage v_C varies the base current. The collector current and thus the current through the motor vary accordingly, so varying the speed of the motor. This is an open loop so speed depends partly on the mechanical load on the motor.*

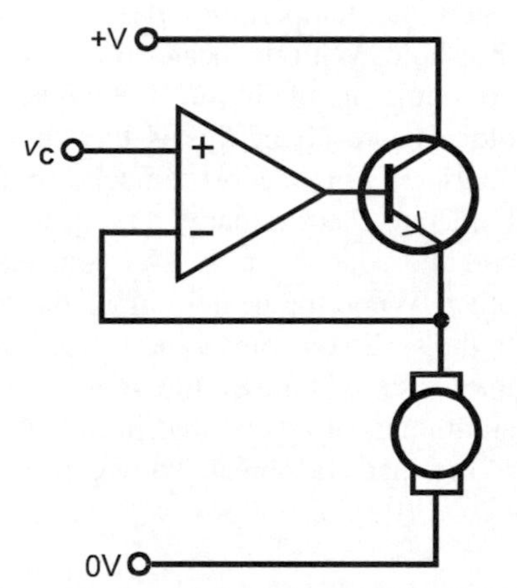

Figure 19.8 *An op amp with negative feed back acts to keep both its inputs at the same voltage. Here it keeps both inputs at v_C and so ensures a constant voltage (v_C) across the motor, and a constant motor speed independently of the mechanical load on the motor. Varying v_C sets the speed of the motor.*

current through the coils, keeping it turning at constant speed, in spite of variations in the load on the motor.

- **Switched mode control:** Instead of controlling the amount of current supplied to the motor the current is supplied in pulses of constant amplitude but with varying mark-space ratio. The speed of the motor depends on the *average* current being delivered to it. In Fig. 19.9 the pulses are generated by a timer ic. The timer runs at 50 Hz or more, producing a stream of pulses. The inertia of the motor keeps it turning at a steady speed during the periods between pulses. Adjusting VR1, varies the pulse length and thus the mark-space ratio of the output to Q1. With the values given in Fig. 19.9, the mark-space ratio ranges from 2 to 50. This gives a 1 to 25 variation in the amount of current supplied to the motor. This technique gives smooth control of motor speed over a wide range. There is less tendency to stall at low speeds because the pulses switch the transistor fully on and the motor is driven with full power. It is possible for the pulses to be generated by a microprocessor for fully automatic speed control. The main disadvantage of the method is that it is an open loop. The actual speed attained depends on the load.

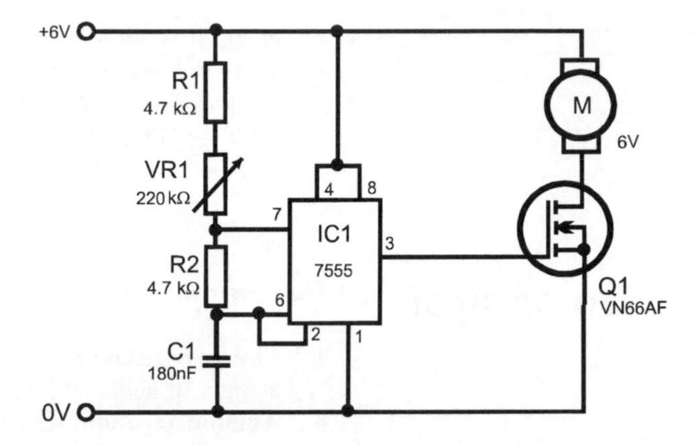

Figure 19.9 *A pulse generator with variable mark-space ratio is used to control the rate of rotation of the motor.*

- **Stepper motor:** A typical stepper motor has four sets of coils, arranged so that the rotor is turned from one position to the next as the coils are energised in a fixed sequence:

Step no.	Coil 1	Coil 2	Coil 3	Coil 4
0	On	Off	On	Off
1	Off	On	On	Off
2	Off	On	Off	On
3	On	Off	Off	On

The sequence then repeats after this. At any step, two coils are on and two are off. To get from step to step, first coils 1 and 2 change state, then 3 and 4 change state, then 1 and 2, and so on. The result of this is to produce clockwise turning of the rotor, 15° at a time. If the sequence is run in reverse, the rotor turns anticlockwise. Six times through the sequence causes the rotor to turn one complete revolution. At any step the rotor is held in a fixed position by halting the sequence. A pulse generator containing the logic to produce four outputs according to the sequence in the table controls the motor. The four outputs are connected to four transistors, which switch the current through each coil on or off. The control circuit may be built from standard logic gates or is available as a special ic, such as the SA1027. The rate at which it runs through the switching sequence depends on the system clock. If this is set to a definite frequency, the motor turns at a fixed rate, one revolution for every 24 pulses. Varying the clock rate varies the speed of the motor. The advantage of the stepper motor is that its speed is controllable with the same precision as the system

Clock: In a logic system, a square-wave oscillator running at a precisely known frequency.

clock. It is not affected by the load on the motor, except perhaps an excessive load, which might completely prevent the motor from turning. The stepper motor may be made to turn 7.5° per step by using a slightly different switching pattern. Motors with a 1.8° step angle are also produced.

Position control

There are two types of position that may be controlled:

- **Linear position:** the location of an object (such as a cutting tool) along a straight path, measured from a point on the path.
- **Angular position:** the direction of a lever-like object (such as an arm of a robot), measured as an angle from some reference direction.

There are dozens of ways in which a part of a machine may be moved along a straight path.

Examples

Worm gears, rack and pinion, and a hydraulic cylinder and piston.

A closed-loop control system needs a sensor to detect the object's position. There are also dozens of types of sensor used for this purpose.

Examples

Broken light-beam sensors, and magnetic proximity sensors.

A simple linear position sensor is illustrated in Fig. 19.10. The output of this is proportional to the position of the object. This is compared with v_C, the control voltage, set by the operator or perhaps generated by a computer.

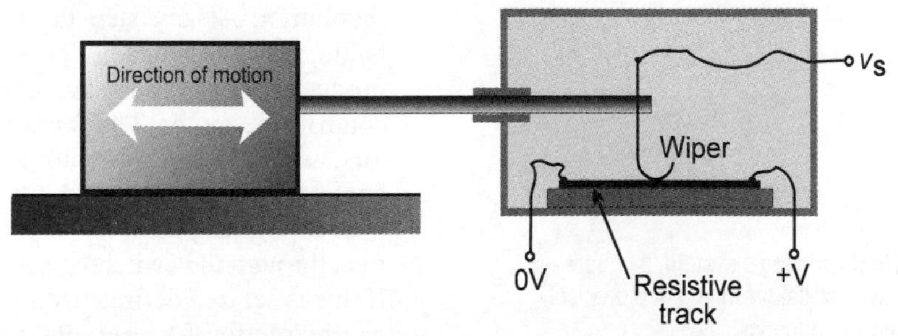

Figure 19.10 *As the object moves, the wiper is moved along the resistive track. The output voltage varies from 0V up to +V as the object moves from left to right.*

There are three situations:

- **$v_C < v_S$:** The output of the comparator is positive. The control circuit supplies current to the motor, so as to increase v_C.
- **$v_C > v_S$:** The output of the comparator is negative. The control circuit supplies current to the motor in the opposite direction, so as to decrease v_C.
- **$v_C = v_S$:** The output of the comparator is zero. The control circuit supplies no current to the motor. The object stays in the same position.

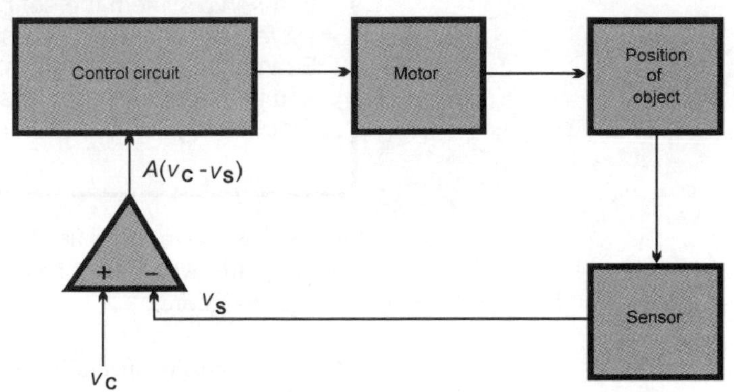

Figure 19.11 *The object is moved by a mechanism connecting it to the motor. The sensor detects the position of the object and produces output v_S. The control circuit controls the direction of rotation of motor, so this determines the direction in which the object moves.*

This is an example of a servo system. It is an on-off system, with the extra feature that the motor is not just switched on and off but may be made to turn in either direction. The control circuit is built from op amps and transistors, or it may be a special ic designed for servo driving. The ZN409C is one such ic. If the object is not at its intended position, the motor rotates one way or the other until $v_C = v_S$. A proportional system has the additional feature that the speed of the motor is proportional to $v_S < v_C$. The motor turns rapidly at first but slows down as the object approaches the intended position.

One of the possible snags of such as system is that the system overshoots. Instead of homing steadily on the intended position, the object is moved *past* that position. It is then moved toward the position from the opposite side, and may overshoot again. This may continue indefinitely, causing the object to oscillate about the position instead of settling there. This is called *hunting*. To avoid hunting, most servo systems include a *dead band* on both sides of the intended position. If, for example, v_C is less than v_S and increasing, the motor is switched off

Regulators and servos

A *regulator* holds the value of a given quantity (temperature, speed, position) at a fixed level, the *set point*. Examples are a thermostat and the system for keeping videotape running through the videoplayer at constant speed. In some regulator systems (for example, an incubator) the set point may be fixed. In others it may be altered occasionally (for example, a room heater).

In a *servo* system the set point is often changed and the system is continually adjusting to new values of the set point. Examples are steering systems in vehicles, in large vessels, and in aeroplanes, and systems for controlling the position of robot arms.

as soon as $v_C - v_S$ becomes less than a given small amount. The object stops a little way short of the intended position but the error is small enough to ignore.

A *servomotor* is specially designed to provide feedback of angular position. It has reduction gearing so that the lever (Fig. 19.12) turns relatively slowly (typically taking 0.2 s to turn 60°) when the motor is running. If required, the angular motion of the lever can be converted to linear or other forms of motion by connecting it to suitable gearing or lever systems. The servomotor has a rotary potentiometer on the same spindle as the lever. If the ends of these are connected to 0 V and +V (compare Fig. 19.10) the output v_S from the wiper of the potentiometer is proportional to the angular position of the lever. As in Fig. 19.11, a control circuit is used to make the motor turn to bring the lever to the intended position. Servomotors are often used in robots and in model aeroplanes and vehicles.

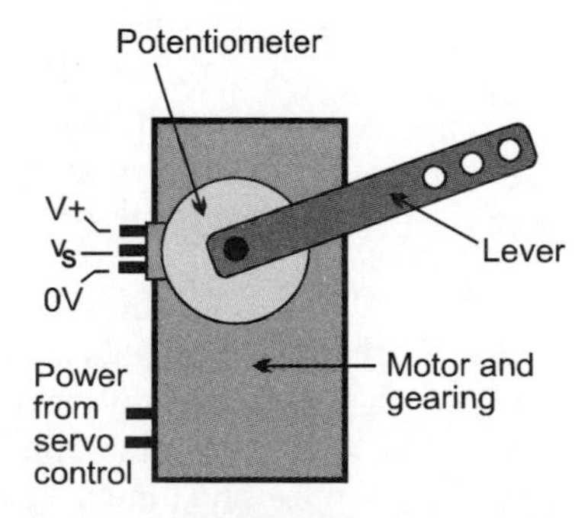

Figure 19.12 *A servomotor has a built-in potentiometer to feed back information about the lever position to the control circuit.*

The rotor of a *stepper motor* (p. 291) can be made to turn to any angle with a precision of 15°, 7.5° or 1.8°. An open-loop digital control system controls the direction in which it turns and supplies the number of pulses needed to turn the rotor to the required position.

Activity 19.1 Control circuits

Use the descriptions and diagrams in this chapter to set up several control circuits. Exact details depend on what sensors and actuators you have available. Set up at least one example of direct control, on-off control, and proportional control. Include at least one thermostat circuit, a motor speed control circuit and a lamp brightness control circuit. These should be powered from batteries of low-voltage PSUs, not from the mains. If you have robotic equipment available, make use of this to demonstrate control circuits. Use a ready-built servo system or a build one from components. Set up a stepper-motor circuit, using the SA1027 or a similar ic, obtaining practical details from the data sheet.

Problems on control systems

1 Explain the difference between on-off control and proportional control.

2 What is the advantage of a close-loop system when compared with an open-loop system?

3 Draw a block diagram of a switched-mode closed-loop circuit for controlling the speed of an electric motor. Describe the action of each part of the system.

4 Explain how an on-off thermostat operates.

5 Give examples to explain what is meant by (a) hysteresis, (b), overshoot, and (c) dead band.

6 Describe a circuit for controlling the speed of a motor by pulses of varying mark-space ratio.

7 What is a stepper motor? How can it be used for controlling (a) rotational speed, and (b) angular position?

20 Microelectronic systems

Summary

Microelectronic systems range from mainframe computers down to simple control systems such as are found in a microwave cooker. All of these systems have many features in common. Data is stored as binary digits, which may be single bits or groups of bits (nybbles, bytes, or words), or larger sets of data reckoned in kilobytes, megabytes or gigabytes. A typical microelectronic system consists of a microprocessor, with clock and memory and a number of peripheral devices for input and output. The address, data and control busses connect the system. The microprocessor comprises a control unit, an arithmetic logic unit, and a number of registers all linked by an internal bus. There are buffers to link this bus to the busses of the microelectronic system. Data is moved within the system under the control of the microprocessor, usually by placing the address of a device on the address bus, then using the control bus to make the device either accept or present data. There are several types of memory, each with its own properties and applications. These types include SRAM, DRAM, ROM, PROM, EPROM and EEROM.

PCs and other microelectronic systems

Personal computers, either the desktop or the laptop versions, are probably the kind of microelectronic system most familiar to most people. They are known as *microcomputers*. There are more complicated systems, such as the *workstations* used by geologists to analyse seismic data when prospecting for oil. These come somewhere between microcomputers and the minicomputers used by medium-scale businesses. The most complex systems are the *mainframe* computers used by the largest businesses such as banks and airlines, or by researchers in highly technological applications such as aircraft design. But the distinctions between types of computers based on size are rapidly becoming out of date. The present day personal computer (PC) has far more power than the mainframes of, say, a decade ago. Also a number of smaller computers can be networked to work together with the effective power of a super-computer.

At the other extreme to the mainframe is the 'computer on a chip', or *microcontroller*. This is often a single ic capable of performing all the functions of a computer but lacking some of the connections to the outside world. It may have no monitor, or only a 16-key keypad instead of a full typewriter-style keyboard. Often it has a fixed program built into it, to perform one particular task. Such microelectronic systems are found in hundreds of different devices, from microwave ovens and smart cards to washing machines and video games machines.

The important thing about all these microelectronic systems is that they all have very similar structure (architecture) and all work in much the same way. Some may lack certain stages and others may have specialised additions, but basically they are all very similar. In the descriptions that follow we refer mainly to PCs, the most familiar and average-sized of the microelectronic systems.

From bits to gigabytes

Some of the terms mentioned in this section have been defined earlier in the book, but these are the fundamental units of data for microelectronic systems so they are repeated here, from the smallest to the currently largest:

- **Bit:** the unit of information. Takes one of two binary values, 0 or 1. In logical terms, these may mean TRUE and FALSE. In positive logic, 1 is TRUE and 0 is FALSE, and this is the system most commonly used. In electronic terms, 0 means low voltage (usually very close to 0 V) and 1 means high voltage (usually close to the supply voltage, which is +5 V in TTL systems.).
- **Nybble:** A group of four bits, or half a byte.
- **Byte:** A group of eight bits. Bytes are used in very many microelectronic systems as the basic unit of data.
- **Word:** the unit of data used in a particular system. In an 8-bit system the word is 8 bits long, so a word is a byte. In a 16-bit system, a word is 16 bits long, or two bytes. In a 32-bit system a word is four bytes.
- **Kilobyte (kB):** According to the metric system, this would be 1000 bytes but in digital electronics this term means 1024 bytes, or 2^{10} bytes. This term is often used when describing the data storage capacity of small devices such as the memory of microcontrollers, and the lengths of typical data files.
- **Megabyte (MB):** This is 1024 kilobytes, or 2^{20} bytes, or 1048576 bytes. It is *just over* a million bytes. This unit is often used in describing the data storage capacity of devices such as floppy diskettes (typically 1.44 MB), memory chips, CD-ROMs and long data files.
- **Gigabyte (GB):** This is 1024 megabytes or 2^{30} bytes. It is just

over a thousand million bytes. The capacity of the hard disk drives of the more recent PCs is reckoned in gigabytes. This is the largest unit in common use at present.

Busses

Fig. 20.1 shows the main features of a microelectronic system. It consists of a number of units connected to each other by busses. A bus may be serial or parallel:

- Serial bus: a single conductor along which the bits are signalled one at a time by making the line voltage high or low.
- Parallel bus: a set of conductors, often 8 or 16 running side by side. At any given moment the buses each carry one bit of a word, so the bus is able to carry a whole byte or double-byte of data.

Test your knowledge 20.1

A system has an address bus that has 8 lines. How many different addresses can this system have?

A serial bus is simpler to wire up or to etch on a printed circuit board. It needs only a two-pin plug and socket for connections (assuming one pin is used for ground). It takes up only one pin of any ic to which it is connected. But a parallel bus transfers data much more quickly.

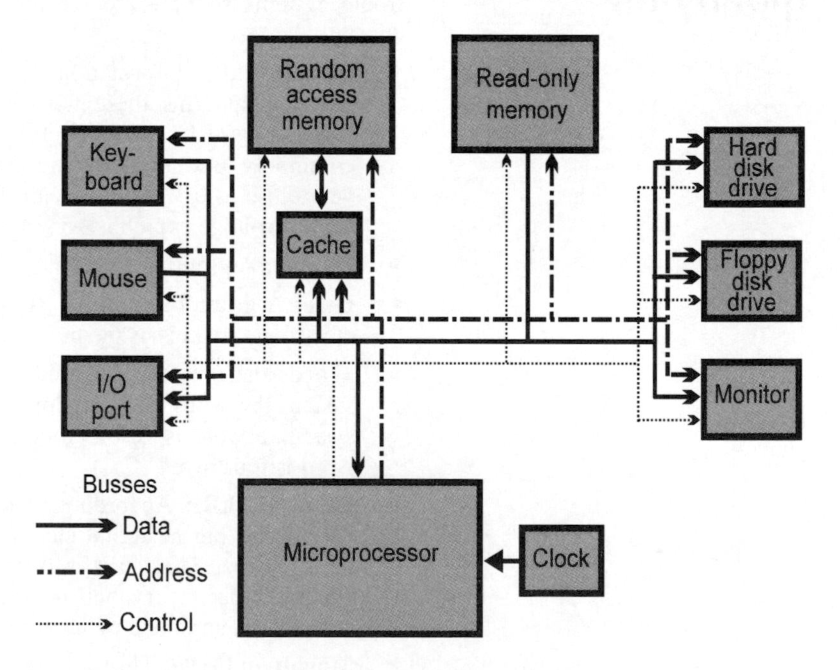

Figure 20.1 *A microelectronics system consists of a microprocessor, a system clock, memory, and a number of devices to interface the system to the outside world. The parts of the system communicate with each other by placing signals on the busses.*

Most busses in a microelectronic system are parallel busses and are divided into three types:

- **Data bus:** used for carrying data between different parts of the system. The data may be numerical values, coded text or coded instructions.
- **Address bus:** used by the microprocessor for carrying addresses of devices (such as the hard disk drive) or locations in memory. When the microprocessor calls a device into action (to receive data, for example) it places the address of that device on the bus. The addressed device then registers the data or instruction that is on the data bus at that time.
- **Control bus:** used by the microprocessor for controlling the other devices in the system.

Only the microprocessor can place signals on the control and address busses, but the data bus can have data put on to it from several sources. These include the microprocessor, the memories, and the keyboard. In such a system there may be times when several devices may be trying to put data on the bus at the same time, is referred to as *bus contention.* This must not be allowed to happen. Each device that may need to put data on the bus has three-state outputs, which are under the control of the microprocessor (through the control bus). The microprocessor allows only *one* device to be *enabled* (to place its data on the bus) at any one time.

Elements of a microelectronic system

The devices that make up a typical microelectronics system are:

- **Microprocessor:** this is a highly complicated VLSI device and is the heart of the system. Its function is to receive data from certain parts of the system, act upon it according to instructions presented to it in a program, and to output the results. It is described in more detail later in this chapter.
- **Clock:** This is a square-wave oscillator that causes the microprocessor to go through one cycle of operation for each pulse. The clock is usually a crystal oscillator and operates at 1 MHz or more. Rates as high as 450 MHz are common in the fastest systems.
- **Read-only memory (ROM):** This is memory that has permanent content, programmed into it during the manufacture of the system. It is used to store programs that instruct the microprocessor to start up the system when it is first switched on. In a microcontroller system intended for a particular purpose (for example, to control a dishwasher) the ROM contains a number of programs and possibly stored data required by the microprocessor in order to run the system. There is more about ROM later in this chapter.

- **Random-access memory (RAM):** This is a temporary data store, used for storing the whole or parts of the program that is currently being run, tables of data, text files, and any other information for which must be registered temporarily. There is more about RAM in Chapter 12.
- **Cache memory:** Certain systems, particularly those required to run at high speed include a small amount (up to 1 MB) of special RAM. This consists of chips that operate at very high speed. It would be too expensive for the whole of RAM to be built from such chips. There may also be a small cache memory built in to the microprocessor. Data that is often required is stored in the cache so that it is quickly accessible. Usually the cache is operated so that data in its way to or from RAM is held there for a while. This data is later overwritten. In some systems it is overwritten if it is infrequently used. In other systems, the overwritten data is that which has been there for the longest time. Either way, the data held in the cache is the data most likely to be needed in a hurry.

Test your knowledge 20.2

A microelectronic system needs a table of fixed numerical data that it needs to refer to every time it is switched on. What type of memory should be used?

To the left and right of Fig. 20.1 are the units that provide input and/or output to the system. These are often referred to as *peripherals*. The figure shows six peripherals that are commonly found in most PCs:

Peripheral: a general term meaning 'on the edge of something'.

- **Hard disk drive:** The key part of this is a disk of non-ferromagnetic metal coated with a magnetic film. There may be more than one disk on the same spindle. There are magnetic heads, one for each surface of the disk or disks, which write data on the surface of the disk or read data from it. The heads are moved by a stepper motor mechanism so that they read or write in one of a number of closely spaced concentric circular tracks. The disk rotates at high speed (about 3600 revolutions per minute) which produces a thin film of moving air close to the disk. The head 'floats' on this film so that it comes very close to the surface of the disk without actually touching it. The high speed of rotation and the narrow gap between head and disk means that a very large number of bits can be stored in a given length of track and that the bits can be written or read at very high speed. It also means that a very large amount of data (rated in gigabytes) may be stored on one drive. In operation, the disk spins continuously (though it may be programmed to switch off if it has not been used for several minutes) and the head is quickly moved to the track that is to receive or that holds the data. The data is laid down in concentric circular tracks. There may be up to 2000 tracks and these are divided into as many as 70 sectors. There are usually 2 disks on a drive but there may be up to 8 disks. Access time is about 20 ms, which is fast enough for data to be written or read during the running of a program without unduly slowing down the operation of the program. When a disk is accessed to write or read data, the

Magnetic recording

The principle of magnetic recording is the same for tape-recorders, hard disk drives, floppy disk drives and the magnetic stripes on bank and other cards. The recording medium is a film containing fine particles of a magnetic substance such as chromium dioxide. The molecules of the particles group themselves into microscopic regions known as *domains*. In a domain all the molecules of the material are aligned with the same magnetic polarity, so that the domain is a very small magnet. But the individual domains are aligned in all directions, at random, so there is no overall magnetic field.

When a signal is recorded, the electromagnetic recording head applies a varying magnetic field to the film, strong enough to align a proportion of the domains in the same direction. As the film moves past the head, the varying signal current in the coils of the head causes a varying field to be applied to the film. This aligns a greater or lesser proportion of the domains in the same direction or the opposite direction (as the field changes polarity). The strength and direction of the magnetic field at successive points along the track is an analogue of the original signal.

On playback, the film passes by the head and the magnetic fields due to the aligned domains induce currents in the head. The strength and direction of the current depends on the proportion of domains aligned and their direction. The current is amplified and reproduces the original signal.

With digital data, the domains representing one bit are all aligned either in one direction or the opposite direction, depending on whether the bit is 0 or 1.

head goes first to one of the outer tracks to find out which track and which sector of the track to go to. A stepper motor takes the head to a particular track and from then on it is matter of timing. A magnetic pattern on the disk marks the 'start' of each track and, with the disk spinning at a known rate the time is calculated which will bring the required track under the head, to begin writing or reading. The gap between the head and the disk is so small that dust and smoke particles must be excluded from the drive. A typical hair, or a particle of smoke or dust has a diameter that is 10 or 20 times the width of the gap between the head and the surface of the disk. For this reason, hard drives are sealed during manufacture in a dust-free factory, and are not openable by the user. A hard disk is used for storage of large amounts of data (including documents) and programs that are often required. These include many short programs such as the 'driver' used for operating the mouse, and short programs used for transferring data from one major program to another. These may remain on the disk for many years as a permanent part of the system. But they can be erased at any time, and possibly replaced by other data.

- **Floppy disk drive:** The magnetic film is coated on both sides of a flexible (floppy) plastic disk enclosed in a stiff plastic case. It has a metal shutter that slides aside automatically when the disk is inserted in the drive. This exposes part of the disk to allow the magnetic read-write head to operate. The amount of data stored is much less than on a hard disk (1.44 MB on the popular 3.5-inch floppy) because the disk rotates more slowly (360 revolutions per minute) and the head can not come as close to the surface of the disk. There are 80 concentric tracks on a 3.5-inch floppy, each divided into 8 sectors. The method of finding the required sector is similar to that described for the hard disk, except that the 'start' of each track is located by detecting the position of a rectangular slot in the metal hub of the disk Typical access time is 200 ms, which is much longer than that of the hard disk. The main advantage of floppy disks is that they are removable, which hard disks are not. Floppy disks bearing data can be stored away from the computer in a safe place, reducing risk of loss of data through fire or theft. They can be sent through the post. A set of program files recorded on one or more floppy disks is a convenient way of marketing software. Various other types of removable disk are

Formatting disks

When a disk (for example, a floppy) is formatted, a number of concentric tracks (usually 80) are laid down on both sides of it. Each track is divided into 8 segments. One track is reserved as an index, to list which segments are being used to store particular batches of data. Then the read/write head is moved to align it with a required track. The start of the segment is found by timing from the instant when an index perforation in the disc passes an optical sensor.

A batch of data (or 'file') may occupy one or more segments. On a new disk a large file is recorded on several adjacent segments. After a disk has been in use for some time and data have been deleted and the segments re-used, it usually happens that few of the vacant segments are adjacent. Then a large file is recorded by skipping from one vacant segment to another, possibly on several different tracks. This increases the time needed to write or read data. The disk has become *fragmented.* At this stage it is possible to use a defragmenter program to reorganise the disk, placing the parts of long files on adjacent segments.

available, with storage capacity from 100 MB to 1 GB. These need their own special drive. They have the additional advantage that they can carry large files (such as high-resolution digitised photographs) which are too long to store on an ordinary floppy. They are also easier to store and handle than the equivalent stack of floppy disks.

- **Monitor:** This is a visual output device for text and graphics, usually in colour. Simple microcontroller systems do not have a monitor but may have a dot matrix LCD display for short messages. An example is the display on a fax machine. Even smaller systems may rely on a few LEDs to provide information about the state of the system. Washing machines often have an array of LEDs to indicate the current stage in the washing cycle and the washing options that have been selected.
- **Keyboard:** A PC usually has a keyboard with over 100 keys, including the alphabet, numerals, punctuation, function keys, a numeric key-pad and several other keys such as Shift, Tab, Escape, and Enter. The keyboard is scanned automatically to detect and store keypresses. Usually a special ic acting as an interface between the key-switches and the data bus does the scanning. It stores keypresses in a buffer memory in the order in which they are made. The contents of the buffer are put on to the data bus when the microprocessor signals on the control bus that it is ready to receive them. Microcontroller systems may have a smaller keyboard or a number of switches of other types, such as the

Handshaking

Signals on the control lines at a port are part of a protocol called *handshaking.* The purpose of this is to speed the efficient transfer of data. When a high level is put on the 'busy' line by a printer (for example), it tells the microprocessor (through the I/O port) that the printer is already occupied printing the data sent previously. As soon as it has finished this, the printer makes the 'busy' line low. Then the microprocessor sends its data to the printer, which is acknowledged by a low level on the 'acknowledge' line. 'Busy' goes high again until this data has been printed. This system ensures that if there is any data waiting to be printed, it is sent as soon as the printer is ready to receive it. It also ensures that no data is lost because of the printer being unable to receive it when busy printing. There are other lines such an 'error' line used by the printer to signal that it is not switched on, or that it is out of paper, or that there is some other error in the printing. On receiving this signal the microprocessor halts the printing and displays an error message on the screen.

push-button or rotary controls of a washing machine.

- **Mouse:** This is a useful input device for moving the cursor around the screen. The mouse has a rubber ball beneath it, resting on the mouse mat. Sensors detect the rotation of the ball as the mouse is moved. Signals from the sensors are interpreted by software and used to position the marker on the screen. A mouse has one or two buttons for selecting objects on the screen, such as the items in a menu.
- **I/O port:** an interface ic connecting the buses (through buffers) to a socket into which leads from external devices are plugged. A port may be a parallel port in which data is received or sent 8 bits at a time. There are additional lines reserved for carrying control signals (for example 'ready to send', 'busy', 'ready to receive', 'buffer full'). A *parallel port* is generally used for connecting a printer to the system. The eight bits of the code for an alphanumeric character is sent. Most computers also have one or more serial ports that send or receive data one bit at a time. A serial port is often used for connecting the computer to a telephone line through a modem. There is more about ports later in this chapter.
- **Data converters:** a microelectronic system may include an analogue to digital converter. This receives analogue data from the outside world, converts into (say) 8-bit digital form and places it on the data bus when enabled. The ADC has three-state outputs. The ADC is controlled by an address decoder so it appears to the microprocessor as a byte of memory. But the contents of the byte change as the analogue input changes. An ADC is not commonly found in a general-purpose microcomputer but can be an important part of a specialised microcontroller system. In a 'smart' electric fan heater, for example, an analogue voltage proportional to air temperature may be converted to digital form so that it can be read by the microprocessor. Some types of microcontroller

Cursor: An arrow-shaped marker displayed on the monitor screen to indicate which 'button' or window is being selected. Also a vertical marker indicating the *insertion point* where newly typed text will appear.

Interface: a circuit, often in a single ic, which processes signals from one part of the system to make them suitable for use by another part of the system or by another system. More complicated interfaces are assembled on circuit-boards (cards) which plug into sockets (slots) inside the computer enclosure. *Examples:* an *I/O port* that interfaces between the system bus and (say) a synthesiser keyboard; a *video card* which interfaces between the microprocessor and the monitor.

Alphanumeric: alphabetic and numeric characters, usually including punctuation marks.

Modem: short for *mod*ulator-*dem*odulator. A device which accepts input from a microelectronic system and produces a modulated signal (usually frequency shift keying) sent along a telephone line to the modem of a distant system. There the signal is demodulated, converted into a serial stream of bits, and received by the system.

Buffers

Buffers are needed at an I/O port for the same reason that line drivers and receivers are used on a transmission line. Although the busses inside the system are relatively short and the leads joining (say) a printer to the parallel port are only a metre or so long, frequencies are high. Pulse shape may become distorted and pulse may tend to merge.The ics used at printer ports contain buffers for each line of the of the data bus and for the relevant lines of the control bus. The buffers square up the distorted pulses as they pass through the output and input ports . At the input ports, the buffers have three-state outputs so that the microprocessor can control exactly which data from outside sources is placed on the bus.

include a built-in ADC. A system may also have a digital-to-analogue converter as one of its output devices. In this way a microcontroller can, for example, control the speed of an electric motor.

- **Real-time clock:** This is not the same thing as the system clock which is really just a pulse-generator used for pacing the microprocessor and keeping other parts of the system synchronised with it. The real-time clock is an integrated circuit with logic very similar to that found in a digital watch. It is regulated by an accurately cut crystal to run with high precision and to register the time of day (hours, minutes, seconds, hundredths of seconds), and also the day of the week, the date, the month, and the year. It automatically gives February a 29th day in a leap year. The real-time clock is connected to the data bus and can be written to (to set the time and to issue instructions), or read from (to find the date and time). As well as general time-keeping, it can act as an alarm clock, interupting the microprocessor at any preset time, daily, or on a particular date. It can also be set to interrupt the microprocessor at regular intervals, such as once a second, or once a minute or once an hour. The microprocessor is programmed to perform a particular action when it is interrupted. In a data logger, for example, the microprocessor could be programmed to measure and record a temperature at 1-minute intervals, but to continue with other tasks in between. Some types of microcontroller have a real-time clock circuit on the same chip.

Many of the elementary microelectronic systems lack some of the units listed above, or have them in simplified form. Microcontrollers very often have built-in RAM and ROM, making separate memory chips unnecessary. In many applications the system includes sensors which put data directly on to the data bus. In a dishwasher, for example, there is a device to measure water temperature. Controllers of industrial machinery have sensors that detect the position of cutting tools, for example, or detect whether safety gates are in position or not. The input from sensors may consists of a single bit placed on one line of the bus.

Many microcomputers have peripherals extra to those shown in Fig. 20.1. These include:

Note the spellings: floppy dis*k*, compact dis*c*.

- **CD-ROM drive:** A compact disc stores data in the form of tiny pits etched into the metallised surface of a plastic disc. The procedure is the same as that used for recording musical CDs. The data is recorded by directing a laser beam at the disc, causing a minute area to melt and become a pit. The track has a string of pits and no-pit gaps, representing 1's and 0's. After being coated with a film of metal the disc can be played back using a laser beam at reduced strength. The beam is reflected from the pits or not reflected at the gaps and this is monitored by a photo-sensor. The signal from the sensor reproduces the original digital data signal. CD-ROMs have a much higher data storage capacity than floppy disks. Typically a CD-ROM stores 600 MB, the equivalent of over 400 of the 3.5-inch floppy disks. Data transfer between disk and data bus is more rapid that that of floppy disks. With the increasing length of programs and with *multimedia* requiring the storage of complicated graphics, including high-resolution photographs and motion pictures, synchronised with music and voice recordings, CD-ROMs are widely used for the distribution of software.

Multimedia: Programs which include materials from many sources, such as text, sound, high-resolution graphics, photographs and motion pictures, in a unified presentation. Very often the program is *interactive*, requiring and responding to contributions from the operator.

- **Tape drive:** Mainframe computers often use reels of tape for data storage. The tape is half an inch wide, with several (usually 7 or 9) tracks recorded side by side, and is up to 3600 feet long. It can store up to 40 MB. Smaller tape drives may be used with minicomputers and microcomputers, including drives which use cassettes of the same type as are used in audio tape recorders. The disadvantage of tape storage is that the data is recorded serially. If the required data is recorded near the end of the tape it is necessary to 'fast-forward' though the tape, searching for the data. This is a relatively slow operation. By contrast the read-write head of a disk drive can be moved to any segment of the disk within a few milliseconds. Tape storage is suitable for back-up of large batches of data, which are generally loaded as a whole.
- **CD-R drive:** A recordable CD used with a special drive allows data to be recorded once only. It can be played back as often as needed, either on the CD-R drive or on an ordinary CD drive.
- **Printer:** Older systems may have a *dot matrix* printer, in which a printing head containing (usually) 9 fine needles moves across the paper, with an inked ribbon between the head and the paper. The needles are 'fired' at the paper under the control of the printer's own microelectronic system, producing a pattern of ink dots on the paper in the form of printed characters. Usually the printer has a ROM containing details of the dot patterns, so the microcomputer only has to send the ASCII code to the printer to tell it which pattern to print. Most printers are also able to accept patterns generated by the computer so that the printer is made to print in various other fonts. Dot matrix printers are commonly used in cash registers, automatic teller machines, and other applications in which their low resolution and slow action are acceptable. *Laser* printers use a fine laser beam to direct a jet of toner (black or colours) at the paper. Laser printing has much finer resolution then dot matrix, and is quicker and quieter. A less expensive technique is *bubble-jet* printing, which is also in black and colours. The cheapest type of printer is the *thermal* printer, often used in fax machines and similar devices where printing quality is not important. Special heat sensitive paper is

Font: Set of type characters with a distinctive style. Examples: the main text of this book is set in Times New Roman. This note is set in Arial Narrow.

ASCII code

This is the *American Standard Code for Information Interchange,* and is the standard way of coding letters, numerals and punctuation and also a number of printer-control commands. It is an 8-bit binary code with a range of values from 0 to 255. The first 32 values are the control characters, such as 'carriage return', 'bell' (produces a 'beep' in modern computers but used to ring an actual bell in the old Teletype machines), and 'end of file'. Code 32 is a space. Codes 48 to 57 are the numerals 0 to 9, then there are some more punctuation marks. Capital letters of the alphabet begin at code 65 and lower-case letters at 97. The codes in between these blocks are used for punctuation.

Examples:
Carriage return is code 13, or 0000 1101.
Number 5 is code 53, or 0011 0101.
A query (?) is code 64, or 0100 0000.
Letter A is code 65, or 0100 0001.
Letter m is code 109, or 0110 1101.

used, which turns black when heated by fine wire filaments. The printing has low resolution and is not permanent.

- **Sound card:** A circuit board that plugs into a socket on the main board of a computer is referred to as a *card*. Different types of card may be used for various functions such as sending and receiving faxes, producing high-quality graphics, and for providing additional memory. A sound card has the circuits required for playing compact discs and for recording sounds in memory. It has ics that generate the sounds of musical instruments under the control of the microprocessor. There are amplifiers to provide high quality audio output to stereo speakers.
- **Scanner:** converts a photograph or other document to digital form, suitable for storing in memory or on disk.
- **Plotter:** Produces diagrams and drawings on paper by moving coloured pens to draw lines under computer control. Architects, design engineers and others whose main requirement is for large, accurate, and highly complicated line drawings often use these.
- **Joystick:** a lever mounted on a base so that it may be pushed forward, backward, and sideward to left or right. The base contains microswitches that close in various combinations as the 'stick' is moved. This gives on-off control of the cursor or other objects on the screen. Some joysticks have variable resistors instead of switches to give proportional control. The joystick may also have one or more 'firing buttons'. Though the function

of the joystick is similar to that of the mouse, joysticks are mostly used when playing those computer games in which action is the main element.

Microprocessors

There are many types of microprocessor, each with its own particular features and architecture. Fig. 20.2 shows the main parts of an imaginary microprocessor that has most of the features typical of all microprocessors. The heart of the microprocessor is the *control unit,* which finds out what the microprocessor must do next and oversees the doing of it. It goes through a repeated cycle of operations, the timing of which is regulated by the system clock. Control lines run from it to all other parts of the microprocessor. It also has connections to certain external parts of the microelectronic system through the control bus.

Architecture: When applied to microprocessors, architecture covers the parts it has, how they are connected together, its inputs and outputs, and the structure of its internal bus.

The microprocessor is also connected to the data and address busses of the system. The *data register* holds outgoing and incoming data and, under the supervision of the control unit, transfers data between the external data bus and the internal bus of the microprocessor. Similarly, the *address register* transfers addresses from the internal bus to the external address bus. It transfers addresses in only one direction, but the data register has two-way transfer.

The *program counter* is a register holding an address that indicates where the microprocessor has got to in memory, following its way

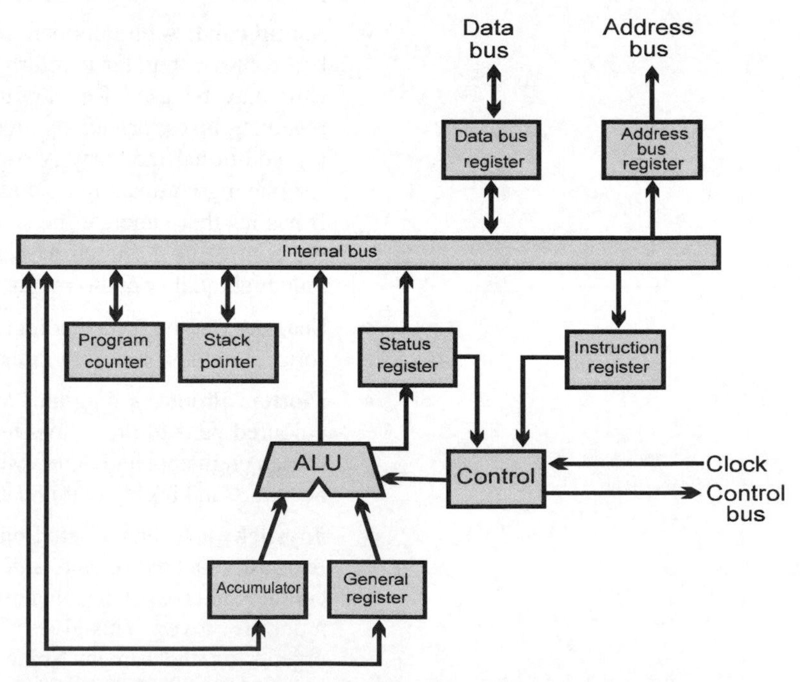

Figure 20.2 *A microprocessor contains the control unit, the ALU and a number of registers (more than shown here) all connected by an internal bus, and all on one silicon chip.*

Pointer: When a register holds an address that is of particular importance, such as the next address to be read, we say that it is a *pointer* to that address, or that it *points to* that address.

through the program that is stored there. In short, it is like a bookmark. At each cycle, the microprocessor reads a byte from the address stored in the program counter. Then the address is automatically increased by 1, so that it 'points to' the next address to be read. Occasionally the microprocessor is made to jump to a different part of memory and continue its reading from a new address. On these occasions the control unit puts the new address in the program counter, to direct the microprocessor to the different part of the program..

The *stack pointer* is another register holding an address. Here the address points to a place in memory where very important data is stored temporarily. The action of the stack is described in Chapter 21.

The *status register* holds essential information about the operation of the microprocessor. It usually has a capacity of one or two bytes, but the information is mainly stored as the separate bits within the bytes. These bits are individually set to 1 or reset to 0 to indicate certain events. Such bits are usually called *flags*. For example, every time a calculation gives a zero result, the *zero flag* is set to 1. If we want to know if the most recent calculation gave a zero result or not, we can look at this particular bit in the status register to see if it is 1 or 0. Another flag may be set when a calculation gives a positive result. A further example of a flag is the *carry flag*, which is set when there is a carry-out from a calculation, that is, when the result is too big to be stored in the register.

The *instruction register* holds the instructions recently read from memory. These tell the control unit what to do next. Closely associated with the control unit is the *arithmetic/logic unit* (ALU) which performs all the calculations and logical operations. Most of these operations involve the *accumulator*. A value placed in the accumulator is operated on in one way or another by the ALU, according to the instructions it has read from memory. Then the result is placed on the internal bus, and is often circulated back to the accumulator to replace the value that was there before. At the same time, one or more of the flags in the status register are set or reset, depending on the result of the operation. Some operations involve two values, one stored in the accumulator and one in a general register. Fig 20.2 shows only one general register, but most microprocessors have several of these to make operations more flexible. The role of the accumulator and the way in which it is concerned in almost everything the microprocessor does is made clearer by examples given in Chapters 21 and 22.

Fig 20.2 and the description above apply to most microprocessors and to most microcontrollers too. The main difference is that microcontrollers may also have memory and other devices such as timers built in to them.

Addressing

It is important for the microprocessor to be able to communicate with any other part of the system. This is done by giving every part of the system an *address*. The address is in binary, a set of low and high voltage levels on the lines of the address bus, but we nearly always write it in hex to make it easier to read and understand. The address bus has a line for every binary digit of the address.

Note: If you find these numbers difficult to understand, turn to the box on 'Number systems' on p. 315.

Example

A system has 16 address lines. The lowest possible address is 0000 0000 0000 0000 in binary. All lines of the address bus carry a low voltage. The highest possible address is 1111 1111 1111 1111 in binary. All lines carry a high voltage. This is the equivalent of addresses from 0 to 65 535 in decimal or from 0 to FFFF in hex. It is a 64 kilobyte system.

The descriptions in this chapter refer to an 8-bit system (data stored and processed in 8-bit units, of 1 byte each), with 16 address lines giving a total memory space of 64 kB. Most PCs run as 16-bit or 32-bit systems, and some have more memory, but the principles are the same.

Sixteen address lines is a common number for microprocessor systems, but many microcontroller systems have only 8 or fewer lines. Some computers need more than 65 566 addresses and so must have more than 16 address lines. Systems with 20 address lines (1 MB of addresses) are common. Another way of increasing address space is to have a relatively small number of lines and to arrange for memory to be switched in or out of the systems in blocks, but occupying the same addresses. Obviously the different blocks of memory can not be used at exactly the same time but memory-managing software can switch it so quickly that there is only a slight delay.

The way that the memory space of a system is allocated to different functions is set out as a *memory map*. A typical memory map of a 64 kB system is shown in Fig 20.3. Different parts of the system are allocated their own addresses within this memory space. At the lowest addresses in memory there is a ROM chip, containing instructions to the microprocessor on what to do when the system is being started up. The microprocessor has 0000 in its program register when it is first powered up so it always begins at address 0000 and goes there first to find out what to do. There it is given instructions to jump to some other address perhaps in another ROM, where it finds instructions for starting up the system. The microprocessor may also go to this low memory when it has received an interrupt (see later), and is then made to jump to one of several interrupt service routines.

Chip: Most semiconductors, from transistors to VLSI integrated circuits, are fabricated on slices of silicon, often referred to as 'chips'. From this comes the popular name 'chip', which is generally used only for the more complex integrated circuits such as memory ics and microprocessors.

The next area of memory may be used for devices such as sound cards and other peripherals. Often these occupy a small range of addresses and the card has logical circuits to decode addresses within its range and to take action accordingly. In general terms, this is referred to as the input/output area of memory space. The devices with addresses within this range are not memory, but they are *memory-mapped*. That is, they are accessed by the microprocessor by using addresses, just as if they were actual memory.

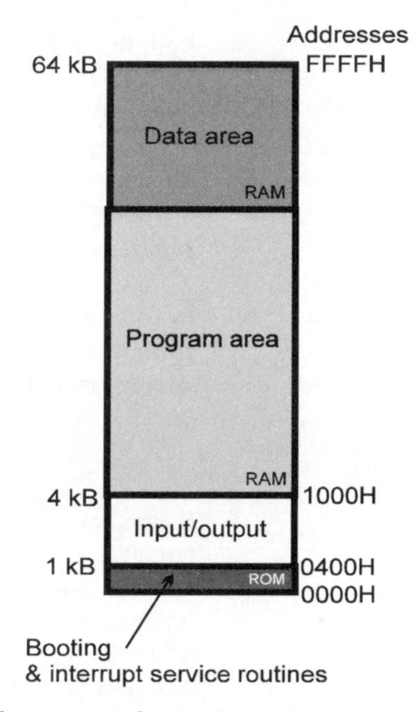

Figure 20.3 *The amount of space in a memory map depends on the number of address lines that the system has. The system shown here has 16 address lines. The memory space is allocated to RAM, ROM and all the input and output devices of the system. Not all of the memory space is necessarily taken up; there are ranges of spare addresses where future additions to the system may be located. The numbers on the right are starting addresses of blocks of memory, the H indicating that they are written in hex.*

RAM for holding programs and data takes up the main part of the address space. Again, not all of the addresses allocated to RAM may actually have chips there. Normally a program and its associated data are stored on the hard disk or on a floppy. It is not possible for the microprocessor to run a program while the data is stored there because the access time is too slow. When a program is to be run, it is copied from the hard disk or floppy into an area of RAM. The microprocessor is then told the address in RAM of the beginning of the program, and jumps to this address to start running the program. In many systems the program or programs are loaded into the lower part of RAM, known as the program area. Data for use by the program (such as graphics, sounds, numerical information, and text) may be stored in the higher addresses. There is no fixed boundary between the data and program areas, which are adjusted in size by the memory manager program to meet the needs of the programs currently being run. Some addresses in RAM are reserved for temporary storage of data. One example is the stack, which is discussed later.

It may happen that a program is too large to fit into the amount of RAM that is available. In this event only part of the program is copied from the disk drive to RAM. When a different part of the program is needed, it is automatically copied into RAM, overwriting the part originally copied there. The disk drive can be heard whirring from time to time as it copies each newly required part of the program into RAM. This may causes a slight delay in the running of the program but often the delay is not noticeable. If this happens frequently it is a sign that additional RAM should be installed.

In some systems the highest memory addresses may be used for I/O ports, drives and other peripherals. There may also be ROM here, storing operating system programs, including interrupt service routines.

Decoding addresses

Every device has connections to the address bus and has a decoder circuit that responds when the address of that device is placed on the bus. The circuit of Fig. 20.4 illustrates the principle of a decoder. This 'system' has only 8 address lines but the logic is the just the same when there are more. An address consists of a number of 0's and 1's. The decoder must recognise the situation when all of those lines that are supposed to be 0 are actually 0, and at the same time all of those lines that are supposed to be 1 are actually 1. This can be done with two logic gates, a NOR gate for the 0's and an AND gate for the 1's. The decoder in Fig. 20.4 is wired to respond to the binary address 101110. The two lines that are supposed to be 0 are connected to the NOR gate. The output of this gate goes high (1) when both inputs are low. Similarly the output of the AND gate goes high (1) when all four inputs are high. As a result of this the output of the NAND goes low (0) when the correct address is on the bus. This output could be connected to the chip select input of a memory chip such as the 2114 shown in Fig. 12.24.

The 2114 chip has 10 address inputs of its own. These are connected to decoding logic within the chip to select which out of the 1 kB memory locations in the 2114 is to be written to or read from. In a 64 kB system

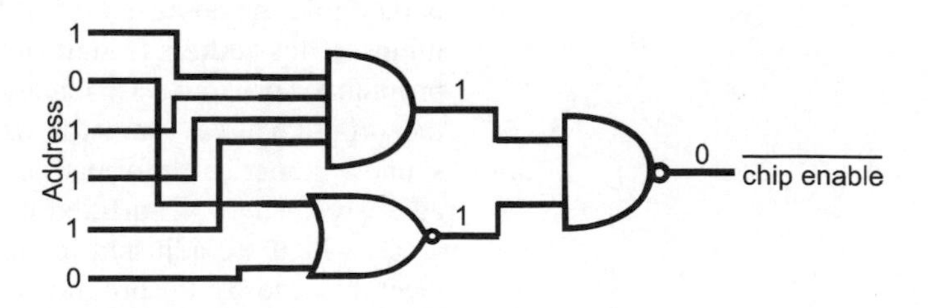

Figure 20.4 *A decoder produces a low output and enables a memory chip or other device when its address is placed on the address bus.*

with 16 address lines, the six inputs to the decoder in Fig. 20.4 would be the top six lines, A10 to A15. Putting 101110 on these lines would select the chip. The levels on the other lines, A0 to A9, would select which address within *that* chip is to be addressed. Thus, the locations in that chip have addresses in the range 1011 1000 0000 0000 (or B800H) to 1011 1011 1111 1111, or BBFFH. A bank of memory could be made up of several such chips each with its own decoder and its own kilobyte of address space. In practice, there would be a single relatively simple decoding circuit for the whole group. There are several other ways of decoding memory chips but there is no space to describe them here.

Booting up: Making the system ready for operation.

Read-only memory (ROM) differs from RAM in that data is stored there permanently. When the system is switched off the data stored in RAM is lost, but the data in ROM remains there and is ready for use as soon as the system is switched on again. A computer does not need a lot of ROM because it is more convenient to store the data on a hard disk or on removable disks. The main need for ROM is for holding the short programs needed by the microprocessor to boot up the system when it is first switched on. This includes checking memory to find out how much there is and if it is working correctly. It also checks to find out what peripherals are attached to the system. In some systems there may be a ROM that holds the program for a high-level language. For example, some microcontroller systems have a ROM that holds BASIC. Instead of having to program the microcontroller in machine code (see Chapter 21) it may be more easily programmed in BASIC. If a system is intended to perform one particular function, there may be a ROM to hold the program. An example is a system intended to act as a word processor. The word-processing program is held in ROM. As soon as the machine is switched on, it is ready to work on text files. But the machine is not able to perform any other task, such as acting as a flight simulator. This contrasts with an ordinary PC, which often has a word-processing program on its hard disk. This is loaded into RAM whenever the operator wants to deal with text files. At other times the operator is able to load and run a wide choice of other programs.

Writing and reading

Storing data in RAM is known as *writing*. It is under the control of the microprocessor and the exact details of the procedure used depend on the type of microprocessor. Typically, a *write cycle* begins with the microprocessor putting on the address bus the address of the memory location in which the data is to be stored (Fig 20.5). At about the same time, it puts on the data bus the data which is to be stored there. Then it makes the write enable ($\overline{WE}$) line of the control bus low, to indicate that this is data to be written into memory. There is then a short low pulse on the chip enable ($\overline{CE}$) input of the memory chip, usually produced by the address decoder. Data is then transferred from the data bus to the addressed location in memory. After that, the chip is disabled and the levels on the other lines change for the next operation.

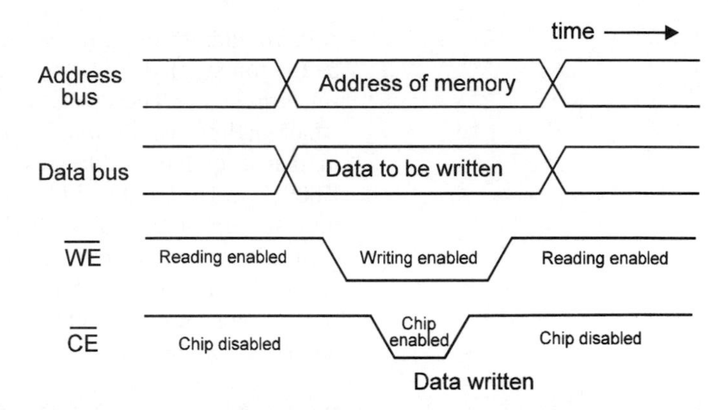

Figure 20.5 *In a memory write cycle the address to be written to and the data to be written are placed on the busses. Then the write-enable line is made low. While this is low the chip-select line is also made low to copy the data to the addressed memory location.*

A *read cycle* puts data stored in RAM or ROM on to the data bus, usually for use by the microprocessor. Note that reading does not destroy the data stored in the chip. It can be re-read as many times as we need to. In the case of RAM it is lost only when it is over-written by new data, or when the computer is switched off. The sequence of the read cycle (Fig 20.6) is similar to that of the write cycle, but the write enable line stays high all the time.

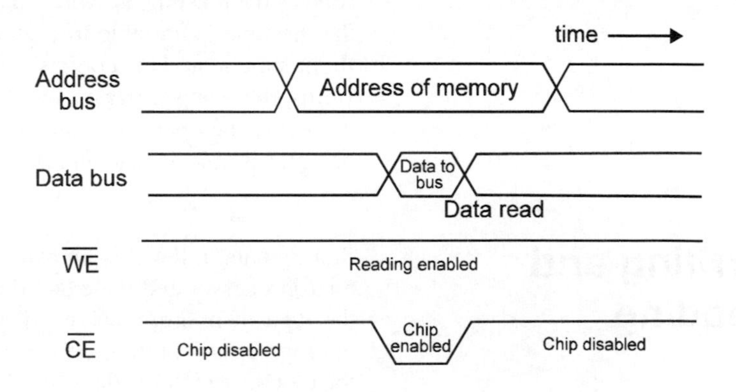

Three-state outputs: the output of the logic gate is at logic high, logic low or high impedance. In the high-impedance state, the output is in effect disconnected from the bus.

Figure 20.6 *In a memory read cycle the address to be read from is placed on the bus. The write-enable line remains high. When the chip enable line is made low, the three-state outputs of the memory chip are made active and the data in memory is copied on to the data bus.*

Test your knowledge 20.3

Express these decimal numbers in binary: 11, 3, 7.

Test your knowledge 20.4

Express these binary numbers in decimal: 1010, 0011, 1110.

Test your knowledge 20.5

Express these binary numbers in hex: 1111, 0110, 1000.

Number systems

Almost all digital electronic systems work in the *binary system.* Flip-flops are either set or reset, voltages are either high or low, and LEDs are either on or off. The situation is two-fold, or *binary*.

Binary numbers are written in ***bi**nary digi**ts***, or *bits*. The value of a bit is either 0 or 1. This is what makes the binary system so convenient for representing numbers in electronic form.

This table lists the first 16 binary numbers and also their equivalents in decimal and hexadecimal numbers.

Decimal	Binary	Hexa-decimal
0	0000	0
1	0001	1
2	0010	2
3	0011	3
4	0100	4
5	0101	5
6	0110	6
7	0111	7
8	1000	8
9	1001	9
10	1010	A
11	1011	B
12	1100	C
13	1101	D
14	1110	E
15	1111	F

A quantity written in binary form uses more digits that one written in decimal. But in logic circuits the two-state binary system of highs and lows lends itself to simple circuitry and reliable action, which more than makes up for the extra digits.

Although a computer can easily handle a long string of binary digits, humans find it confusing. This leads to mistakes. To help programmers and software engineers, we have adopted another number system, known as *hexadecimal*, or *hex* for short. This is based on sixteens. Counting runs through the digits 0 to 9 in the usual way, then proceeds to letters of the alphabet from A to F.

Techniques for converting one sort of number into another sort are explained in the Supplement.

Types of ROM

There are different kinds of ROM:

- **Mask-programmed ROM:** Special masks are used when the chip is made, so that the content of every memory location is fixed from the very beginning. It can never be altered. It is a special kind of logic device in which each combination of input levels (the address, plus low on the chip-enable input) produces a particular set of output levels (the data). Making special masks is very expensive so this technique is only used when a large number of identical ROMs are required. The ROMs for a popular model of washing machine might be made in this way because it is known in advance that thousands of such machines will be made.
- **PROM**: This is *programmable ROM*, the principle of which is illustrated in Fig. 20.7. The input from the three lines of the address bus is decoded and *one* of the eight lines representing memory location is made high. There are four data output lines,

With 3 address lines, there are 8 possible binary addresses, 000 to 111.

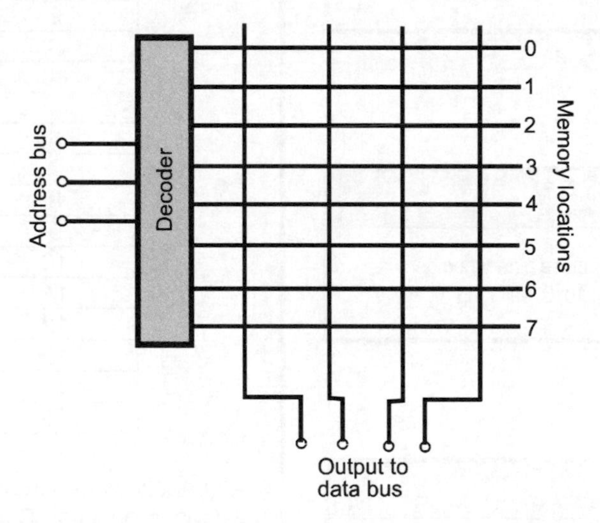

Figure 20.7 *A PROM consists of an address decoder, a line for each memory location and a set of data output lines. Where they cross there may by a link between the memory line and the data line.*

which run across the memory locations but are not necessarily connected to them. At each place where the two sets of lines cross there is a transistor connected as in Fig. 20.8. The emitter of this has a fusible link connected to it. This is a thin connection that can be 'blown' by passing a fairly large current through it. When the ROM is being programmed the link is either blown or not blown. To blow it we connect the data output line to 0 V and then apply a high level to the memory location line. If V+ is made higher than usual and large current flows through the

Addresses

If a block of memory contains 1 kB of locations, this is 1024 bytes.

Numbering begins with address zero so the last address is 1024 – 1 = 1023. This is 3FF in hex. This is 1111111111 in binary. Counting the number of digits tells us that we need 10 address lines to be able to set up all the addresses from 0 (all 0's) to 1023 (all 1's).

In general, if the number of address lines is n, the address block it can cover is 2^n. If the lowest address is zero, the highest address is $2^n - 1$.

Test your knowledge 20.6

How many addresses can be covered by an address bus of 12 lines? If the lowest address is zero, what is the highest address in hex?

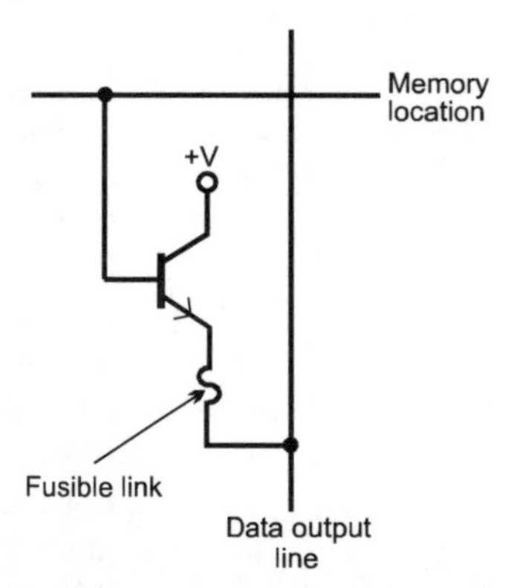

Figure 20.8 *There is a transistor at the every crossing point in Fig. 20.7, linking the memory location to the data line through a fusible link. This link may be blown when the PROM is being programmed.*

transistor, and makes the fusible link very hot so that it melts, or fuses. All of this can be done automatically with the PROM plugged into a socket on a special PROM programmer. This is loaded with the data that is to be blown into the PROM and it works through the memory locations one at a time, either blowing the link or leaving it unblown. Fig 20.9 shows the links of memory location at address 011 (3 in decimal). Two have been blown and two left unblown. After the PROM has been programmed it is ready for use as ROM in a microelectronic system. The voltage +V used for reading the ROM is not as high as it was for programming. Turning on the transistor now does not blow the unblown links. When a memory line is addressed it goes high and turns on the transistors that have unblown links.

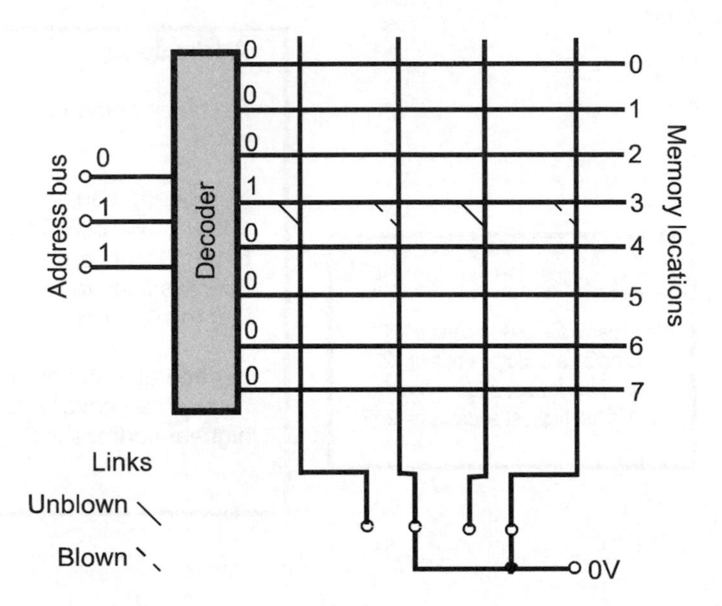

Figure 20.9 *The address 011 is on the address bus and the line for memory location 3 is high. With the some of the data outputs connected to 0 V as here, some of the links are blown and some are not. The data stored is 1010.*

Currents flow through these to the data bus and are equivalent to a stored '1'. No current flows where a link is blown and is equivalent to a stored '0'. In Fig 20.9, the blown and unblown links produce a nybble of data equivalent to 1010. Blowing the links destroys them permanently, so a PROM can be programmed only once. If a mistake is made in programming, or the data needs to be altered, another PROM must be programmed with the corrected data. PROMS are programmed one at a time so are unsuitable for mass production. However, once a correct version of the program has been developed, the process is automatic and suitable for applications for which mask-programmed ROM would be too expensive.

- **EPROM:** This is *erasable programmable ROM.* The ROM consists of an array of MOSFETs, one for each bit. During programming, a charge is pulsed on to the gates of those MOSFETs that are to be set to '1'. The principle is similar to that of DRAM except that the charge remains for very much longer, usually for several years. EPROMs have a quartz window though which the chip can be exposed to ultraviolet light if the data is to be erased. Ultraviolet is an ionising radiation and the ions produced in the chip discharge the gates and erase the data. An EPROM can be reprogrammed after erasing, which makes it very suitable for use when a system is at the development stage and frequent revision of the program and data are needed.

- **EEROM:** This is *electrically erasable ROM*, which can be programmed and reprogrammed while still connected in its circuit. The ROM of many microcontrollers is of this kind. Although this type of ROM could be used for temporary storage in a way similar to RAM, writing new data into EEROM is too slow for this to be practicable.

Ports

Ports connect a microelectronic system to its peripherals and to other circuits in the outside world. A port is connected on one side to the data bus and on the other side to a plug or the bus of the peripheral. Ports that connect to a plug are generally located on the main circuit-board of the system. Those that interface the system to a circuit card are generally located on the card. In a microcontroller system the microcontroller may have one or more ports built into it.

A *parallel port* may have 8 or more input lines and 8 or more output lines. Often there are 8 lines connected to the data bus and a few lines connected to the control bus. These allow the port to be controlled by the microprocessor, so that data can be read from it or into it as required. There may be a connection from the port to an interrupt input of the microprocessor, so that the program may be interrupted when there is new data waiting at the port.

Parallel ports are of three main types, output ports, input ports, and bidirectional ports. An output port is connected directly to the data bus and, when it is enabled, it places the levels of the data bus on to its outputs. An input port has three-state outputs connecting it to the data bus so that it places data on the bus only when the microprocessor is ready to read it. A bidirectional port can pass data in either direction. It may include a *data direction register* which must be set before data is passed through the report. Each line of the port is set individually either as an output or an input by setting the appropriate bit in the data direction register to 0 or 1. This makes it easy to send data in either direction through the port at high speed.

A *serial port* may have only two lines, one for data, the other for ground (0 V). Some have a third line for handshaking. Serial ports are cheaper to build and simpler to operate. But they are slower than parallel ports as they transfer only one bit at a time.

Activity 20.1 Microprocessor systems

The topics in this chapter are best studied by using a simple microprocessor development system or tutor. Decide on a topic to investigate (memory, bus structure, keyboard, input, output and others) look up the topic in the system manual and carry out suggested practical work.

Activity 20.2 Microcomputers

Examine the interior of a microcomputer, and identify the mother board, with power supply unit, cooling fan, microprocessor, RAM, ROM, input ports, output ports, slots for audio and video cards, disk drives, CD-ROM drive.

Problems on microelectronic systems

1 Name four kinds of microelectronic system, state who uses them and what they are used for.

2 Explain what bit, nybble, byte and word mean.

3 Describe the three types of bus found in a microelectronic system and the type of information they carry.

4 What is meant by *bus contention* and how may it be avoided?

5 List the differences in properties and uses between RAM and ROM.

6 Draw a diagram of a typical microelectronic system and explain briefly what each part does.

7 Compare the properties and uses of a floppy disk drive with a hard disk drive.

8 What methods are used for the storing data magnetically? Describe their main features, their advantages and disadvantages.

9 Name three devices that are found as peripherals to a microelectronic system. Briefly describe what each is used for and outline the way it works.

10 Describe the functions and uses of a sound card.

11 Name a multimedia program that you have seen and describe the features that you found most interesting, and most useful.

12 Draw a diagram of a typical microprocessor and explain briefly what each part does.

13 Draw the memory map of a typical PC and explain the functions of the different areas.

14 Draw a circuit for a simple 4-line decoder and explain how it works.

15 Describe the different types of ROM and the purposes for which they are used.

Multiple choice questions

1 The unit of information is a:

A byte.
B bit.
C nybble.
D word.

2 The number of bits in a byte is:

A 4.
B 8.
C 16.
D 32.

3 A kilobyte is:

A 1000 bytes.
B 1024 bits.
C 1024 bytes.
D 1000 bits.

4 The kind of memory used in a microcomputer for storing the booting up program is:

A cache.
B SRAM.
C DRAM.
D ROM.

5 Of those listed below, the device with the fastest access time is:

A ROM.
B DRAM.
C a hard disk drive.
D a floppy disk drive.

6 Of the peripherals listed below, the one that is most likely to be useful to a nature photographer is a:

A plotter.
B joystick.
C modem.
D scanner.

7 Of the output peripherals listed below, the one which does not allow data to be changed is:

A hard disk drive.
B tape drive.
C floppy disk drive.
D CD-R drive.

8 The control unit of a microprocessor:

A reads data from the data bus.
B tells the ALU what to do.
C controls the rate at which the clock oscillates.
D writes addresses on the address bus.

9 The binary number 11 0100 1110 1011 is represented in hex by:

A BE43.
B 34EB.
C 13547.
D D3A3.

10 A ROM can hold 1 kB. Its top memory address, in hex, is:

A 400.
B 3FF.
C 1024.
D 3E8.

11 A system is designed to have an address space from zero to 3FFF. The number of address lines required is:

A 16 383.
B 16 384.
C 14.
D 15

12 ROM is required in a microcontroller system that is being designed for a racing car, which is presently undergoing track trials. The most suitable type of ROM is:

A EPROM.
B Mask-programmed ROM.
C PROM.
D EEROM.

21 Programming

Summary

A microprocessor understands only machine code, but it is easier for the programmer to use an assembler. It is easier still, and quicker, to use a high-level language such as BASIC. A microprocessor operates by a series of fetch-execute cycles, in which it fetches instructions and data from memory and acts upon them. Instructions are stored in binary form, the hex form of which is called an opcode. Programs written in opcodes are known as machine code, but we more often use an assembler program so that we can work with mnemonics instead. A microprocessor works by following a sequence of very small steps, but it executes them very quickly. The instruction set consists of instructions to carry out data transfers (loading and storing), arithmetic and logical operations, for branching to other parts of the program as the result of testing flags and sundry other operations. Interrupts as a result of errors or on demand from peripheral devices cause the microprocessor to halt what it is doing, jump to a routine to service the interrupt and then jump back to resume what it was doing before the interrupt. Interrupts from various sources may be ranked in order of priority.

Languages

A microprocessor must be told what to do, but we must give it its instructions in a language it understands. The only 'language' that a microprocessor understands is a set of high and low voltage levels on its data input lines. For our own reference, we can write these down as a set of 1's and 0's, in other words, as values in the binary system. But binary values need many bits, seldom fewer than eight, and we prefer to convert them into hex to help us remember them and understand them.

Program: any complete sequence of instructions, whether in machine code, assembler or high-level language.

It is possible to write a program in hex, that is, to write out a list of coded instructions, the *opcodes*. The microprocessor will understand these when they are converted into binary voltage levels. We load them into RAM and finally make the microprocessor read them, in turn, and execute them. As well as the opcodes, there will be various numerical values also in hex. This completely coded way of giving instructions is known as *machine code.*

Assembler

Writing programs in machine code is possible but difficult and is very subject to human mistakes. Although using hex makes it easier to write out the opcodes in a compact form, a microprocessor may respond to a hundred or more different opcodes. With such a large number it becomes difficult to remember which is which. The solution to this problem is to write the program in an *assembler* language. One of the advantages of assembler is that it replaces opcodes with *mnemonics*. Mnemonics are a set of abbreviations, usually of 2 or 3 letters, one mnemonic for each opcode. They are intended to help the human programmer distinguish the different operations and remember what each operation does.

Examples

The instruction 0100 1100 on the data bus, equivalent to 4C in hex, tells the microprocessor to add 1 to the contents of the accumulator. Instead of 4C, we use the mnemonic INC, which is short for '**Inc**rement the accumulator'. When using the assembler program, we type in 'INC' and the assembler converts this to 0100 1100 as it assembles the program ready for the microprocessor.

Another advantage of an assembler is that lets us use *labels*. Remembering where a particular program routine begins in RAM means remembering the exact address of its first byte. Labels are used to mark the entry points of the routines, and in general as an easy and memorable way of specifying addresses of locations in memory.

Example

A routine for waiting for one of the keys on the keyboard to be pressed begins at memory location D350H. Instead of having to remember this (along with the addresses of a dozen or more other routines) we can define the label KEYB to mean address D350H. Using the assembler program, we type in 'KEYB' and the assembler converts this to the binary address 1101 0111 0101 0000.

Note that we type in opcodes in hex and define labels in hex but everything ends up in binary, ready for the microprocessor. This includes addresses of ports, the sound card, the printer and various other peripheral devices that are memory-mapped.

An assembler allows us to run the whole or parts of the program, and to go through the program one step at a time. At any stage it displays the contents of the accumulator, the flags and all the registers of the microprocessor so that we can see exactly what is happening. We can also put a temporary break in a program so that it stops at a given point to allow us to examine the contents of all the registers at that point. When the program is finished and debugged, we can download it into PROM or save it on tape or disk for future use.

Programming in assembler is a slow matter because we have to issue the microprocessor with instructions one at a time. It usually takes several instructions to perform an elementary action such as adding two numbers together. If we want to add numbers that are larger than 255, or have many more significant figures, or that are negative, or to add *three or more* numbers together, the routines become even more complicated when written as single steps. In addition to this, there are dozens of routines that may be used hundreds of times in a program. These include adding numbers, conditional jumps, loops, displaying a letter on the screen, beeping a buzzer, and many more. It is inefficient to have to write them out in assembler every time they occur, and to write them all over again in the next program.

High-level languages

Although assembler is often used for writing fairly limited programs for microcontrollers, it is less often used when writing programs for computers. Most programmers use a *high-level language* for this purpose. A high-level language is one stage removed from assembler because a single command in the high-level language is often the equivalent of many tens of assembler instructions.

Example

In the popular high-level language known as BASIC, we could add two numbers together and divide them by a third number by typing:

30 LET total = (price + 75)/100

Source code: the form in which a program is *written*, which may be machine code, assembler or high-level language.

This is a typical line from a BASIC source program. It means 'take the value at present stored as a variable named 'price', add 75 to it, divide the sum by 100 and store the result as a variable named 'total''. The '30' is the line number, because the instructions are numbered and executed in numerical order in most versions of BASIC. A BASIC statement or command looks very much like an ordinary sentence written in English. The meaning of this instruction is clear, even to a person who does not know BASIC. It is much clearer and much shorter than the machine code into which is has to be converted before the microprocessor can get to work on it.

To be able to create a BASIC *source* program, you need a BASIC *language* program in the computer to accept the commands you type in, to display them on the screen, and to allow you to edit it, to test it, to check for programming errors and to save it to tape or disk. Many of the early home computers had their BASIC language program stored in ROM, so that you could start programming as soon as the computer was switched on. But present-day computers such as a PC do not have any built-in language programs. If you want to program in BASIC, you buy a disk with the BASIC language program on it, save it on your hard disk and run it whenever you want to write a BASIC source program.

A BASIC *source* program is stored in memory or saved on to tape or disk as a series of bytes, which are coded to represent the words and numbers in the program lines. Numbers and text are represented by their ASCII codes, and the BASIC commands are represesented by bytes of code. In this form it is meaningless to a microprocessor. It must be *interpreted*. When you 'Run' the source program, a section of the BASIC language program known as an *interpreter* takes over. It goes through your source program line by line, and command by command and turns it into machine code, which it then sends to the microprocessor. Actually the microprocessor has to do several things at once. It has to read a BASIC source program line, use the interpreter to find out what it means, turn it into machine code, and then obey the instructions of the machine code.

Because the microprocessor has to run both the source program and the interpreter in the language program *and then* obey its machine code form, the rate at which things get done is relatively slow. Even so, with a BASIC program running on a modern PC, most things seem to happen very fast. But if there is a lot of processing to be done, this may not be fast enough. Also it is wasteful for the microprocessor to have to interpret every line every time it is used. Imagine a loop that repeats an action, say, 100 times. The lines within the loop are interpreted 100 times, always giving the same result. It would be far more efficient and far quicker if the lines were interpreted just once and for all. This is the task of a *compiler*.

Compilers

A compiler is a program that takes a program written in a high-level language and converts it directly into machine code. It does this in just one session and produces a program in machine code that can be stored in memory or saved to disk. When the compiled program is run, the microprocessor has only to read in the machine code instructions and obey them. For this reason a compiled program runs very much faster than an interpreted one. A compiler works by using a set of rules to convert the high-level language into machine code. If the rules are well thought out, the resulting code is short and runs very quickly. But there are often instances in which a human programmer, using assembler, can devise a neater, quicker way of doing things. Compiled programs are fast but programs written using an assembler are often slightly faster.

Compilers are widely used by professional programmers working with high-level languages such as Pascal, Fortran and C. Compiling is a complicated process that takes the computer an appreciable amount of time. The program can not be tested until compiling is complete. If any errors are found, the programmer has to go back to the source program in the high-level language, correct the errors and then re-compile the program. Then it has to be tested again it to see if the errors have been corrected satisfactorily. It may take several trials and several compiling sessions before all errors are eliminated. This contrasts with the quickness of BASIC in which source program lines can be edited in a few seconds and the amended version takes effect as soon as the program is run.

Although high-level languages have many advantages, particularly because they have so many ready-to-use routines, the user tends to lose touch with what is actually happening in the microprocessor. For this reason we shall say no more about high-level languages. We return to a study of source code written in opcodes and in assembler.

Instructions

A microprocessor can do nothing without instructions. The instructions are stored in memory as a sequence of op codes, some with numeric values or addresses stored immediately after them. When the system is booted up, the microprocessor first looks for instructions in the small ROM where the booting instructions are stored. Then it goes on to read instructions in RAM, or perhaps in other ROMs containing special programs. Each time it reads an instruction, the microprocessor steps through the following sequence of operations:

- the control unit takes the address stored in the program counter
- it puts that address on the address bus
- the addressed unit of memory puts its stored data on the data bus
- the control unit loads the data into the instruction register
- the control unit does what the instruction says, usually with the help of the ALU and the accumulator

This is known as the *fetch-execute cycle.* It is timed by the system clock. The program counter is incremented by 1 every time an instruction has been executed. So the microprocessor goes to the next address to load the next instruction. The microprocessor works its way through the program, instruction by instruction, from beginning to end.

Once an instruction is in the instruction register it may be executed straight away or its execution may require further fetch-excute cycles. The control register may need more information. Here are some

Simplified instruction set

The microprocessor referred to in this chapter is a model one with a simplified set of instructions (see p.332). They resemble the instructions of several real microprocessors so they will help you understand how a microprocessor works and how they are programmed.

Most microprocessors have many more instructions than we describe here. Some have two hundred or more, compared with the 26 instructions of our model microprocessor. But some real microprocessors too have relatively few instructions. These are called RISC processors (Reduced Instruction Set Computer). They operate very quickly because they have few different kinds of instruction to deal with. In some applications, which use fairly basic operations, this is a big advantage. On the other hand, they need more program steps to execute some types of operation, which slows down execution and makes the programs longer.

examples of ways of loading the accumulator using 1, 2 or 3 cycles.

Examples

Using 1 cycle: The opcode 4C (mnemonic INC) tells the control unit to increment the accumulator. This instruction is complete in itself and does not need any further fetch-execute cycle.

Address	Op code or data
47B2	Program continues
47B1	4C
Program counter	47B1
Instruction register	4C
Accumulator	25→26

The table shows what happens when this instruction is read from address 47B1, for example. The top of the table represents two consecutive addresses in the program area of RAM (remember that the addresses run *upward* in the table). The lower part represents registers in the microprocessor. The program counter points to 47B1, so its contents, 4C, are fetched to the instruction register. Code 4C is the opcode for INC, which is then executed. The accumulator is incremented from its existing value, 25, to a new value, 26.

Using 2 cycles: The opcode CA (mnemonic LDI) tells the control unit to load the accumulator with a particular binary value, but it does not say *what* value. To find out what value to load, the microprocessor requires a *second* fetch-execute cycle. After the program counter has been increased by 1, to point it to the next address (47B2), the required value 2E (say) is found in that address and placed in the accumulator. The complete instruction is CA2E, stored at two consecutive addresses.

Address	Op code or data
47B3	Program continues
47B2	2E
47B1	CA
Program counter	47B1 to 47B2
Instruction register	CA 2E
Accumulator	2E

The table has the same layout as before. Instruction CA 2E is loaded from two consecutive addresses in two fetch-execute cycles. Then the complete instruction is executed and the value 2E is placed in the accumulator.

Using 3 cycles (see p. 330): The opcode CE (mnemonic LDA) tells the control unit to load the accumulator with a binary value that is at present stored in an address somewhere else in memory. Because it takes two bytes to specify an address, it takes two more fetch-execute cycles to obtain it.

First the address is obtained by reading it from the next two memory locations, taking two cycles. The complete instruction is CE D102. The address D102 is a location in the data area of memory where the required value is stored. In a third cycle, the address D102 is placed on the address bus. The addressed location puts the stored value (54) on the data bus, and this value is loaded into the accumulator.

Reading the data from the destination address does not alter what is stored there. It can be read as many times as we like.

Address	Op code or data
D102	54
Above is data area	
Below is program area	
47B4	Program continues
47B3	02
47B2	D1
47B1	CE
Program counter	47B1 to 47B3
Instruction register	CE D102
Accumulator	54

Test your knowledge 21.1

1 What is the opcode for 'load the accumulator with 65 (decimal)'.
2 Express this as a mnemonic.
3 What is the meaning of CE 035AH?

Programming

A sequence of instructions such as those described is used to program the microprocessor to perform several operations one after another.

Example

To program the microprocessor to take a value from address D102 (see table opposite), increment it, and store it in address A204, we would use assembler and write:

```
LDA D102H
INC
STA A204H
```

The assembler would convert this to:

```
CE D1 02 4C DE A2 04
```

This sequence of bytes is stored in 7 consecutive memory addresses. Here we have used another mnemonic, STA, to tell the microprocessor to 'Store the contents of the accumulator at the following address'. Its opcode is DE. The table opposite shows the program stored in a block of memory, starting from 47B1. Address D102 holds the value to be loaded, and address A204 awaits the result on the calculation. Eventually the value 55 is stored at this address.

Test your knowledge 21.2

1 Write the assembler for 'Load accumulator with 99 (decimal) and store the value at address 2F56H'.
2 What is the meaning of LDA 750EH DEC STA 5FFFH?

Address	Opcode or data
D102	54
A204	?
Above is data area	
Below is program area	
47B7	Program continues
47B6	04
47B5	A2
47B4	4C
47B3	02
47B2	D1
47B1	CE
Program counter	47B1 to 47B7
Instruction register	See listing
Accumulator	54→55

The short piece of program, or *source code*, in the example above illustrates the fact that a program proceeds in very small steps. Microprocessors can do only very simple things. They often need to take many steps to do something that we humans can do in our heads in (apparently) a single step. But they can perform millions of steps in a second, so they perform calculations much more quickly than we do.

Instruction set

The table opposite lists the *instruction set* of our simple microprocessor. This has an accumulator with an 8-bit register, and another general-purpose register, register X, which also has 8 bits. Most microprocessors have several such registers for holding data, including storing the intermediate results of multi-stage calculations. Some registers may hold 16 or more bits.

The model computer has up to 65536 locations in its memory map, so addresses can be written in hex with four digits.

Data transfer instructions

Of the data transfer instructions in the table, we have already described LDA, LDI, and STA. Of the remainder, LDX is the mnemonic for ‘Load register X with the following byte of data’.

Type	Mnemonic	Opcode +	Operation
Data transfer	LDA	hhll	Load A with *contents* of hhll.
	LDA	(hhll)	Load A with contents of *address* in hhll.
	LDI	dd	Load A immediate with dd.
	LDX	hhll	Load X with *contents* of hhll.
	STA	hhll	Store A at hhll.
	STA	(hhll)	Store A at *address* in hhll.
	PSH	-	Push A on to stack and increment stack pointer.
	POP	-	Pop A from stack and decrement stack pointer.
Arithmetic and logic	INC	-	Increment A.
	DEC	-	Decrement A.
	INX	-	Increment X.
	DEX	-	Decrement X.
	ADD	hhll	Add the *contents* of hhll to A.
	ADX	-	Add X to A.
	SUB	dd	Subtract dd from A.
	CMP	dd	Compare A with dd.
	ANA	dd	AND A bitwise with dd.
	ORA	dd	OR A bitwise with dd.
	NEG	-	Form 2's complement of A.
Test and Branch	JMP	hhll	Set PC to hhll.
	JMR	dd	Add dd to PC.
	JZ	dd	Add dd to PC if zero flag = 1.
	JNZ	dd	Add dd to PC if zero flag = 0.
	JSR	hhll	Push PC, A, X and status on stack; set PC to hhll.
	RTS	-	Pop status, X, A and PC from stack.
	HLT	-	Halt action, awaiting possible interrupt.

In the third column, *hhll* is an address of four hex digits, the high byte followed by the low byte. Letters *dd* represent a numerical value of two hex digits (or 1 byte) .

PSH and POP are a pair of operations involving the stack. The stack is a small region of RAM set aside for the temporary storage of data. It works in a 'last-in-first-out' manner, like the plate dispenser sometimes seen in canteens. The stack pointer holds the address of the top of the stack.

Example

This table shows a section of RAM and the data stored there:

Address	Stored data
3B2C	0
3B2B	0
3B2A	48
3B29	0E
Stack pointer	3B2A
Accumulator	65

The most recently stored data is 48. Data 0E is left over from a previous operation. The stack pointer is pointing to the top of the stack, at 3B2A, at present storing 48. Now the PSH opcode occurs in the program and is executed. The data stored is:

Address	Stored data
3B2C	0
3B2B	65
3B2A	48
3B29	0E
Stack pointer	3B2B
Accumulator	65

The value 65 has been pushed on to the top of the stack from the accumulator. For the moment the accumulator still holds 65. The stack pointer points at the top of the stack, now at 3B2B. The microprocessor may now be engaged in other operations not involving the stack, which remains unchanged. Eventually, with (say) A2 in the accumulator, the POP opcode occurs in the program and is executed:

Address	Stored data
3B2C	0
3B2B	65
3B2A	48
3B29	0E
Stack pointer	3B2A
Accumulator	65

All questions in this chapter refer to the model microprocessor and its instruction set as listed on p. 332.

The original value 65 has been restored to the accumulator and the stack pointer is decremented to 3B2A. Then POP occurs again and is executed:

Address	Stored data
3B2C	0
3B2B	65
3B2A	48
3B29	0E
Stack pointer	3B29
Accumulator	48

Test your knowledge 21.3

If, in the stack example in the main text, the accumulator holds B5 and there is a third PSH operation, where is the data stored?

The new 'top of the stack' is loaded into the accumulator and the stack pointer again decremented.

The stack is used for temporarily storing an intermediate result in a calculation. With a simple PSH, it is quickly pushed from the accumulator on to the stack and may be recovered at a later stage with a POP instruction. The stack is also used for registering the state of operations when a jump to subroutine instruction is executed (see table), or when the microprocessor is responding to an interrupt. As soon as the microprocessor returns from the subroutine or interrupt it recovers all the essential information from the stack. It then continues operating at the same point in the program that it had reached before it jumped or was interrupted. There is more about this later.

Arithmetic and logic instructions

In this group we have already looked at INC. DEC performs the reverse of this. INX and DEX perform the same operations on the contents of the X register. Addition is performed by ADD, which adds the contents of the specified address to whatever value is currently in the accumulator. The contents of the accumulator are then replaced by the result of addition. ADX adds the contents of the X register to the accumulator, placing the result in the accumulator and leaving X unchanged. SUB subtracts a given value from the accumulator. Note that ADD and SUB are not the reverse of each other in this limited set of instructions. If the person writing the program decides to use twos complement arithmetic (see box on p. 335), then the ADD and ADX instructions can also be used for subtraction.

CMP subtracts a specified value from the contents of the accumulator, *leaving the accumulator unchanged.* The zero flag is set (=1) if the result is zero (that is if the accumulator is *equal* to the specified value). The zero flag is reset (=0) if the result is not zero. This instruction is useful in counting and looping routines.

Test your knowledge 21.4

Find the twos complements of 0, –1, –2, –4, –5 and –6, when expressed as 4-bit numbers plus the sign bit.

Test your knowledge 21.5

Use the twos complement technique to evaluate (a) –4+1, (b) –5+5, (c) –3+6, and (d) –2-4.

Negative numbers

We usually represent a negative decimal number by writing the negation sign (–) in front of it. This does not work in a computer because the registers contain 0's or 1's, and there is no '–'. There are several ways of representing negative numbers in binary without using the negation sign. The most useful of these ways is the *twos complement.*

We must first decide the number of digits in which we are working. This includes a *sign digit* on the left. We will work in four digits plus the sign digit. To form the twos complement of a number, write the positive number using 4 bits. Then write a 0 on the left to represent a positive sign. Next write the ones complement by writing 1 for every 0 and 0 for every 1. Form the twos complement by adding 1 to this number. Ignore any carry digits. This is the binary equivalent of the *negative* of the original number.

Example

Find the 4-bit equivalent of –3.

Write +3 in binary	0011
Write the sign digit for +	00011
Find the ones complement	11100
Add 1	1
Result is twos complement	11101

11101 is the equivalent of –3, the 1 on the left meaning that it is negative.

Twos complement may seem an arbitrary process, but it can be shown to work when two negative numbers are added together.

Example

Add –2 to –3

Twos complement of –3	11101
Twos complement of –2	11110
Add	[1]11011

Ignoring the carry digit in brackets, this is the twos complement of –5, the expected result.

Twos complement arithmetic can be used when adding a positive number to a negative number, to get a positive result.

More on negative numbers

Example

Add +4 to –3.

Twos complement of –3	11101
Binary equivalent of +4	00100
Add	[1]00001

The result is +1.

Adding positive to negative may also produce a negative result.

Example

Add +2 to –6.

Twos complement of –6	11010
Binary equivalent +2	00010
Add	11100

The result is –4.

The technique also works with two positive numbers.

Example

Add +3 to +2.

Binary equivalent of +3	00011
Binary equivalent of +2	00010
Add	00101

The result is +5.

Our set includes two logical instructions, ANA and ORA. These are bitwise operations, meaning that the corresponding bits in the accumulator and in the data *dd* are ANDed or ORed individually.

Examples

The accumulator holds 2DH and the instruction is ANA CBH. To follow what happens we must expand this into binary form:

Accumulator	0 0 1 0 1 1 0 1
Data	1 1 0 0 1 0 1 1
Result	0 0 0 0 1 0 0 1

The result at any given place contains a 1 only when both accumulator AND data contain a 1 at that place. The result is 09 in hex.

Given the same values, the result of ORA CBH is:

Accumulator	0 0 1 0 1 1 0 1
Data	1 1 0 0 1 0 1 1
Result	1 1 1 0 1 1 1 1

The result contains a 1 when either the accumulator OR data contain a 1 at that place. The result is EF in hex.

Test your knowledge 21.6

What is the result of ANDing B3H with 10H bitwise?

Test your knowledge 21.7

What is the result of ORing C5H with 56H?

Test and branch instructions

These operations allow the microprocessor to jump to other parts of memory and to continue operating there. The first instruction is JMP, which is an *unconditional* jump. When it reaches this instruction in memory the program counter is always set to the specified address. With a new address in the program counter, the next fetch-execute cycle automatically fetches from the new address and execution continues from that address. Note that JMP can be used to transfer the operation back to an earlier address. The JMR instruction has a similar action except that the place to jump to is specified in a different way. JMP gives the actual address to jump to. We call this *absolute* addressing. JMR gives the address relative to the present address, and is known as *relative* addressing. This is useful for short jumps within a routine. It often happens when a program is being revised that the routines need to be shifted around in memory. The jump always skips *the same number of bytes*. There is no need to alter the length of the jump because of the changing location of the routine. The JZ and JNZ instructions also use relative addressing.

JSR and RTS are concerned with *subroutines*. A subroutine is a small self-contained part of a program that performs a frequently required action.

Example

A program needs to calculate square roots of different numbers at several stages during a calculation. Instead of writing out the code for this operation every time a square root is required, it saves programming space (and hence memory) to include the code once, as a distinct part of the program and to send the microcontroller to it whenever a square root is needed. The square root routine is usually placed at the end of the main program, where it is known as a *subroutine*. Every time a square root is needed, the value to be rooted is stored in memory.

Then the microcomputer is made to jump to the first byte of the square root subroutine by issuing the JSR instruction followed by its address. When it gets to this address the microprocessor operates on the number in memory accumulator according to the subroutine, finally storing the result in memory. At the end of the subroutine the instruction RTS makes the microprocessor 'return from subroutine' back to the same point in the main program that it originally came from.

Test your knowledge 21.8

A microprocessor has 45H in its accumulator, FEH in its X register, 03H in its status register, and 650AH in its program counter. It is interrupted and pushes its accumulator, X register, status register and program counter on to the stack. List the contents of the stack, from the top byte downward.

The microprocessor is always able to jump back to the place it came from because it pushes all the relevant data on to the stack before jumping to the subroutine. The data includes values stored in the accumulator, X register, status register and program counter. After performing the subroutine, it recovers these values by popping them from the stack. The value then in the program counter tells it where it came from. It returns to that place with all the values intact and continues from there.

Input and output between the system and its peripherals may be handled in two different ways. In some systems the input and output ports are controlled by a decoder so that they appear to be locations in memory. Input data is read from a port by using instructions such as LDA. Output data is sent to a port by using instructions such as STA. In some systems with bidirectional ports it may first be necessary to set the direction of each line of the port by writing a pattern of bits into the data direction register. This register has its own address in memory and it written to using STA. In other systems, the ports are numbered and there are instructions (and opcodes) such as IN p and OUT p, to input or output data through port *p*.

Interrupts

A microprocessor is always busy handling the program, proceeding from one instruction to the next unless told to stop (HLT). Events sometimes occur which interrupt this steady progress of the microprocessor. An immediate response is essential. Such events include:

- **Arithmetic errors:** a typical example is 'division by zero'. It may happen that a particular expression is the divisor in an equation. If this expression occasionally evaluates to zero, the division can not be performed. This should not happen in a correctly written program but may sometimes occur during the early stages of development of a new program. This error would be detected inside the microprocessor and causes it to interrupt itself. Instead of going on to the next instruction, it jumps to a special *interrupt service routine* (ISR). This might cause a message to be displayed saying 'Division by zero error'. The program is then halted and the programmer checks to find why such an error occurred. The programmer *debugs* the program to find all such possible errors in the program and re-writes the program to eliminate them. Failure to find arithmetic errors may cause the program to 'hang' or to 'crash' when the final version is being used. Debugging is a very essential, and often time-consuming, part of programming.

- **Logical errors:** these may have effects as serious as arithmetic errors.
- **Clocked interrupts:** These are described in the previous section.
- **Peripherals:** The keyboard buffer may be holding key-presses which the microprocessor needs to know about, so the keyboard-buffer sends an interrupt signal (usually a low voltage level on a control line) to the interrupt input pin of the microprocessor. This then enables the buffer to allow it to place the keypresses on the data bus one at a time. The result is further action by the microprocessor, perhaps storing the ASCII codes in memory, or displaying characters on the monitor screen. A printer is another device that frequently interrupts to tell the microprocessor that it is waiting to receive data for printing.

Because the microprocessor can not handle two interrupts at the same time, interrupts by different devices are dealt with according to their priority. An interrupt from the keyboard, for example, may have a higher priority than one from the printer. If the printer interrupts while the microprocessor is dealing with an interrupt from the keyboard, the printer is ignored until the keyboard has been dealt with.

This raises the problem of how the microprocessor finds out which device is interrupting. With some microprocessors there are several interrupt input pins, ranked in order of priority. Devices are connected to these pins according to their priority. Priority is fixed by the wiring. Other microprocessors may have only one interrupt line, which is part of the control bus. The microprocessor knows there is an interrupt but has to find out its source. In some systems there is a line in the control bus called *interrupt query*. When the microprocessor puts a low signal on this line the signal automatically goes to all the devices that might possibly be interrupting. The one that is interrupting puts its own device code on the data bus so that the microprocessor knows what action to take.

As an alternative to interrupts, a system may rely on *polling* the individual devices. Each device has a register in which there is an interrupt flag. In between its main activities, the microprocessor interrogates each device to find out which one or more of them has its interrupt flag set. It then takes action. This system is the slowest of all, as a device has to wait to be interrogated and, by then, it may be too late to avoid a program crash or a loss of data.

It may be that the microprocessor is in the middle of a complicated routine which might fail if interrupted. With some microprocessors there is an opcode which causes it to ignore all interrupts, until interrupts are enabled again by using another opcode. We use these instructions to *mask* (shut out) all interruptions while a difficult routine is in progress. When to use these instructions is a matter for the programmer to decide. But this is not the highest level of control. There may be high-priority interrupts that are *non-maskable interrupts*, even by instructions in the program.

Problems on programming

1 Describe what happens during a fetch-execute cycle.

2 What is the difference between an opcode and a mnemonic?

3 Give three examples of instructions written in assembler and explain what they mean.

4 What does the status register of a microprocessor hold? Give one example and explain a way in which it is used in programming.

5 Describe how values are stored when an interrupt is called, and how they are recovered afterward.

6 Explain how to obtain the twos complement of an 8-bit number. Give an example of adding a negative number to a positive number when both are in twos complement form.

7 What is a subroutine? Why are subroutines an important feature of many programs?

Multiple choice questions

1 A relative jump instruction makes the microprocessor branch to an address:

A stated after the jump opcode.
B stored in the program counter.
C a given number of bytes away.
D stored in the stack.

2 A conditional jump is ignored if:

A the microprocessor is interrupted.
B the condition is not true.
C the zero flag is set.
D the accumulator holds zero.

3 If B2H is bitwise ANDed with 37H, the result is:

A E9H.
B B7H.
C 11H.
D 32H.

4 A bit that shows whether a certain condition is true or not is called a:

A nybble.
B opcode.
C flag.
D mnemonic.

5 The twos complement form of –9 is:

A 10111.
B 10110.
C 11001.
D 00110.

6 The accumulator holds 07H. The next instruction is CMP 03H. After that the accumulator holds:

A 03H.
B 07H.
C 04H.
D 1CH.

7 The result of bitwise ORing B3H with 10H is:

A B3H.
B 10H.
C C3H.
D A3H.

8 A routine that is jumped to several times during the running of a program is called a:

A loop.
B interrupt service routine.
C subroutine.
D relative jump.

9 The interrupt query control line is used:

A to tell the microprocessor that there is an interrupt.
B when there is a non-maskable interrupt.
C to ask an interrupting device to puts its code on the databus.
D to interrupt the microprocessor.

10 A name that is given to an address where a variable is stored is called:

A an opcode.
B indirect addressing.
C a mnemonic.
D a label.

11 A program that produces a machine code program from source code written in a high-level language is called:

A an assembler.
B an interpreter.
C a compiler.
D BASIC.

22 Computer routines

Summary

The best way to understand how to program a computer is to write and test routines for performing simple tasks. In this chapter we look at a range of short routines illustrating a variety of computing functions.

Decision box: A stage in a a flow-chart at which a decision has to be made. It is usually represented by a diamond-shaped box.

The best way to become familiar with the instruction set of a microprocessor or microcontroller is to program it to perform a specified task. In some cases it may be helpful to begin with a *flow-chart* (Fig. 22.1), which sets out all the stages of an operation and the way the computer is to get from one stage to the next. A flow-chart is most useful when a routine involves many *decision boxes* so there are several different paths through the routine. In other cases a flow-chart is not particularly helpful. Instead it may be better to try a *dry run* on paper before keying the program into the computer for testing. The best approach in any given situation is often a personal choice.

This chapter describes some program segments or routines that illustrate what the various machine code instructions do. In some of the routines there may be sections that are optional in content. The place where the instructions go is indicated by ¤.

The segments are all written in an imaginary assembler language matched to the instruction set of the imaginary microprocessor described in Chapter 21. The complete instruction set is listed in the table on p. 332.

Most assemblers allow the user to add comments to explain what is going on. This is a very useful and important facility because it is so easy to come back to a program that one has written only a few weeks earlier, and be quite unable to remember how it works. Comments are essential, but they *must* be brief and to the point. Otherwise the comments dominate the program listing, which then becomes difficult to follow. Comments are on the right in our listings, separated from the instructions by a semi-colon (;). Remember that the comments are intended for people reading the program and are completely ignored by the microprocessor.

One further concept illustrated in these examples is the *label*. This is a name given to a particular point in the program. In reality, this point is identified by its address, the address in memory at which that particular opcode is stored or at which a routine begins. But names are easier to remember than hex addresses, particularly if they have some meaning relevant to the action of the program. Once a label has been defined in an assembler program its name may be used instead of the address. If, later, the program is moved to a different part of memory, the addresses change but the labels do not. Labels are printed in CAPITAL letter in the listings in this chapter. In the descriptions the accumulator is referred to as A.

The routines are divided into five main sections: arithmetic and logical routines, timing routines, input routines, and output routines. Because they are arranged by function, a few of the more difficult routines come early in the list. Skip over these the first time through.

Arithmetic and logical routines

1) Add two bytes stored at adjacent memory locations and place their sum in the next location. The bytes are stored at 9000H and 9001H. Their sum is to be stored at 9002H. The values in 9000H and 9001H are to be preserved.

```
LDA 9000H  ; Get first byte.
STA 7000H  ; 7000H is an address in an area of RAM being used
           ; like a sheet of scrap paper for jotting down
           ; intermediate results.
LDA 9001H  ; Get second byte.
ADD 7000H  ; Add first byte to second byte, sum now in accumulator.
STA 9002H  ; Store sum at 9002H.
```

Even a simple action may require several stages. By using twos complement arithmetic, this routine may also be used for subtraction. It would be possible to write this program in assembler using labels instead of the addresses 7000, 9000, 9001 and 9002. This program does not include any routine to deal with the possibility of *overflow*. If the sum of the numbers is greater than the value of a single byte (255) there will be an error in the result. Similar remarks apply also to the multiplication and division routines described below.

2) Repeat an action a specified number of times. The number of times is stored in 9000H, which acts as a *loop counter*.

```
         LDA 9000H  ; T9he number of repeats loaded into A.
BEGIN:   PSH        ; BEGIN is a label for the start of the loop.
                    ; Store  the loop count on the stack. This
                    ; frees the accumulator for further use.
         ¤          ;The required action goes here. If anything
         ¤          ; is pushed on to the stack it must be
         ¤          ; popped off again by the end of this part
         ¤          ; of the program.
         POP        ; Recover the stored number.
         DEC        ; Count down.
         JNZ BEGIN  ; Jump to the BEGIN label if the number
                    ; is not zero.
         ¤          ; The program drops through to here when
         ¤          ; the stored number is zero and the action
         ¤          ; has been repeated the specified number
         ¤          ; of times.
```

The structure of this program is a *loop*. It holds a routine that is to be repeated the number of times set by the initial value in 9000H. The routine within the loop could be a simple addition such as:

```
         LDI 03     ; Put 3 in the accumulator.
         ADD 9001H  ; Add 3 to the contents of 9001H
```

Each time round the loop we add 3 to the contents of 9001H. If 9000H contains 9, and 9001H starts at zero, the content of 9001H is 27 (3×9) after the last time round the loop. This is a simple technique for *multiplication*, though it is not the only one used with microprocessors which have more instructions in their set.

3) Divide one number by another. The numbers are stored at two locations labelled DEND and DSOR (dividend and divisor). The loop counts the number of times that the divisor may be subtracted from the dividend. The result is stored in QUOTIENT. Note that, from now on in this section describing routines, we are using labels for memory locations, as well as for destinations for jumps.

```
         LDA DSOR   ; The divisor
         NEG        ; Form its twos complement.
         STA DSOR   ; Store it.
         LDX DSOR   ; The negated divisor in X.
         LDA DEND
BEGIN:   ADX        ; Adding a negative is equivalent to
                    ; subtracting.
         PSH        ; Temporary store for A. A still holds the
                    ; result.
```

```
LDA QUOT  ; Assuming QUOT(ient) stored zero at the
          ; start.
INC       ; Counting number of subtractions.
STA QUOT  ; Result in QUOT
POP       ; Recover A.
ANA 8000H ; Gives 1 if A is negative, 0 if positive.
JZ BEGIN  ; If A still positive.
¤         ; Program continues.
```

The use of ANA for picking out certain bits from a number is explained below in routine (6). Note that we have written the value to be ANDed in hex. This is better for logical operations as it is easy to understand that 'ANA 8000H' means 'AND with 1000 0000'. If we wrote the decimal equivalent 'ANA 32768', the meaning would be much less clear.

Although an assembler program allows you to single-step through a routine and examine the contents of registers and memory at each stage, a short routine like this one can also be tested by a 'dry run' on paper. Allocate a column for each register or memory location. In this example, 23 is divided by 7 (all values shown in decimal):

Instr	DSOR	DEND	QUOT	A	X	Stack	Zero
	7	23	0	0	0	0	0
LDA DSOR				7			
NEG				–7			
STA DSOR	–7						
LDX DSOR					–7		0
LDA DEND				23			
ADX				16			0
PSH						16	
LDA QUOT				0			1
INC				1			0
STA QUOT			1				
POP				16			0
ANA 8000H				0			1
JZ BEGIN	Zero flag is 1, so jump						
ADX				9			0
PSH						9	

Values are written in the cells only when a new value is put there. There is still a positive amount left in the accumulator, so the routine is not yet completed. The use of twos complement arithmetic restricts the maximum values in a byte to ±127.

4) Move a table of data to a different place in memory. Each time round the loop we need to be able to change the addresses to load from and to store in, so we need *indirect addressing* (see box). We are given OLD, which stores the original address of the first byte of the table, NEW, which stores the new start address, and LEN which stores the length of the table, in bytes.

```
        LDX LEN    ; Get the number of bytes.
START:  LDA (OLD)  ; Load the data in the old address.
        STA (NEW)  ; Store the data in the new address.
        LDA OLD    ; Load the old address.
        INC        ; Point to the next old address.
        STA OLD    ; Store next old address for next loop.
        LDA NEW    ; Load the new address.
        INC        ; Point to the next new address.
        STA NEW    ; Store the next new address.
        DEX        ; Reduce number of bytes to move.
        JZ START   ; Continue until all bytes moved.
```

This routine works if the two storage regions do not overlap, and also if they partly overlap but the new addresses are lower than the old ones.

5) Conditional jump. Perform one action if today is Sunday but a different action if today is a weekday. The day of the week is stored in a register of the real-time clock. The address of this register is at DAY. The register holds values 1 to 7, according to the day of the week, beginning with 1 for Sunday.

```
        LDA DAY    ; Find the day of the week.
        CMP 01     ; Is it Sunday? If it is, zero flag = 1.
        JZ: SUN    ; Jump to Sunday routine if zero flag = 1.
        ¤          ; Weekday routine.
        HLT        ; Prevents microcontroller from
                   ; continuing on to the Sunday routine.
SUN:    ¤          ; Sunday routine.
```

This routine demonstrates the important of HLT, or perhaps an unconditional JMP to skip over the Sunday routine and continue with the weekday program if today is not Sunday.

6) ANDing to mask out unwanted bits. In routine (3) we need to find out if DEND is positive or negative. If it is negative, its most significant bit is 1 in twos complement arithmetic. To isolate this bit we mask the byte with another byte that has a 1 as its MSB.

A holds the latest value of DEND (say 9)	0 0 0 0 1 0 0 1
ANDed with 8000H	1 0 0 0 0 0 0 0
Bitwise ANDing gives	0 0 0 0 0 0 0 0

This sets the zero flag and there is a jump back to BEGIN.

Target data: the data we are aiming to find and load.

Addressing modes

Microprocessors have several different ways or *modes* of specifying addresses, though different microprocessors differ in which modes they are able to use. The four modes most often found are:

Direct addressing: The opcode is followed by the actual memory address where the target data is stored.

> *Example :* LDA 9E63, means load accumulator with the data stored at 9E63.

This is the mode most often used.

Indirect addressing: The opcode is followed by an address at which *the address* of the target data is stored.

> *Example*: The data is stored at 9E63. This address is stored at 0214. The opcode LDA (0214) tells the microprocessor to 'load indirect' 1402. It goes to 0214, finds the address 9E63 there, then goes to 9E63, finds the data and loads it.

Example (4) illustrates the difference between direct and indirect addressing.

Indexed addressing: There is an *index register* in the microprocessor. The opcode is followed by a value that is added to the contents of the index register to obtain the address at which the data is stored.

> *Example:* The opcode is LIXA 45, the index register already holds 3200. The data is found at 3245.

Indexed addressing is useful for look-up tables. The index register holds the address of the first byte in the table. The data is simply located by quoting its position number after the opcode.

Relative addressing

The target data is stored at an address a given number of bytes further on (or further back).

> *Example*: LRA 35 means add 35 to the address at present in the program counter and find the data there. The JMR instruction in our model uses relative addressing.

After a few more subtractions, DEND might become –5. Then:

A holds the latest value of DEND (–5)	1 1 1 1 1 0 1 1
ANDed with 8000H	1 0 0 0 0 0 0 0
Bitwise ANDing gives	1 0 0 0 0 0 0 0

This resets the zero flag to 0, there is no jump, and the routine ends.

This is a useful routine for our model microprocessor. Many real microprocessors would not use this routine because they have positive and negative flags in their status registers. These are set or not set according to the latest value in A, and there are conditional jumps which use these flags.

Timing routines

7) Delay. The simplest technique is to make the microprocesor repeat the same action many times. In other words, let it run round a loop a given number of times. Routine (2) can be used, with no task inside the loop. The maximum value that can be stored in the loop counter in a 16-bit system is $2^{16} - 1 = 65535$. In a system with a slow clock, the loop might take 1 ms to run, so the total time is 65 s. In a computer system with a clock rated in hundreds of megahertz, the loop runs much more quickly, and this simple routine may be impracticable. One way to extend it is to place a second inner loop inside the first outer loop. The counts for each loop are stored at OUTLOOP and INLOOP.

```
          LDA OUTLOOP     ; Loop count for outer loop.
BEGIN:
          LDX INLOOP      ; Loop count for innner loop.
     REPEAT:   DEX        ; Count down inner loop.
               JNZ REPEAT ; Jump to REPEAT if the number
                          ; is not zero.
          DEC             ; Count down outer loop.
          JNZ BEGIN       ; Jump to BEGIN if the number
                          ; is not zero.
```

There is no need for PSH and POP in this routine because the accumulator is used only in the outer loop. The X register counts the inner loop. Note that the values stored in OUTLOOP and INLOOP remain unchanged, so the routine can be used again without the need to reset the values. The structure of this routine is a little more complicated than usual so it is illustrated by a flowchart in Fig. 22.1. For every time round the outer loop, the microprocessor must go many times round the inner loop.

The times obtained with delay routines such as these are not precise. They depend on the rate of the system clock, which is not usually of high precision. Also there may be interrupts, which increase the delay.

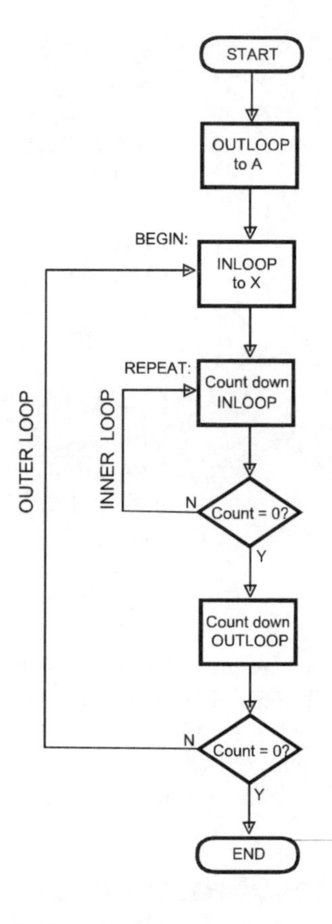

Figure 22.1 *The flowchart of a program with two nested loops. The program segment begins and ends with the oval boxes. Rectangular boxes contain instructions for each step. Decisions are required where there is a diamond-shaped box.*

8) Timing an operation. This routine gives more precise timing as it uses a special peripheral ic, a real-time clock. At the beginning of the operation, MINS and SECS (locations in RAM) store the minutes and seconds readings of the real-time clock, taken from the clock's MINUTES and SECONDS registers. At the end MINS and SECS store the time taken. The time taken is the starting time subtracted from the finishing time. There would be no difficulty if both the minutes and the seconds values were bigger at the end. But often the seconds at the end is more than the seconds at the beginning. In this case we need to 'borrow' a minute before subtracting the seconds.

```
LDA MINUTES     ; Get starting minutes.
NEG             ; Form its twos complement.
STA MINS        ; Store it.
LDA SECONDS     ; Get starting seconds.
NEG             ; Form its twos complement.
STA SECS        ; Store it.
¤               ; The operation to be timed.
¤
¤
LDA SECONDS     ; Get finishing seconds.
```

```
        LDX SECS      ; Get negative starting seconds.
        ADX           ; Find difference of seconds.
        STA SECS      ; Store it.
        ANA 8000H     ; Positive or negative?
        JZ POS        ; OK if positive, jump to deal with
                      ;  minutes.
        LDA 60        ; Borrow 60 seconds.
        LDX SECS      ; Get negative value.
        ADX           ; Subtract from 60.
        STA SECS      ; Store the difference (now
                      ; positive).
        LDA MINS      ; Load negative of starting minutes.
        DEC           ; Compensate for borrowing.
        STA MINS
POS:    LDA MINUTES   ; Deal with minutes.
        LDX MINS
        ADX
        STA MINS
```

We assume that the finishing minutes are greater than the starting minutes, but they may not be. If they are not, the routine should continue by borrowing 60 minutes and subtracting from that. This routine illustrates how essential it is to take into account all the possible outcomes that can occur.

Input routines

9) Registering a keypress. Routines for taking input from a keyboard are too complicated to be considered here. Instead consider an input circuit connected to the line of the lowest bit of an input port. The circuit is operated by a key or switch, or is the output of a logic gate. It produces a high voltage (usually +5 V, relative to system ground) for logic high, and a low voltage (0 V) for logic low. The input port is memory-mapped at PORTA.

```
        LDA PORTA     ; Read the input data lines.
        ANA 01        ; Bitwise AND all lines with
                      ; 0000 0001.
        JNZ  HIGH     ; If input is high.
        ¤             ; Routine if input is low.
        ¤
        ¤
        JMP CONT
HIGH:   ¤             ; Routine if input is high.
        ¤
        ¤
CONT:   ¤             ; Continue with the program.
```

The JMP CONT prevents the program from running on to the high input routine immediately after the low input routine.

10) Sensing proportional input. An analogue voltage is fed to an ADC, which puts its output on the data bus when its address, ANALOG, is on the address bus. The reading is to be stored in LEVEL, and the microprocessor is to wait until the reading is between 100 and 102.

ADC: analogue to digital converter.

Waiting generally requires a looping routine.

```
START:   LDA ANALOG    ; Read input.
         STA LEVEL     ; Store it.
         CMP 100       ; Is it 100?
         JZ DONE       ; If so, continue with program.
         CMP 101       ; Is it 101?
         JZ DONE       ; If so, continue with program.
         CMP 102       ; Is it 102?
         JZ DONE       ; If so, continue with program.
         JMP START     ; Not in range 100-102, so go
                       ; back to take another reading.
DONE:    ¤             ; Continue with program.
```

CMP leaves the accumulator unchanged so we can use it to test the value three times. This is too long a routine if there are many acceptable values in the range. In that case we compare the reading with the smallest and greatest values in the range. It must be equal to OR greater than the smallest AND equal to OR less than the greatest. This requires operations not available in our model microprocessor.

Output routines

11) Flash an LED once, indefinitely, or a given number of times. The LED is controlled by the lowest line of the outport port. This can be made low or high. A high level may be used to drive a MOSFET switch or a BJT switch. These routines can be used to drive LEDs lamps, motors, relays and many other electrically powered devices. This is the program to flash the LED once:

```
LDI 01         ; load accumulator with 0000 0001.
STA PORTB      ; Port B is an 8-bit output port. Turns LED on.
JSR DELAY      ; use one of the delay routines as a subroutine.
LDI 00         ; has 0 in the least significant position.
STA PORTB      ; turn LED off with 0000 0000.
```

For indefinite flashing the routine needs another delay while the LED is off, before it loops back to flash the LED again:

```
START:    LDI 01
          STA PORTB      ; turn LED on.
          JSR DELAY      ; to delay subroutine.
          LDI 00
          STA PORTB      ; turn LED off.
          JSR DELAY      ; to delay subroutine.
          JMP START      ; repeat the flash.
```

This assumes that the LED is to be on and off for equal lengths of time. Otherwise, use different delays, or perhaps go to the same subroutine more than once while on or off (see next routine).

Flashing a given number of times requires a loop with a loop counter, FLASH. Here the LED is off for twice as long as it is on:

```
          LDI 00
          STA FLASH      ; Set loop counter to zero.
START:    LDI 01
          STA PORTB      ; Turn LED on.
          JSR DELAY      ; To delay subroutine.
          LDI 00
          STA PORTB      ; Turn LED off.
          JSR DELAY      ; To delay subroutine.
          JSR DELAY      ; To delay subroutine again.
          LDA FLASH      ; Get loop count
          INC
          STA FLASH
          CMP 10         ; Compare with 10.
          JNZ START      ; Repeat for 10 flashes.
```

12) Switch on a lamp if a passive infrared (PIR) detector senses a person approaching AND it is dark. This could be used to switch on a porch lamp automatically, but only at night. We need two sensors, the PIR detector and a light sensor. The PIR detector has logic high output when it has sensed a person. The light sensor circuit has a high output when it is dark. These are connected to the two lowest lines of the input port, PORT. The relay controlling the lamp is controlled by the lowest line of the output port.

```
START:    LDA PORTA      ; Get state of inputs from sensors.
          AND 03         ; Select inputs from PIR and
                         ; light sensor.
          CMP 03         ; The value of A if both outputs are
                         ; high.
          JZ LAMP        ; Both are high, so jump to switch
                         ; on lamp.
          JMP START      ; Only one or neither are high, so
                         ; read inputs again.
```

```
LAMP:      LDA 01          ; Lowest bit for lamp relay.
           STA PORTB       ; Switch lamp relay on.
```

This routine needs to be expanded to switch the lamp off after a short period, then to repeat. The routine operates independently of other devices being connected to other lines (bits) of the ports.

13) Turn on a siren if a PIR sensor detects a human OR a security loop broken. When the security loop (doors, windows) is broken the output from the circuit goes high.

```
START:     LDA PORTA       ; Get state of inputs from sensors.
           AND 03          ; Get inputs from PIR sensor and
                           ; loop circuit.
           CMP 00          ; The value of A if both outputs are
                           ; low.
           JNZ  START      ; Both are low, so read inputs
                           ; again.
           LDA 01          ; Lowest bit for siren switch.
           STA PORTB       ; Switch siren on.
```

14) Drive a 7-segment display using a lookup table. Instead of sending just 1 bit to the output port, we send 7 bits, one for each segment. The LEDs that need to be switched on to display any given figure (Fig. 13.2) can not be calculated by using a formula. Instead we store the bit pattern in a block ten consecutive bytes of memory. If the bits represent segments *a* to *g*, reading from right to left, the first three bit patterns are:

For '0'	00111111
For '1'	00000110
For '2'	01011011

The variables are:

BASE	the first address in the bit-pattern block (the address where the pattern for figure 0 is stored).
FIG	the address where the figure to be displayed is stored.
POINT	holds the address in the bit-pattern block where the pattern for the required figure is stored.
PORTB	the LEDs *a* to *g* (Fig. 13.1) are switched by the outputs from the lowest 7 lines.

Here is the routine:

```
LDA BASE         ; Get base address.
ADD FIG          ; Figure value added to BASE to point to
                 ; figure pattern.
```

```
STA POINT        ; Storing the address of the required figure
                 ; pattern  in POINT.
LDA (POINT)      ; Using indirect addressing, get the figure
                 ; pattern, which is stored at the address
                 ; stored in POINT.
STA PORTB        ; Switch on the segments.
```

This routine is better understood after a dry run on paper or by single-stepping the assembler.

15) Control 8 parallel output lines to produce a sequence of 8-bit outputs with delays between. Routine (14) is able to control 8 output lines, and we can use a lookup table to specify the action at each stage of the sequence. There can be any reasonable number of bytes in the lookup table, storing the bit-patterns for a long sequence of actions if necessary. In this example we assume that there are 15 bit-patterns in the table. The routine requires us to increment FIG each time round the loop. FIG determines which bit-pattern is used at each stage, and also acts as a loop counter.

```
        LDI 00      ; Clears the accumulator.
        STA FIG     ; FIG is set to zero.
START:  ¤           ; Routine (14) goes here to output the
        ¤           ; pattern selected by the value in FIG.
        INC
        STA FIG     ; FIG incremented to point to next pattern.
        CMP 15      ; End of table?
        JNZ START   ; Repeat if FIG less than 15.
```

The 15 bit-patterns in the lookup table are put on the output lines once only, in sequence.

Problems on computer routines

1 What is a lookup table and how is it used?

2 Explain the difference between direct addressing and indirect addressing. Give an example of how to use each mode.

3 Write a routine with the same function as routine (1) above, but without the requirement to preserve the values in 9000H and 9001H.

4 Complete the dry run of routine (3). Repeat it with another pair of values for DEND and DSOR.

5 Do a dry run of routine (4) for a block of 4 bytes, start address 40H, being moved to a new start address 30H.

6 Adapt routine (4) for moving a table of data when the areas of memory occupied by the old and new tables partly overlap, and the new addresses are higher in memory (start at a later address) than the old ones.

7 Adapt routine (5) to perform one action on a Sunday, a different action on a Wednesday and Saturday, and a third action on the other weekdays.

8 Complete routine (8) to allow for the finishing minutes being less than starting minutes.

9 In routine (9), assume there are two keys connected to the two lowest data input lines of PORTA. Write a routine to go to two different routines according to which key is pressed, or to skip both routines if neither is pressed. If both are pressed, give priority to the lowest input line.

10 Write a routine to flash two LEDs alternately, like the red lights on a railway crossing.

11 Write a routine to control three LEDs (or lamps), red, yellow and green, in the sequence of traffic lights.

12 Write routines to control a stepper motor to (a) make single steps in either direction, (b) to make fixed number of steps in one direction, and (c) to make steps at specified rate.

13 Write routines to control 8 output lines connected to a digital-to-analogue converter so that it produces (a) square waves, (b) triangular waves, and (c) sawtooth waves.

23 Systems with faults

Summary

A system may be faulty because of its design, the way it is handled and used, and defects in its components. In addition, a logical system may be faulty because of its programming. The first stage in finding the cause of a fault is a *written* report describing the exact nature of the fault – what happens, or what does *not* happen, and when. Next follows a general examination of the system, and a written report on the findings. In particular, check that there is a positive supply to all the parts that need it. Check that the ground (0 V) connection is intact for all parts. Examine the circuit board and soldering for defects. Also note what parts could not be examined at this stage, and why not. If the cause of the fault has not been discovered by this stage, more detailed testing is required. In analogue systems this may involve using an oscilloscope, a signal generator and a signal tracer. In digital systems a logic probe and a logic pulser are used to observe circuit operation. A logic analyser is required for the more complicated logic circuits. Other techniques for checking logic circuits include in-circuit emulation and signature analysis.

Causes

A system may develop faults because of:

- Design faults
- Handling or usage faults
- Component faults

Most systems work properly for most of the time. But even a system that has been properly designed may show faults when it is operated under unusual or extreme conditions. Possibly the designer did not take these into account when the system was in the design stage.

Example

An amplifier normally works correctly, but becomes unstable

when the ambient temperature exceeds 35°C. The heat sinks are too small or the ventilation of the enclosure is not adequate.

Ambient temperature: the temperature of the surroundings.

With a commercially built system there may be little that can be done to correct designer faults, except to complain to the manufacturers. With a system you have built yourself, it is a matter of 'back to the drawing board'.

Handling or usage faults occur when the system is used in a way that was not intended by the designer.

Examples

- A circuit board is connected to a supply voltage of 12 V when it is intended to operate on a maximum of 5 V.
- A circuit is connected to a supply of the reverse polarity.
- This type of fault can also occur accidentally, as when there is a spike on the mains supply as a result of a lightning strike.

Defective components are rarely the *cause* of a fault. Provided that new components have been purchased from a reputable supplier, and are connected into a well-designed system, they should seldom fail. If components are not new but have been used in other circuits, they may have been damaged previously by overvoltage or excessive current, by incorrect polarity, or they may have been damaged mechanically. Unless precautions are taken, CMOS and similar devices may be damaged by static charges while being handled, prior to being use in a circuit. This is not a defect of the component as such, but is the result of mishandling the component.

Although it is unlikely to be the cause of a fault, a defective component may often be the *result* of a circuit fault of another kind.

Examples

- An inductive load switched by a transistor does not have a protective diode. Consequently, the high current induced when the load is switched off eventually damages the transistor. The basic cause of the defect is faulty circuit design.
- An excessive current is drawn from an output pin of a microcontroller by connecting a low-impedance load to it. The output circuit is damaged and no further output is obtainable from that pin. The basic cause is incorrect usage.

Fault report

The first stage in diagnosing a fault is to state exactly what is wrong with the system. It is not sufficient to report simply that 'It does not work'. The initial report should give answers to the question:

What does it NOT do that it should do?

This can be answered by one or more statements such as:

- The pilot LED does not come on when power is switched on.
- There is no sound from the loudspeaker.
- There is no picture on the screen.
- The heater does not switch off when the set temperature is exceeded.
- Nothing happens when a control button is pressed.

To be able to provide such answers, the engineer must be very clear about what the system should do. Read the User Manual and consult it often.

With some systems it may also be informative to answer this question:

What does it do that it should NOT do?

Typical answers could be:

- There is a smell of scorching.
- The indicator LED flickers irregularly.
- There is a loud buzz from the loudspeaker.
- The servomotor never settles to a fixed position.

Another question that should be asked is:

WHEN does it show the fault?

It may be necessary to run the system for several hours to answer this fully, but typical answers are:

- All the time.
- Not often. It becomes faulty and then recovers.
- Not often, but after the fault develops it stays.
- When it is first switched on, but not later.
- Not when it is first switched on, but starts about 10 minutes later.
- Only when the gain is set high.

Inspection

With the fault report in mind, the next stage is to inspect the circuit, looking for obvious faults, such as wires that have broken away from the circuit board or the terminals of panel-mounted components, signs of mechanical damage to the circuit board and components, signs or smells of scorching. These visible faults should be written down. At the same time it is important to note what parts of the system can not be inspected and why. In some systems there are complicated components (such as a microcontroller, or a RAM chip) that can not be tested without special equipment. These could be looked at later if the initial inspection fails to deliver any solution to the problem.

Mains supply

Turn off the mains supply and remove the plug from the socket before inspecting or testing mains-powered equipment.

Unless you are fully experienced you should never test or operate mains-powered equipment with the case open. You should never attempt to test the circuit while it is plugged into a mains socket.

Preliminary diagnosis

When the inspection is complete it may already be possible to decide what the cause of the fault may be. The next stage is to carry out some elementary practical tests. It is not possible to provide a set of routines because so much depends on the type of system and the nature of the fault. Some manufacturers provide checklists of faults and remedies applicable to their equipment. If such help is available it should be followed.

In the absence of advice from the manufacturer, or even when advice is available, it is essential that the person looking for the fault fully *understands* the system and how it works. A schematic diagram is almost essential, preferably marked with typical voltage levels at key test points. With circuit diagrams and other data to hand, and with a good understanding of the system, it is usually possible to eliminate certain possibilities and to concentrate on locating particular kinds of defect.

Examples

- A total failure of the equipment to do anything at all suggests that there may be a failure of its power supply. With mains equipment, first check that the mains socket provides mains current when switched on (try it with some other piece of equipment plugged in).

- With a battery-powered circuit, check the battery voltage, remembering that the voltage from a nickel-cadmium rechargeable battery drops sharply as it reaches full discharge. It might have been correct yesterday, but is it still delivering full voltage today?
- From the circuit diagram, pick out points in the circuit that should be at full positive supply voltage, and check these with a meter. Such points include the positive supply pins of all integrated circuits and all other pins (such as unused reset pins) that are wired permanently to the positive supply.
- If the supply voltage is low at any of these points, place an ammeter in series with the power supply and the equipment and measure the current it is taking. If the current is excessive, it suggests that there may be a partial or total short-circuit between the positive supply rail and the ground (0 V) rail. This could be due to a defective component.
- Use a continuity tester to check that all points that should be connected directly to the 0 V rail are in fact connected. This includes 0 V pins of all ics and all other pins that are wired permanently to 0 V.

Even if a list of expected voltage levels is not available, it is often possible to work out what they should be.

Example

If the base-emitter voltage of a BJT is more than about 0.6 V the transistor should be on. This means that its collector voltage should be relatively low, perhaps not much greater than its base voltage. Conversely, if the base-emitter voltage is less than 0.6 V, the transistor should be off and its collector voltage may be equal to the positive supply voltage.

Capacitors

In many types of circuit, the capacitors retain their charge for a long time, perhaps for several hours, *after* the power supply has been switched off. They can deliver a very unpleasant and possibly lethal shock.

Treat all capacitors with respect and make sure they are discharged before touching them or their terminal wires. Capacitors of large capacity should always be stored with their leads twisted together.

Common faults

There are some very basic faults that occur more often than faults of other kinds. It is best to look for these first. Several of them are related to the circuit board and construction techniques. Quality control should eliminate most such faults from commercially built systems, but self-built systems should be checked *before* power is first switched on:

- There are short-circuits caused by threads of solder bridging the gaps between adjacent tracks on the pcb. A hair-thin thread can result in a fault, so it is *essential* to use a magnifier when inspecting a circuit for this kind of fault.
- There are hairline gaps in the tracks, perhaps caused by an actual hair present on the negative when the board was exposed to light in the early stage of etching. Alternatively, breaks in tracks may result from flexing due to mechanical stress in mounting the board. These gaps are difficult to see and it may happen that they only open up after the board has been warmed from some time. Faults of this kind may develop after the circuit has been switched on for several minutes.
- Cold-soldered joints are the result of a lead or the pad or both being too cold when the solder was applied. It may also be the result of a greasy or corroded surface. The molten solder does not wet both surfaces properly. Often the solder gathers into a ball on the track or lead and the fact that the joint is dry can be seen under a magnifier. Suspected cold-soldered joints should be resoldered.
- Components may be soldered on to the board with the wrong polarity. This applies particularly to diodes, LEDs, electrolytic capacitors and tantalum capacitors. It is also possible (and damaging) to insert an ic the wrong way round in its socket.
- It is possible that a wrong ic has been used. Check all type numbers.
- An ic may be inserted in its socket so that one or more pins is bent under the ic and does not enter the socket. This type of fault is hard to see. The only method is to remove the ic from the socket and inspect its pins.
- Resistors and capacitors may be mechanically damaged by rough handling so that leads lose their electrical connection with the device without actually becoming detached. With ics, excess pressure on the ic may flex the silicon chip inside so that connections are broken.

Devices such as diodes and transistors may be totally damaged or their characteristics may be altered by excessive heat. This is usually happens when the device is being soldered in, but may also be the result of excess current. The resistance of resistors may also be altered by excessive heat.

Analogue circuits

The items listed above may lead directly to the cause of the fault, which can then be rectified. If this fails, a more detailed inspection is required. Suggestions for digital circuits are listed in the next section. Analogue equipment is checked by further use of a testmeter and with more elaborate equipment such as an oscilloscope. Signal generator and signal tracers are often used with audio equipment. Detailed instructions for using these are found in their Users' Manuals.

The probe of an oscilloscope has high input impedance so the instrument can be used in the same way as a voltmeter for checking voltage levels at key points. In addition, it is possible to detect rapid changes in voltage to discover, for example, than an apparently steady voltage is in fact alternating at a high rate. In other words, the circuit is oscillating. In other systems we may discover that a circuit which is supposed to be oscillating is static. Clues such as these, combined with the understanding of circuit operation, may point to the cause of the fault. In particular, the switching of transistors may be examined, to see that the base (or gate) voltages go through their intended cycles and that the

Test meters

There are two main types of testmeter, or multimeter: the *analogue* type with a needle moving over a scale, and the *digital* type with an LCD numeric scale. The digital type is usually preferred because it is more precise and it draws very little current from the circuit being tested. If the circuit has components of high resistance in it, an analogue meter may draw so much current that voltage levels are affected and a seriously incorrect reading is obtained. On the other hand, assuming that the circuit is able to provide the meter with enough current, an analogue meter is preferred for monitoring *changing* voltages. The motion of the needle is easier to interpret than the ever-changing readings of a digital display. Some models of digital meter also have an analogue display, and thus combine the advantages of both types.

Typical analogue meters have an input resistance of 20 kW per volt. For example, when set to its 10 V DC scale, the resistance of the meter is 200 kW, so that it takes 50 mA when measuring a voltage of 10 V. If such a meter is being used to measure the voltage at the base of a biased transistor, the current it takes may pull the base voltage down, so that the transistor becomes switched off. It is better to use a digital meter for such measurements.

collector and emitter (or source and drain) voltages respond accordingly. A dual-trace oscilloscope is invaluable for checking transistor operation.

A signal generator may be used for introducing an audio-frequency signal into an audio system. In a properly functioning system, a tone is then heard from the loudspeaker. Beginning at the power output stage, we work backward through the system, applying the signal at each stage and noting whether it appears at the loudspeaker and whether it is noticeably distorted. An oscilloscope can be used to check the waveform, particularly to look for clipping, ringing and other distortions. If we get back to a stage from which no signal reaches the speaker, or distortion is severe, the fault must be located in this stage. Then we examine this stage more closely. Alternatively, we can work in the opposite direction. We supply the input of the system with an audio signal from a radio tuner or disc player. Then we use a signal tracer (a simple audio-frequency amplifier with a speaker) to follow the signal through the system from input to output. Again we note the stage at which the signal is lost and investigate this stage in more detail. Sometimes it is helpful to use both techniques, working in from both ends of the system to arrive at the site of the trouble.

Logic circuits

The preliminary techniques, such as checking power supplies and circuit boards, are the same for logic circuits as for analogue circuits. Cold soldered joints or cracks in the circuit board tracks can cause a loss of logic signal between the output of one gate and the input of the next. A common fault is for an input to be unconnected to the power rails or to a logic output. This can happen because of incorrect wiring, a bent ic pin, or a dry solder joint. An unconnected CMOS input will then 'float', staying in one state for some time but occasionally changing polarity. Output from the ic will vary according to whether the disconnected input is temporarily high or low. With TTL, an unconnected input usually acts as if it is receiving a high logic level. If outputs of a logic ic are not firmly low or high it may be because of lack of connection on the ic power input pins to the power lines. If there is no connection to the 0 V line, output voltages tend to be high or intermediate in level. If there is no connection to the positive line, the ic may still appear to be working because it is able to obtain power through one of its other input terminals which has high logic level applied to it. But output logic levels will not be firmly high or low, and the performance of the ic will be affected. This situation can lead to damage to the ic.

The more detailed checks require the use of special equipment. A *logic probe* is the simplest and most often used of these. It is usually a hand-held device, with a metal probe at one end for touching against terminals or the pins of the logic ics. The probe is usually powered from the positive and 0 V lines of the circuit under test. In this way the probe is working on the same voltage as the test circuit and is able to recognise the logic levels correctly. The simplest logic probes have three LEDs to indicate logic high, logic low and pulsing. The third function is very important because the level at a given pin may be rapidly alternating between high and low. With an ordinary testmeter the reading obtained is somewhere between the two levels. We can not be certain whether this indicates a steady voltage at some intermediate level (which probably indicates a fault) or whether it is rapidly alternating between high and low logical levels (which probably means that it is functioning correctly). With a logic probe, the 'high' and 'low' LEDs may glow slightly, but this again could mean either an intermediate level or rapid alternation. The 'pulsing' LED indicates clearly that the voltage is alternating. Some probes also have a *pulse stretcher* LED. Many circuits rely on high or low triggering pulses that are far too short to produce a visible flash on the 'high' or 'low' LEDs. The pulse stretcher flashes for an appreciable time (say 0.5 s) whenever the probe detects a pulse, even if the actual pulse it is only a few milliseconds long.

A related device is the *logic pulser*, which produces very short pulses that effectively override the input from other logic gates. They can be used in the similar way to an audio signal generator to follow the passage through a system of a change in logic level. A *logic analyser* is used for analysing faults in complex logic systems. These can accept input from several (often up to 32) points in the system and display them in various combinations and on a range of time-scales. It can detect and display *glitches* in the system as short as 5 ns duration. The existence of a glitch indicates that there is an unexpected *race* in the system, producing incorrect logic levels. There are also input circuits triggered by various input combinations, making it possible to identify particular events in the system and examine circuit behaviour as they occur. A logic analyser requires expertise on the part of the operator, both in using the equipment and in interpreting its results.

Glitch: An unintended short pulse resulting from timing errors in a logic circuit.

Race: when two or more signals are passing through a logical system, one or more of them may be slightly delayed, perhaps because a certain device has a longer delay time than the others. Logic levels do not change at precisely expected times and the result is often a glitch.

Signature analysis is a technique for checking that each part of a logical circuit is acting correctly. When the system has been assembled and found to operate correctly, its inputs are fed with a repeating cycle of input signals. The voltage levels at each key point in the system are monitored and recorded for each stage in the cycle. These are the *signatures* for each point. If the system develops a fault, its inputs are put through the same cycle and the signatures monitored at each key point. If the inputs of a device have the correct signatures but its outputs show incorrect signatures it indicates that the device is faulty. This technique is one that can be made automatic, so that circuit boards may be tested by computer. Very complicated systems can be put through the testing cycle in a very short time. But the input sequence must be expertly planned, or there may be errors that the test fails to reveal.

When a system is controlled by a microprocessor or a microcontroller is may happen that the fault lies not in the hardware but in the software. The tests described above fail to show anything wrong with the power supplies, the circuit board, the wiring or the individual components, but still the system operates wrongly. It may be that there is a bug in the program, even one that was there from the beginning. In many programs there are routines that are seldom used and it is relatively easy for a bug to lurk there while the rest of the program functions correctly. Eventually the faulty routine is called and the bug has its effect. The same thing can happen as a result of faulty copying of a program or if one of the elements in a RAM is faulty. It is necessary only for a single '0' to change to a '1' or the other way about. Then the machine code is altered, with possibly dramatic effects.

Bug: A software fault, which causes the microprocessor to operate incorrectly.

Thorough testing of programs at the development stage is essential if all bugs in a program are to be found and eliminated. Although programs can be developed using assembler or a high-level language on a computer, the only way that it can be completely tested is to run it on the system itself. This is often done by using an *in-circuit emulator.* The emulator is connected into the system by plugging its connector into the socket that the system's microprocessor would normally occupy. The emulator acts in the same way as the microprocessor would act, but it has the advantage that it can easily be reprogrammed if the program does not control the system correctly. The program can be run and stopped at any stage to allow readings to be taken or to examine the content of memory and of registers in the microprocessor. When the program has been perfected, it is burnt into a PROM for use in the system, and the actual microprocessor chip replaces the emulator. An emulator is an expensive piece of equipment so its use is limited to industries in which relatively complex logical systems are being produced on the large scale. Emulators are usually able to simulate a number of different popular microprocessors, which saves having to have a different one for every processor used. But when a new processor is produced, there may not be an emulator to use for testing circuits based on the new processor. This limits the use of in-circuit emulation to the well-tried processors.

Problems on systems with faults

1 Select one of the circuits described below and list all the things that can possibly go wrong with it. For each possible fault, describe the symptoms you would expect the circuit to show, state what tests you would perform to confirm that the fault exists, and briefly state how you would attempt to cure it.

(a) common-emitter amplifier.

(b) common-drain amplifier.

(c) thermostat circuit using a thermistor, a BJT and a relay.

(d) BJT differential amplifier.

(e) tuned amplifier.
(f) Hartley oscillator.
(g) Phase shift oscillator.
(h) non-inverting op amp circuit.
(i) op amp adder.
(j) op amp sample and hold circuit.
(k) precision half-wave rectifier.
(l) BJT push-pull power amplifier.
(m) triac lamp dimmer.
(n) half-adder.
(o) magnitude comparator.
(p) pulse-generator based on NAND gates.
(q) astable based on 7555 ic.
(r) microcontroller receiving input from a light-dependent resistor, and sending output to a MOSFET, which switches on a lamp in dark conditions.

2 What are the main questions to ask when presented with a faulty circuit or system?

3 List the common faults that should be looked for first when checking a faulty circuit. Explain how to test for them.

4 What simple items of equipment can be used to test an analogue circuit? Explain how to use each item.

5 What simple items of equipment can be used to look for faults in a digital circuit? Explain how to use each item.

6 Explain briefly (a) the main uses of an oscilloscope, (b) signature analysis, and (c) in-circuit emulation.

Supplement A

Useful definitions and equations

This is a summary of facts and formulae that you will find useful when studying electronics and when building and testing circuits on the workbench or simulator.

Electrical quantities and units

Current: This is the basic electrical quantity. Its unit, the *ampere*, is defined by reference to the mechanical force a current produces between two parallel conductors.

Symbol for the quantity, current: I.
Symbol for the unit, ampere: A.

Charge: The unit of charge is the *coulomb,* defined as the amount of charge carried past a point in a circuit when a current of 1 A flows for 1 s.

Symbols for quantities are printed in italics (sloping) type. Symbols for units are printed in roman (upright) type.

Symbol for the quantity, charge: Q.
Symbol for the unit, coulomb: C.

Equation: $Q = It$.

Potential: The unit of potential is the *volt*. The potential at any point is defined as the work done in bringing a unit charge to that point from an infinite distance. This is rather an abstract definition. In electronics, we are more concerned with *potential difference,* the difference in potential between two points relatively close together, such as the two terminals of an electric cell. Potential difference is usually measured from a convenient reference point, such as the Earth (or 'ground') or from the so-called negative terminal of a battery. The reference point is taken to be at 0 V. Potential difference is often called *voltage*.

There is often confusion between V and V. Some books use U for the quantity instead of V to avoid this.

Symbol for the quantity, potential: V.
Symbol for the unit, volt: V.

Electromotive force: The unit of emf is the *volt*. It is a measure of the ability of a source of potential to produce a current *before* it is connected into a circuit The emf of an electric cell depends on the chemicals it is made of. The emf of a generator depends on such factors as the number of turns in its coils, the magnetic field and the rate of rotation. When the battery or generator is unconnected, the potential difference between its terminals is equal to its emf. When the battery or generator is actually connected to a circuit and current flows into the circuit, the potential difference between its terminals falls because of internal resistance, even though its emf is unaltered.

Symbol for the quantity, electromotive force: E.
Symbol for the unit, volt: V.

Resistance: The unit of resistance is the *ohm.* When a current I flows through a conductor there is a fall in potential along it. The potential difference between its ends (or, in other words, the voltage across it) is V, and its resistance R is defined as:

$$R = V/I$$

This equation expresses Ohm's Law (but this is not the form in which Ohm originally stated it). Other useful versions of the equation are:

$$V = IR \qquad I = V/R$$

Symbol for the quantity, resistance: R.
Symbol for the unit, ohm: Ω.

The ohm is also the unit of reactance (symbol, X) and impedance (Z), both of which vary with frequency.

Conductance: The unit of conductance is the *siemens.* It is defined as the reciprocal of the resistance. If a conductor has a voltage V across it and a current I flowing through it, its conductance G is:

$$G = I/V$$

This is another source of confusion. The symbol for siemens is a capital S, but the symbol for second is a small s.

Symbol for the quantity, conductance: G.
Symbol for the unit, siemens: S.

Power: The unit of power is the *watt.* In electrical terms power P dissipated in an electrical device is proportional to the current flowing through it and the voltage across it:

$$P = IV = I^2R = V^2/R$$

Symbol for the quantity, power: P.
Symbol for the unit, watt: W.

Capacitance: The unit of capacitance is the *farad*. If the charge on the plates of a capacitor is Q coulombs and the potential difference between them is V volts, the capacitance C in farads is given by:

The definition of the farad is easy to understand but it produces a unit that is much too large for most practical purposes. In electronics we more often use the submultiples, microfarad (μF), nanofarad (nF) and picofarad (pF).

$$C = Q/V$$

Symbol for the quantity, capacitance: C.
Symbol for the unit, farad: F.

Inductance: The unit of inductance is the *henry*. The inductance of a coil in which the *rate of change* of current is 1 A per second is numerically equal to the emf produced by the change of current. This definition applies to a single coil as is more correctly known as *self-inductance*. If two coils are magnetically linked, their *mutual inductance* in henries is numerically equal to the emf produced in one coil when the current through the other coil changes at the rate of 1 A per second.

Symbol for the quantity, self-inductance: L.
Symbol for the quantity, mutual inductance: M.
Symbol for the unit of either, henry: H.

Frequency: The unit of frequency is the *hertz*. It is defined as 1 repetition of an event per second.

Symbol for the quantity, frequency: f.
Symbol for the unit: Hz.

Networks

Resistances in series and parallel

When two or more resistances are connected in series (Fig. A.1a), their total resistance is equal to the sum of their individual resistances. When two or more resistances are connected in parallel (Fig. A.1b) their total resistance R_T is given by the equation:

$$\frac{1}{R_T} = \frac{1}{R_1} + \frac{1}{R_2} + \frac{1}{R_3} + \ldots + \frac{1}{R_n}$$

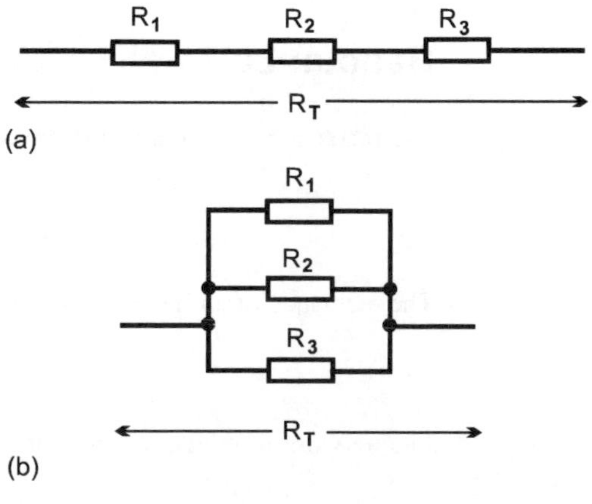

R1 R2 R3
RT
(a)

R1
R2
R3
RT
(b)

Figure A.1 *Two or more resistors may be connected (a) in series, or (b) in parallel.*

For two resistances in parallel, this simplifies to:

$$R_T = \frac{R_1 \times R_2}{R_1 + R_2}$$

Capacitances in series and parallel

When two or more capacitances are connected in series (Fig. A2.a), their total capacitance C_T is given by the equation:

$$C_T = \frac{1}{C_1} + \frac{1}{C_2} + \frac{1}{C_3} + \ldots + \frac{1}{C_n}$$

When two or more capacitances are connected in parallel their total capacitance is equal to the sum of their individual capacitances (Fig. A.2b).

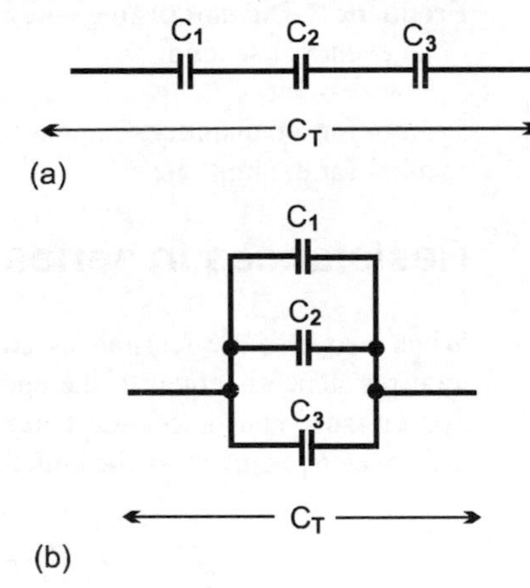

Figure A.2 *An uncommon capacitance value can often be made up by wiring two or more capacitances in parallel.*

Reactance

The reactance of a capacitor decreases with frequency:

$$X_C = \frac{1}{2\pi fC}$$

The reactance of an inductor increases with frequency:

$$X_L = 2\pi fL$$

In a network in which a capacitor and an inductor are connected in series or in parallel, the resonant frequency of the circuit occurs when $X_C = X_L$, giving:

$$f = \frac{1}{2\pi\sqrt{LC}}$$

If the capacitor and inductor are in parallel, the impedance of the network is a maximum at the resonant frequency. If the capacitor and inductor are in series, the impedance is a minimum.

Charging a capacitance

If a capacitance C is charged from a *constant* voltage supply V_C, through a resistance R, the *time constant* of the circuit is defined as RC. The graph of Fig. A.3 is plotted for a circuit in which $V_C = 10$ V, $R = 100\ \Omega$, and $C = 1$ μF. The time constant is:

$$RC = 100\ \mu\text{s}$$

With any pair of values of R and C, v_{OUT} rises to $0.63V_C$ after RC seconds (1 time constant). The graph is plotted for 5 time constants after the voltage step and it can be seen that the curve levels out *almost* completely during this time. In theory, the capacitor never reaches full charge but, for all practical purposes, we can say that it takes 5 time constants to fully charge the capacitor.

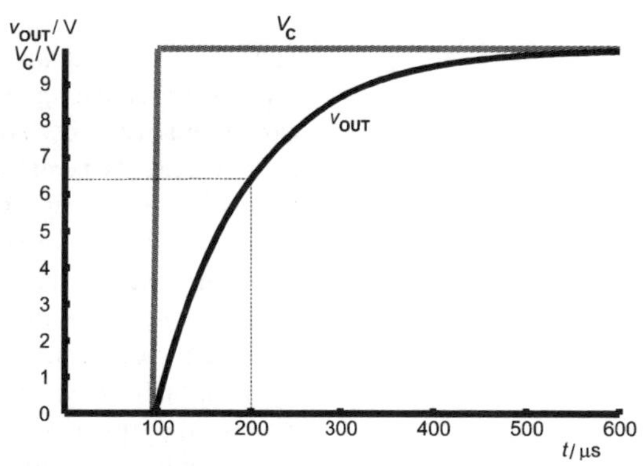

Figure A.3 *If* V_C *= 10 V,* R *= 100 Ω, and* C *= 1μF, it takes 500 μs (5 time constants) to fully charge the capacitor.*

If a capacitance C is already charged to V_C and is discharged through a resistance R, the voltage across the capacitor falls as in Fig A.4. In 1 time constant v_{OUT} falls to 3.7 V (that is, the voltage across it *falls* by $0.63V_C$). After 5 time constants, the capacitor is completely discharged.

The shape of the curves in Figs. A.3 and A.4 is described as *exponential*. This is because the equations for the curves contain the *exponential constant*, e.

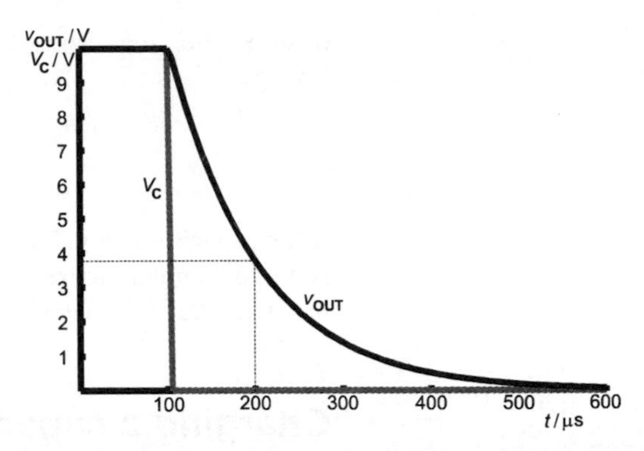

Figure A.4 When the capacitor discharges, the voltage across it falls to 0.37 times the original voltage across it in 1 time constant.

Alternating current

An alternating current is any current that periodically changes direction. The most familiar alternating current is the *sinusoid* or *sine wave,* which has this equation:

$$i = I \sin 2\pi ft$$

I is its amplitude (maximum deviation above and below 0 V), f is its frequency (in hertz) and t is the instant at which it has value i.

The *root mean square* of an alternating current or voltage is a way of expressing its average value over an interval of time, the interval being a whole number of periods. Because the signal is symmetrical about the 0 V or 0 A axis, its positive values are exactly cancelled out by its negative values so its *average* value is zero. This means that taking the average value is of no use for specifying the size of the signal. This is why we use the root mean square. If we first square the values, both positive and negative, *all* values become positive. Then we can add them and divide by the number of values to obtain their average. This is an average of their squares, their *mean square.* To return to the original units, volts or amps, we take the square root of the mean square, the *root mean square*. This is usually abbreviated to r.m.s.

The idea of r.m.s. applies to any alternating current or voltage, but the r.m.s. of a sinusoid is of special interest. If a voltage sinusoid is:

$$v = V \sin 2\pi ft$$

The r.m.s. value is given by:

$$v_{\text{r.m.s.}} = \frac{V}{\sqrt{2}}$$

A more practical version of this equation is:

$$v_{\text{r.m.s.}} = 0.7071V$$

Because the power dissipated in a device is proportional to the *square* of the voltage across it, the $v_{\text{r.m.s}}$ is equal to the constant (DC) voltage that causes the same power to be dissipated as the sinusoidal voltage, amplitude V. For example, a DC voltage of 7.07 V generates the same amount of heat in a resistance as a sinusoidal voltage, amplitude 10 V.

Transformers

In an ideal (no-loss) transformer that has n_p turns in the primary coil and n_s turns in the secondary coil, the ratio between the r.m.s. voltage v_p across its primary coil and the r.m.s. voltage v_s across its secondary coil is:

$$\frac{v_p}{v_s} = \frac{n_p}{n_s}$$

The inverse relationship applies to rms currents:

$$\frac{i_p}{i_s} = \frac{n_s}{n_p}$$

Logic

Logical identities

These are useful for solving logical equations when designing combinational logic circuits. In these identities, the symbol • means AND, and can be omitted when writing out the identity and when solving equations. The symbol + means OR. A bar over a symbol means NOT.

The order of ANDed and ORed terms makes no difference:

A•B = B•A
A + B = B + A

Brackets can be inserted and removed just as in ordinary arithmetic:

A•B•C = (A•B)• C = A•(B•C)
A + B + C = (A + B) + C = A + (B + C)
A•(B + C) = A•B + A•C

ANDing A with itself and with true and false:

A•A = A
A•1 = A
A•0 = 0

ORing A with itself and with true and false:

$A + A = A$
$A + 1 = 1$
$A + 0 = A$

Negation:

$A + \overline{A} = 1$
$A \bullet \overline{A} = 0$
$\overline{\overline{A}} = A$

De Morgan's Laws:

$\overline{A + B} = \overline{A} \bullet \overline{B}$
$\overline{A \bullet B} = \overline{A} + \overline{B}$

Miscellaneous:

$A + A \bullet B = A$
$A + B \bullet C = (A + B) \bullet (A + C)$
$A + \overline{A} \bullet B = A + B$

Logic gates

The symbols used in this book are those defined by the American National Standard Institute (ANSI). Fig. A.5 shows the ANSI symbols, and also the corresponding symbols defined by the British Standards Institute in BS3939.

Converting number systems

Converting binary to decimal

1) Evaluate each binary digit.
2) Sum their values.

Example

Convert 1101 0101 to decimal

Starting with the least significant digit (on the right):

Any number to the power zero is equal to 1.

$1 \Rightarrow 1 \times 2^0 = 1$
$0 \Rightarrow 0 \times 2^1 = 0$
$1 \Rightarrow 1 \times 2^2 = 4$
$0 \Rightarrow 0 \times 2^3 = 0$
$1 \Rightarrow 1 \times 2^4 = 16$
$0 \Rightarrow 0 \times 2^5 = 0$

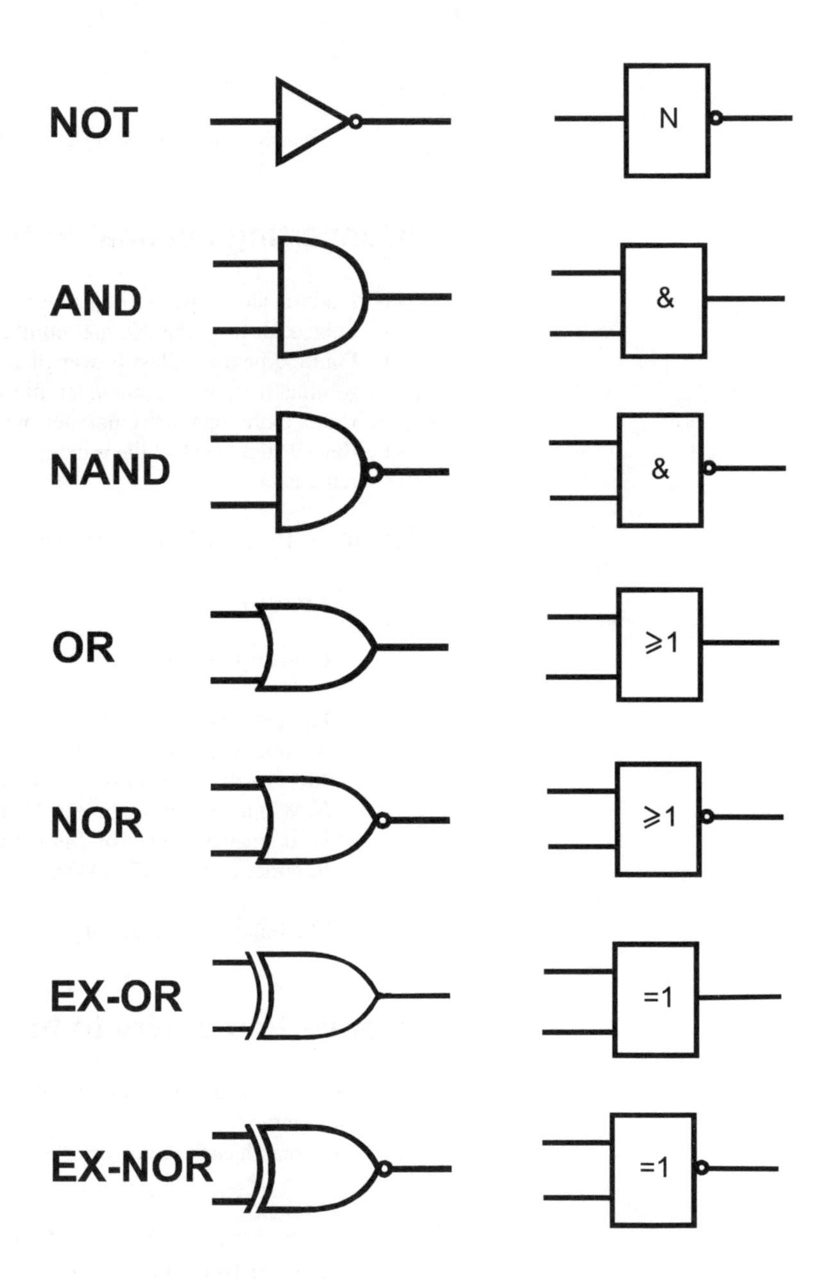

Figure A.5 *These are the two-input versions of the symbols, except for NOT which, by definition, has only one input. The truth tables for all of the two-input logic gates are summarised on p. 154.*

$1 \Rightarrow 1 \times 2^6 = \quad 32$
$1 \Rightarrow 1 \times 2^7 = \quad \underline{64}$
Total 117

The decimal equivalent of 1101 0101 in binary is 117.

Converting decimal to binary

1) Find the largest power of 2 that is less than the decimal number.
2) Subtract it from the decimal number. Write down '1'.
3) Find the next smallest power of 2. If this is less than the remainder, subtract it from the remainder and write down '1'.
   If it is more than the remainder, write down '0'
4) Repeat step (3) until the remainder is 0 or 1. Write down the remainder.

The 0's or 1's are written right to left.

Example

Convert 145 to binary.

Largest power of 2 is 128	Write 1
Subtracted from 145 leaves 17 remainder	
Next smallest power of 2 is 64; more than 17	Write 0
Next smallest power of 2 is 32; more than 17	Write 0
Next smallest power of 2 is 16; less than 17	Write 1
Subtract 16 from 17; leaves 1.	Write 1

The binary equivalent of 145 in decimal is 1 1001.

Converting binary to hex

1) Write the binary number with the digits in groups of 4, starting from the right.
2) Underneath each group write its hex equivalent (see table on p. 315).

Example

Convert 10 1011 0001 1101 to hex.

Binary number in fours	10	1011	0001	1101
Hex equivalents	2	B	1	D

The hex equivalent of 10 1011 0001 1101 in binary is 2B1D.

Converting hex to binary

1) Write out the hex number, spacing the digits widely.
2) Underneath each digit write its 4-bit binary equivalent (see table on p. 315).

Example

Convert *C42E* to binary.

Spaced hex digits	C	4	2	E
Binary equivalents	1100	0100	0010	1110

The binary equivalent of C42E in binary is 1100 0100 0010 1110.

Supplement B

Answers to questions

Chapter 1

Test your knowledge

1.1 (1) 125 mS, (2) 0.1875 V.
1.2 220 kΩ, 820 kΩ.
1.3 289 kΩ.
1.4 66 pF, nearest 68 pF.
1.5 60 Ω, nearest 62 Ω.
1.6 (1) 8.75 dB, (2) 19.1 dB, 904.4 mV.

Multiple choice questions

1 C 2 A 3 C 4 A 5 B 6 D

Chapter 2

Test your knowledge

2.1 2.41.
2.2 3 kΩ.
2.5 Input resistance = 29 kΩ, C_1 = 5.49 pF.
2.6 −300 (assuming r_e = 25 Ω).
2.7 common-drain.

Multiple choice questions

1 C 2 B 3 D 4 C 5 D 6 C

Chapter 3

Test your knowledge

3.1 Common-source.
3.2 0.89 Ω.
3.3 It decreases.
3.4 It is more than 10 kΩ.
3.5 4.33 A, next highest standard rating is 5 A.

Multiple choice questions

1 A 2 C 3 C 4 C 5 B 6 C

Chapter 4

Test your knowledge

4.1 10 000.
4.2 480.
4.3 X_C = 2022 Ω, 2006 Ω, 1981 Ω.
X_L = 1991 Ω, 2006 Ω, 2032 Ω.
4.4 31.8 kHz.

Multiple choice questions

1 B 2 A 3 A 4 B 5 B

Chapter 5

Test your knowledge

5.1 633 mH.
5.2 29.4 Hz.

Multiple choice questions

1 C 2 B 3 C 4 D 5 B 6 D 7 A

Chapter 6

Test your knowledge

6.1 To the (1) positive, (2) negative supply voltage.
6.2 373.
6.3 2.2 kΩ.
6.4 2.19 kΩ.
6.5 (1) 27 Ω. (2) 1 TΩ. (3) 29.4 kΩ.
6.6 2.42 V.
6.7 55°C.

Multiple choice questions

1 D 2 B 3 A 4 B 5 B 6 A 7 D 8 C

Chapter 7

Test your knowledge

7.1 13.4 mA.
7.2 1.07 kΩ
7.3 −106 mV.
7.4 UTV = 8.49 V, LTV = 6.51 V, hysteresis = 1.30 V.
7.5 UTV = 12.5 V, LTV = 2.5 V.
7.6 154 kHz.

Multiple choice questions

1 A 2 C 3 D 4 A 5 A 6 D
7 B 8 A 9 C 10 D 11 C 12 B

Chapter 8

Test your knowledge

8.1 CC = emitter follower, CD = source follower.
8.2 (1) 2.4 °C/W, (2) 44 °C.
8.3 4.66 W.

Problems

6 The maximum ambient temperature is 23.3°C.

Multiple choice questions

1 C 2 B 3 A 4 D

Chapter 9

Test your knowledge

9.1 From anode to cathode.
9.2 4 V.
9.3 Holes.
9.4 4.55 V.

Multiple choice questions

1 D 2 A 3 A 4 D 5 A 6 B

Chapter 10

Test your knowledge

10.1 500 Ω; take next highest E12 value, 560Ω.
10.2 (a) Unlimited in practice, nominally 50.
(b) 1.

Multiple choice questions

1 D 2 B 3 D 4 A 5 D

Chapter 11

Test your knowledge

11.1 0, 0.

11.2

D	*C*	*B*	*A*	*Z*
0	0	0	0	1
all 15 other rows				0

11.2 The circuit is equivalent to exclusive-NOR.

11.5 (a) 0, (b) 1.

11.6

Inputs			*Outputs*	
3	*2*	*1*	*Z1*	*Z2*
0	0	0	0	0
0	0	1	0	1
0	1	0	1	0
1	0	0	1	1

11.7 3.

Multiple choice questions

1 C 2 D 3 D 4 D 5 A 6 A 7 A 8 B

Chapter 12

Test your knowledge

12.4 3.63 μs.
12.5 103 kΩ (nearest is 100 kΩ).
12.6 407 Hz, 2.47.
12.8 63.
12.9 5.

Multiple choice questions

1 A 2 D 3 D 4 A 5 D 6 B
7 D 8 A

Chapter 13

Test your knowledge

13.1 293 Ω (nearest 300 Ω).

Multiple choice questions

1 A 2 B 3 C

Chapter 14

Test your knowledge

14.1 –3.4375 V. 14.2 –2.8125 V.

Multiple choice questions

1 C 2 A 3 A 4 D

Chapter 15

Multiple choice questions

1 B 2 A 3 A 4 B 5 C

Chapter 16

Test your knowledge

16.1 Increases.
16.2 High frequencies are absorbed by inductors and shorted through capacitors. Low frequencies pass theough inductors but not through capacitors.
16.3 Slew rate is the rate at which the output can be driven from its maximum high level to its minimum low level.
16.4 $n = 1.47$.
16.5 There are no majority carriers so the region has very low conductivity and is in effect a dielectric.
16.6 10 km.

Multiple choice questions.

1 B 2 C 3 D 4 D 5 C 6 B 7 A 8 A

Chapter 17

Problems

4 SNR = 46.5 dB.

Test your knowledge

17.1 147 dB.
17.2 9.4 dB (one of the advantages of working in decibels is that they can be added to calculate their total effect).

Multiple choice questions

1 A 2 B 3 A 4 A 5 D

Chapter 18

Test your knowledge

18.1 15.9 Ω.
18.2 188 kΩ.
18.3 (a) 25.455 MHz, (b) 455 kHz.

Multiple choice questions

1 C 2 C 3 A 4 A 5 B

Chapter 20

Test your knowledge

20.1 $2^8 - 1 = 255$.
20.2 ROM.
20.3 1011, 0011, 0111.
20.4 10, 3, 14.
20.5 F, 6, 8.
20.6 4096, FFF.

Multiple choice questions

1 B 2 B 3 C 4 D 5 B 6 D
7 C 8 B 9 B 10 B 11 C 12 A

Chapter 21

Test your knowledge

21.1 (1) CA 41H, (2) LDI 41H, (3) Load accumulator with the data stored at 035AH.
21.2 (1) LDI 63H STA 2F56H (2) Load accumulator with the data stored at 750EH, subtract 1 from it, and store the result at 5FFFH.
21.3 At 3B2CH.
21.4 00000, 11111, 11110, 11110, 11101, 11100, 11011, 11010.
21.6 10H.
21.7 D7H.
21.3 0A 65 03 FE 45.

Multiple choice questions

1 C 2 B 3 D 4 C 5 A 6 B
7 A 8 C 9 C 10 D 11 C

Index